THE MOTHS OF AMERICA NORTH OF MEXICO

The Moths of America North of Mexico

INCLUDING GREENLAND

FASCICLE 26.1

NOCTUOIDEA

NOCTUIDAE (PART)
CUCULLIINAE, STIRIINAE
PSAPHIDINAE (PART)

ROBERT W. POOLE
SYSTEMATIC ENTOMOLOGY LABORATORY
UNITED STATES DEPARTMENT OF AGRICULTURE

COLOR PHOTOGRAPHS BY
CHIP CLARK
SMITHSONIAN INSTITUTION

MONOCHROME PHOTOGRAPHS BY
ROBERT W. POOLE AND PATRICIA GENTILI

LINE DRAWINGS BY
CHARLES L. HOGUE (STIRIINAE)
BRIT GRISWOLD

1995
WASHINGTON

THE WEDGE ENTOMOLOGICAL RESEARCH FOUNDATION

THE MOTHS OF NORTH AMERICA

This work is to be cited as
Poole, Robert W., 1994. Noctuoidea, Noctuidae (part)
in Dominick, R. B., et al.,
The Moths of America North of Mexico, fasc. 26.1

DISTRIBUTORS

THE WEDGE ENTOMOLOGICAL RESEARCH FOUNDATION
% National Museum of Natural History,
MRC-127, Washington, D.C. 20560, U.S.A.

E. W. CLASSEY LTD
P.O. Box 93, Faringdon, Oxfordshire SN7 7DR, England

BIOQUIP PRODUCTS
17893 LaSalle Avenue
Gardena, California 90248, U.S.A.

ENTOMOLOGICAL REPRINT SPECIALISTS
P.O. Box 77224, Dockweiler Station
Los Angeles, California 90007, U.S.A.

ISBN 0-933003-07-2

LIBRARY OF CONGRESS CATALOG NUMBER 94-61667

Printed in the United States of America by Allen Press, Inc., Lawrence, Kansas

Color plates by Lanman Progressive Co., Washington, D.C.

ABSTRACT

The North American species of the subfamilies Cuculliinae and Stiriinae, and the two tribes Psaphidini and Feraliini of the subfamily Psaphidinae are revised. Forty species of the Cuculliinae are recognized including the following new species: *Cucullia styx* (Mexico and Arizona), *Cucullia lethe* (eastern and central Texas), *Cucullia charon* (western Texas and southern Arizona), and *Cucullia eurekae* (Utah). One new generic name is proposed in the Cuculliinae, *Dolocucullia* (type species: *Xylina dentilinea* Smith). Forty species of the subfamily Stiriinae are recognized including four new species: *Plagiomimicus heitzmani* (Illinois, Missouri, and Kansas), *Plagiomimicus mimica* (southwestern United States and northern Mexico), *Basilodes straminea* (southern Arizona), and *Stiria satana* (southern Arizona). The following new generic names are proposed in the Stiriinae: *Lineostriastiria* (type species: *Antaplaga hachita* Barnes), *Hoplolythrodes* (type species: *Lythrodes arivaca* Barnes), *Angulostiria* (type species: *Basilodes chryseochilus* Dyar), and *Argentostiria* (type species: *Antaplaga koebelei* Riley). Thirteen species are placed in the Psaphidini and six species in Feraliini. No new genera or species are proposed for either tribe. Adults of all species are illustrated in color. The male and female genitalia of each species are illustrated by drawings or photographs. Distribution maps of specimens examined for each species are provided. Keys are provided for each genus or subfamily depending on which is more appropriate. A new classification of the trifid Noctuidae is proposed and some characters are discussed that justify it. A phylogenetic analysis of the Stiriinae is included.

Dedicated to the memory of
CHARLES L. HOGUE
Entomologist
Natural History Museum of Los Angeles County

CONTENTS

PREFACE

Among the so-called "trifid" noctuids of the Noctuidae are several groups of species previously placed in the "subfamilies" Cuculliinae and Amphipyrinae that appear to be the scattered remnants of some of the more primitive lineages of the "trifid" noctuids. There groups of species often are characterized by a lack of shared, derived morphological characters.

This fascicle deals with three subfamilies (as newly defined in this work): the Cuculliinae, Stiriinae, and Psaphidinae (tribes Psaphidini and Feraliini). Only two tribes of the Psaphidinae are covered; the remaining tribes will be treated in a following fascicle. Some genera cannot be adequately placed anywhere in the "trifid" noctuids, but they have a general similarity to genera in the Stiriinae. Thus, an *incertae sedis* group has been placed after the section on the Stiriinae.

I do not accept the subspecies concept, and subspecific names are not used in this fascicle. However, geographically distinct populations, if present, are discussed. If an available name exists for the population, it is mentioned. My feelings about forms, varieties, and aberrational names are considerably stronger. If a name is available in the sense of the *International Code of Zoological Nomenclature,* it is included. If a name is unavailable in the sense of the Code, it is not included. Infrasubspecific names have been listed in some previous fascicles even though they are nomenclatorially unavailable. I feel form, variety, and aberration names represent bad science, and bad science should not be dignified by listing these names.

The distribution maps were prepared with the computer program MAPINFO. The locality and other information for all specimens examined was recorded in an xbase (in this case FOXPRO 2.5) database. MAPINFO reads these database files and plots the localities on a map according to the latitude and longitude of the locality.

All wing measurements are in millimeters and taken from the base of the forewing to its apex. Sample sizes of 10 usually were used unless fewer specimens were available. Sample sizes and standard deviations also are given.

AUTHOR'S ACKNOWLEDGEMENTS

Many individuals have provided important help and support in the preparation of this fascicle. I particularly want to acknowledge the late Charles L. Hogue who very generously allowed me to use his excellent figures of the structural morphology of selected genera of the Stiriinae. I also wish to thank J. D. Lafontaine of Agriculture Canada specifically. Many of the ideas expressed in the discussion of the phylogeny of the "trifid" noctuids are partially his and represent collaboration between the two of us. He also reviewed the manuscript extremely thoroughly. As always, the managing editor R. W. Hodges made significant improvements to the manuscript and is responsible for the high quality and continuance of the *Moths of America North of Mexico* series in general. John G. Franclemont reviewed the manuscript, provided significant information for the fascicle, and loaned me material from his collection. Eric Metzler reviewed the manuscript and supplied both specimens and records of all of the relevant species from the Ohio Lepidoptera Survey. I am also grateful for the reviews of D. C. Ferguson and R. Leuschner.

I acknowledge the help of the following individuals for lending material in their care or helping me on visits to their institutions: the late A. Blanchard (Houston, Texas), D. Bowers (Museum of Comparative Zoology, Cambridge), V. Brou (Abita Springs, Louisiana), A. Brower (Augusta, Maine), C. Covell (University of Louisville), T. Dickel (Homestead, Florida), J. P. Donahue (Natural History Museum, Los Angeles County), M. Honey and the late A. Hayes (The Natural History Museum, London), E. Knudson (Houston, Texas), J. D. Lafontaine (Biosystematics Research Centre, Agriculture Canada, Ottawa), B. Mather (Clinton, Mississippi), S. McKown (Boise, Idaho), J. Miller and F. Rindge (American Museum of Natural History, New York), P. Opler (Fort Collins, Colorado), J. Powell (University of California at Berkeley, California), J. Rawlins (Carnegie Museum of Natural History, Pittsburgh), R. Robertson (Santa Rosa, California), and F. Stehr (Michigan State University, East Lansing).

I thank R. Stanger and P. Gentili for technical support in dissecting male and female genitalia. The color plates of the adults were photographed by C. Clark of the Smithsonian Institution. Finally I would like to express my gratitude in general to the members of the Systematic Entomology Laboratory, USDA and the Department of Entomology, Smithsonian Institution.

THE MOTHS OF AMERICA NORTH OF MEXICO

SUPERFAMILY
NOCTUOIDEA (continued)

FAMILY
Noctuidae (continued)

A PRELIMINARY PHYLOGENY FOR THE "TRIFID" NOCTUIDAE. This is the third fascicle on the family Noctuidae to appear in the *Moths of America North of Mexico*. No explicit phylogeny or sequence of groups for the fascicles has yet been proposed for the family, although some of the conclusions discussed below were presaged by comments made in Lafontaine and Poole (1991). The classification of the Noctuidae historically has been based on a system developed by Hampson in his famous series *Catalogue of the Lepidoptera Phalaenae in the Collection of The British Museum*. This system is largely artificial. However, the scheme works so well, the family is so large, and the interpretation of characters in the Noctuidae is so difficult that no one has tried to construct a natural classification reflecting the phylogeny of the family.

This state of affairs eventually must cease. Over the years J. D. Lafontaine and I have discussed alternative classifications for the Noctuidae, concentrating primarily on the so-called "trifid" noctuids, i.e., noctuid groups with vein M_2 in the hindwing reduced in strength versus the "quadrifids" with vein M_2 fully developed. This concentration of effort on the "trifid" noctuids reflects our area of interest and expertise in the family. The phylogenetic tree presented below attempts to present a new and more nearly natural classification of the "trifid" noctuids. Errors and inconsistencies are solely my responsibility, but whatever value the classification may have is the result of discussions with J. D. Lafontaine and many others, both in print and in informal conversation. This classification is not the result of a formal analysis, a project that would take several years and is beyond the scope of this paper. My sole purpose is to provide a tentative classification that will serve as a framework of groups of the "trifid" noctuids for the *Moths of America North of Mexico*. The tree is presented almost without justification although a few characters are discussed to give it some foundation. A sturdy classification of the Noctuidae is unlikely to result from any one person's single publication or study. A secondary purpose of this discussion is to provide an explicit preliminary hypothetical phylogeny to serve as a target for future studies on noctuid phylogeny.

The classification is based almost exclusively on the North American elements of the "trifid" noctuids. Although most of the groups making up the "trifid" Noctuidae are represented in North America, several Old World or Neotropical groups are not represented in either the United States or Canada and consequently are absent from the tree. Secondly, the classification deals only with the "trifid" noctuids and ignores the larger problem of the phylogeny of the family Noctuidae as a whole. Finally, the largest percentage of the trifid noctuids falls in the subfamily Noctuinae (as here defined), and the magnitude of the subfamily is hidden by its representation as a single branch on the tree. The delineation of tribes within the Noctuinae and their relationship to each other are outstanding problems to be tackled in later fascicles.

A few comments on the distinction between "trifid" and "quadrifid" noctuids are in order. In a technical sense the "trifid" state refers to species with the hindwing vein M_2 absent (actually reduced in strength relative to the other median veins). In contrast the "quadrifid" state refers to the presence of vein M_2 (vein M_2 of equal strength to the other median veins). On an informal basis "trifid" noc-

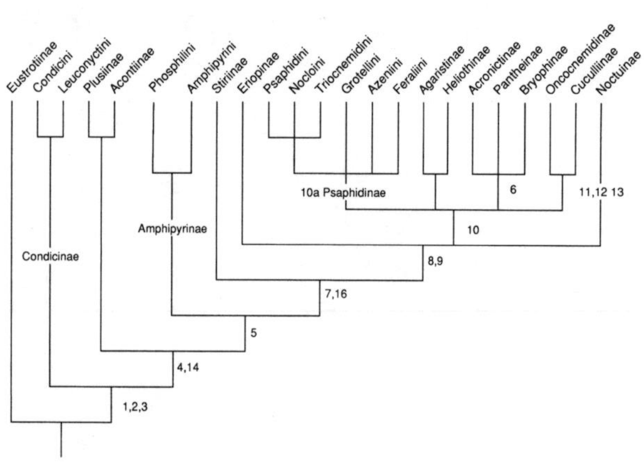

FIGURE 1: POSTULATED PHYLOGENY OF THE TRIFID NOCTUIDAE

tuids are those species with the characteristic "miller moth" shape and appearance while "quadrifid" refers to "the rest of them." The informal definition of "trifid" has become more important, in many ways, than the technical version. For example, the subfamilies Stiriinae and Condicinae have a fully developed vein M_2 in the hindwing but have always been placed in the "trifid" noctuids because they look like "trifid"noctuids. The tree is intended to represent the "trifid" noctuids, but the state of M_2 in the hindwing varies from well developed to almost absent, with the general trend being from strong to weak (primitive to derived), i.e., from the Plusiinae and Condicinae, which are clearly "quadrifid," to the Noctuinae, which are usually (but not always) clearly "trifid." This character, as so many others in the Noctuidae, must be interpreted with great care and is very susceptible to reversals, losses, and reappearances. In addition, the character is not an "either-or" character, and its interpretation is usually subjective to some degree.

Another comment on my operating principles is needed. It is my belief that in the Noctuidae (at least) the loss of a character usually refers to its loss of expression but not its genetic loss. I believe that in the Noctuidae (and probably everywhere) characters are easily lost and as easily regained because the loss of a character commonly represents a loss of gene expression, not the loss of the genetic information necessary to produce the character. Not a single character discussed below is consistent throughout all the groups on the tree. I believe it is common in the Noctuidae for characters to be lost and subsequently regained, sometimes repeatedly throughout the evolution of the family.

CHARACTERS ON THE TREE. The following characters were used in constructing the tree, and their appearance or the appearance of the derived state is marked with the appropriate number on the tree in text figure 1.

1. Pocket IV of the internal tympanic structure: Pocket IV is almost always completely enclosed in the "trifid" noctuids. Pocket IV is usually dorsoventrally divided in the "quadrifid" groups. An undivided pocket IV is considered to be a derived state based on its divided condition in the outgroup families Arctiidae and Lymantriidae. The internal tympanic structure of a species of *Cucullia* is shown in text figure 2 *f.* Pocket IV is intimately fused to the air sac in many groups of the "trifid" noctuids. The reader should refer to Richards (1933) for a discussion of the tympanic structures of the Noctuoidea and for illustrations and definitions. The structure of Pocket IV shows considerable variability and needs to be examined in more detail for the "trifid" noctuids. The Pocket IV criterion places the Plusiinae and the Nolinae among the "trifid" noctuids. Other characters indicate a closer relationship of the Nolinae with the Sarrothripinae and Chloephorinae in the "quadrifid" noctuids. The Sarrothripinae and Chloephorinae are possibly basal in the Noctuidae and may have given rise to the "quadrifid" and "trifid" lines of the family. Indeed, the barshaped retinaculum found in the Sarrothripinae and Chloephorinae is shared with the outgroup family Arctiidae.

2. Position of the silk pore in the larval spinneret: The silk pore of the spinneret in "quadrifid" noctuids is concealed in a medial depression near the apex of the spinneret, and this depressed area is margined by lateral lobes. The silk pore of "trifid" noctuids is open and either apical or subapical. The pore is distinctly present in most "trifid" noctuids although in some groups of Noctuinae the opening is a transverse or U-shaped slit, seldom visible except in properly prepared material.

3. Sclerotized apodemes from the basolateral corners of the eighth sternum of the male abdomen present (derived) or absent (primitive): The function of these apodemes is unknown, but probably they are muscle attachments. The number "3" in the tree (text figure 1) represents the first known appearance of this structure in the Noctuidae. This structure is most consistently developed in the Condicinae, Heliothinae, and Cuculliinae but is found in other groups, usually in some reduced form. The structure is apparently easily lost and is very inconsistent. For example, the apodemes are well developed in the Condicinae but are absent in all subsequent branches of the tree (except the Phosphilini of the Amphipyrinae) until the Heliothinae, Cuculliinae, and a few species of Nocloini. The basal placement of the Condicinae requires the loss of these apodemes in all of these subfamilies and their reacquisition in the Phosphilini, Nocloini, Heliothinae, and Cuculliinae. The subfamily Condicinae may be misplaced. However other characters, in particular the 3 setae in the SV setal group on larval abdominal segment one and the absence of a free pleurite in the male genitalia, strongly suggest a basal position for the Condicinae.

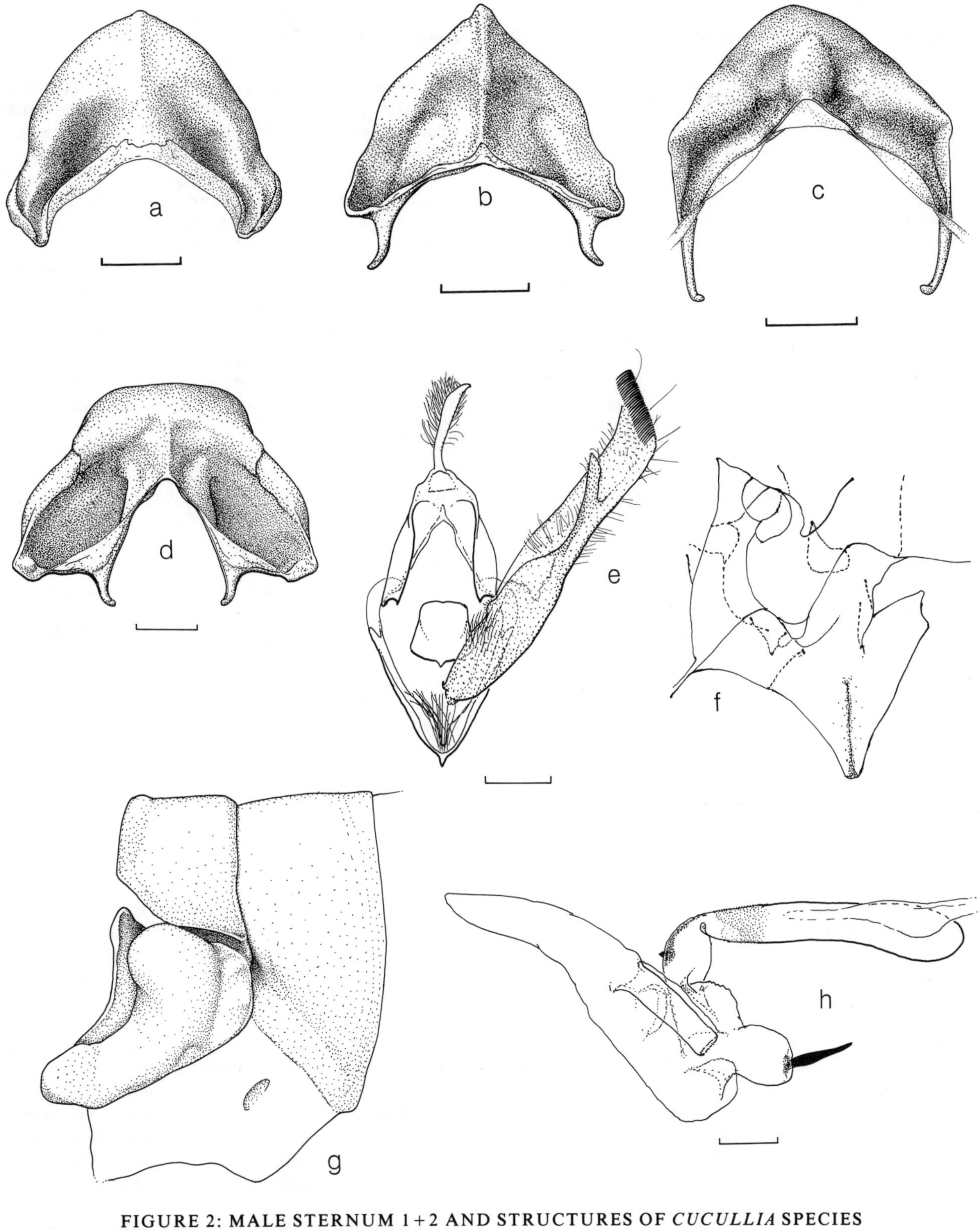

FIGURE 2: MALE STERNUM 1+2 AND STRUCTURES OF *CUCULLIA* SPECIES

a. Male sternum 1+2 of *Halysidota tessellaris. b.* Male sternum 1+2 of *Acronicta lepusculina. c.* Male sternum 1+2 of *Apamea lignicolora. d.* Male sternum 1+2 of *Cucullia serraticornis. e.* Male genitalia of *Cucullia omissa. f.* Internal tympanic structure of *Cucullia asteroides. g.* Lateral view of 1st and 2nd abdominal segments in *Cucullia asteroides. h.* Aedoeagus of *Cucullia omissa.*

4. Seta SV2 absent on larval abdominal segment 1: Three setae are present in the SV group on abdominal segments one and two in the primitive condition. Seta SV2 is absent on segment 1 and present on segment 2 in the derived state. For example, SV2 is found on segment A1 in the Condicinae and Eustrotiinae and in most "quadrifid" groups, but it is absent (as far as I have looked) in all groups beyond this break point in the tree except in some genera of the Plusiinae. The character, however, can be variable. For example SV2 is absent on A1 in most genera of the Plusiinae, but it is present in *Syngrapha* Hübner, *Plusia* Ochsenheimer, and some species of *Autographa* Hübner. I consider the occurrence of SV2 in these genera to be a reversal to the primitive 3 setae state. The two character states are illustrated in Lafontaine and Poole (1991). In addition, SV2 is absent in the "quadrifid" subfamily Sarrothripinae. The Sarrothripinae are considered to be "quadrifids" because of the "quadrifid" larval spinneret (a concealed silk pore and lateral lobes); the absence of SV2 in the Sarrothripinae is considered to be an independent loss.

5. Double, eversible hair pencils exist in the male eighth abdominal sternum of the Plusiinae, Acontiinae, and Leuconyctini (Condicinae) (plate A, figure 1): These double hair pencils are considered primitive for the "trifid" noctuids because they occur sporadically throughout the "quadrifid" subfamilies and in the Arctiidae as well. The hair pencils are lost and usually replaced with a single row of long hairs in the more derived "trifid" noctuids. The presence of the double hair pencils in the Acontiinae argues for a placement near the base of the trifid noctuid tree. The presence of a strong corona on the valve of the male genitalia and the "larvae liberae" (see following) condition of the larva argue for the placement of the Acontiinae within the "larvae liberae" group of subfamilies. The conflict is unresolved, but the Acontiinae are placed with the Plusiinae in this particular hypothetical phylogeny. In some genera of the Noctuinae (e.g., *Leucania* Ochsenheimer and *Aletia* Hübner of the old subfamily Hadeninae) the single row of hairs found in the eighth sternum of males of the higher trifids has been greatly enlarged, although not in the form of a double hair pencil. The existence of this partial redevelopment of the eighth abdominal hair pencils in an unrelated group argues that a reversal to the primitive condition in the Acontiinae is plausible.

6. Secondary setae in the larvae present (derived) or absent (primitive): The presence of secondary setae is hypothesized to be a reversal to a primitively hair larva as is found in the Arctiidae, Lymantriidae, and Nolinae (Noctuidae). The adults of the Acronictinae, Pantheinae, and Bryophilinae are clearly "trifid" noctuids on the basis of other larval characters and the structure of the tympanal region.

7. Flower and seed feeding larval habits: This is a particularly weak character. This node in the phylogeny represents the first group of the Noctuidae with larvae feeding on the flowers and fruits of their host plants. For example, flower feeding is found in the Stiriinae, Heliothinae, Cuculliinae, and Oncocnemidinae. The habit is postulated to have been secondarily lost and replaced by other life histories in the remaining groups above this break point in the tree, e.g., in the Noctuinae, Acronictinae, Agaristinae, and parts of the Psaphidinae. Species groups exist within the same genus exhibiting both flower feeding and other types of life histories. In the genus *Oncocnemis* (Oncocnemidinae) most species appear to feed on flowers of the Schrophulariaceae, but a few species have become foliage feeders on trees and shrubs.

8. Basal hair pencils on the male abdominal sternum 1+2: This character is a suite of structures illustrated in plate A, figures 2, 3. First paired lateral hair pencils (plate A, figure 2) are attached to the distal sides of the sternum by long apodemes. When not in use, they are contained in pockets (plate A, figure 3) located in the intersegmental membrane of the fourth abdominal segment. Paired glands (Stobbe's glands) (text figure 2 *a–d*) are in the basal to medial part of sternum 1+2. They consist of small, ovate invaginations with large glandular cells each subtending a long hair through the opening of the gland. These structures appear to be important defining characters for this branch of the tree. The most "primitive" state is found in species of the Eriopinae and some unplaced South American genera such as *Cropia* Walker and *Paratrachea* Hampson. Occurrence of these hair pencils, however, is very sporadic. This character is apparently very easily lost (reversal) and probably as easily regained (reversal of the reversal). For example, some species of a genus may have the structures, and others not. Various reduced states also are commonly seen. The potential presence of hair pencils (i.e., the genetic information is present but the character is unexpressed) is usually determinable from the structure of sternum 1+2, which is generally narrower proximodistally and wider side to side in groups with hair pairs pencils or the potential for hair pencils than in groups that do not have the structure (actually or potentially). Sternum 1+2 is approximately as wide as long in its primitive state, wider than long as in the derived state. The basal processes of sternum 1+2 are usually long in their hair pencil group, but absent or short and bifurcate in those groups without hair pencils or the genetic information for hair pencils. For example, compare text figure 2 *a* of an arctiid (the outgroup) without hair pencils or their potential with text figure 2 *b–d* of three species within this branch of the tree, all of which lack hair pencil structures but presumably have the genetic potential for them.

9. Digitus present (plate A, figure 4) in the male genitalia: The absence of a digitus is assumed to be primitive, and its presence (unless secondarily lost) is derived. The placement of this character on the tree is very difficult. The current position of derivation on the tree assumes that the digitus was lost in all the Psaphidinae, Agaristinae, and Heliothinae: a hypothesis I find very weak. A weak digitus is present in some species of the Oncocnemidinae and Cuculliinae, but it is absent in most. The presence of a digitus is most indicative of the species in the

Noctuinae. If it were restricted to the Noctuinae, a digitus would be a good synapomorphy for the group. However, its presence in the Cuculliinae and Oncocnemidinae greatly confuses matters. Other interpretations are possible, including its independent evolution in the Noctuinae line and the "larvae liberae" line (see following). Alternatively, the Noctuinae might have evolved as part of the "larvae liberae" line but regained the "larvae confluentae" structure of the tenth abdominal sternum.

10. Larvae liberae vs. larvae confluentae: This character was used by Crumb (1956) to separate a large portion of the "trifid" Noctuidae. Crumb's description and figures of the character are confusing. Godfrey and Stehr (1985) recently redefined the character and gave SEM photographs of the two conditions. Readers are referred to their figures. The venter of larval abdominal segment 10 in the primitive state is grooved posteriorly, and the posterior margin of the segment is medially impressed. The venter of segment 10 is not grooved posteriorly in the derived state, and the posterior margin is truncate or convex and projects (ventrally) as a quadrate plate below and anterior to the anus. Those groups without the groove and with a projecting quadrate plate (the derived state) represent Crumb's "larvae liberae" and those with the groove (the primitive state) are his "larvae confluentae." The "larvae confluentae" condition is considered to be primitive because it is found in the "quadrifid" groups of the Noctuidae and in the outgroup family Arctiidae. This character can be very difficult to interpret at times, particularly if the anus has been everted or the larvae inflated. For example, Crumb includes the two subfamilies Bagisarinae and Euteliinae in the "larvae liberae" group even though both subfamilies are "quadrifids" on the basis of other characters, most notably the structure of the larval spinneret. Species of the Psaphidinae show an intermediate condition with a partial groove toward the posterior margin of the venter of the 10th abdominal segment but without a groove anteriorly. This partial state is treated as an intermediate state (10a) and not a partial reversion.

11. Orientation of clasper + differentiation of cucullus and sacculus: First, the clasper of the male valve of the Noctuinae (plate A, figure 5) runs along the proximodistal axis of the value in the derived state. In the primitive state the clasper runs along the dorsoventral axis of the valve (plate A, figure 6). Second, a general tendency exists in the Noctuinae for the cucullus and sacculus to be distinctly differentiated. Neither character is absolutely reliable. In many genera of the Noctuinae (perhaps the more primitive ones) the cucullus is not distinctly differentiated from the sacculus. Also, the axis of the clasper appears to have reversed to a dorsoventral orientation in some genera of the Noctuinae.

12. Capped tegumen: The tegumen has a distinctive double appearance (plate A, figure 7) in most genera of the Noctuinae (as here defined). Specifically, the distal part of the tegumen near its junction with the uncus is set off from the larger basal part, with a distinct juncture between the two parts. In the rest of the "trifid" noctuids the tegumen (on each side) is a single structure (plate A, figure 8). The double structure is considered to be derived. The structure apparently is easily lost or weakly developed in many genera placed in the Noctuinae. I have not exhaustively surveyed the noctuine tribes to see whether the character typifies the entire subfamily, but list it as a character to examine for a more intensive phylogenetic analysis.

13. Posterior position of the larval subspiracular line: The larval subspiracular line, when it is present or defined, runs down the long axis of the anal prolegs in the subfamily Noctuinae. The subspiracular line, when present, runs into the anal shield or ventral to the anal shield in other groups of the "trifid" noctuids. The condition in the Noctuinae is assumed to be derived.

14. Free pleurite present between the tegumen and vinculum in the male genitalia: a roughly rectangular, free pleurite articulated to the end of the vinculum and running along the outer side of the lower end of the tegumen (plate A, figure 9) is present in most groups of the "trifid" Noctuidae. This free pleurite is further modified in some members of the Noctuinae (plate A, figure 10) and may be twisted or have a plate for a muscle attachment. Possibly intermediate conditions exist with the pleurite present, but fused with the end of the vinculum. The presence of this pleurite is considered to be a derived state. A free pleurite is absent (plate A, figure 11) in the primitive condition. The pleurite is absent in the Arctiidae, the "quadrifid" Noctuidae, and within the "trifid" subfamilies Condicinae and Eustrotiinae. The Condicinae may represent an intermediate condition because the end of the vinculum is attenuated and extends up the side of the tegumen; however, this strap is clearly part of the vinculum and not free from it.

15. Ventral half of the front of the head naked: This character is not marked on the phylogeny because it seems to be a derived state found only in some of the "quadrifid" groups, in particular the Ophiderinae (Catocalinae in North American lists), and Hypeninae. The lower half of the frons is scaled in all the "trifid" groups, the Sarrothripinae, and the Arctiidae and is assumed to be the primitive state.

16. Corona present: A row of strong spines, a corona (plate A, figure 12), is present along the apex of the male valve in most of the "trifid" Noctuidae. A corona is absent in the primitive state; namely, in the Eustrotiinae, Condicinae, Plusiinae, and the Amphipyrinae. The character must be interpreted with caution because the corona can be lost, particularly in small species, or those species with highly modified valvae. The apex of the valve may be massively spiny and the corona obscured by the general cover of large spines in parts of the Noctuinae and Agaristinae (e.g., the genus *Apamea* Ochsenheimer. The corona is irregularly developed in the Stiriinae and may represent an intermediate condition.

SUMMARY OF CHANGES. This phylogeny presents several significant changes from the classical view of the "trifid" Noctuidae.

1. The Eustrotiinae and Condicinae are placed as the most primitive of the "trifid" noctuid groups. Indeed the Eustrotiinae might be treated as "quadrifids." The Condicinae consist, in part, of the genera *Condica* Walker (*Platysenta* Grote of the *Check List*, Franclemont & Todd, 1983), *Ogdoconta* Butler, *Draudtia* Barnes & Benjamin, and *Stibaera* Walker. Its total composition is not known, particularly in the New and Old World tropics. In the *Check List* these genera are placed in the Amphipyrinae. The basal placement of the Condicinae results from the presence of three setae in the SV group on larval abdominal segment one, the absence of a free pleurite between the tegumen and vinculum, and the absence of a corona. The hindwing venation is strongly "quadrifid." The condition of the larval spinneret cannot be determined because the apex of the spinneret is absent and the remainder of the spinneret is reduced to a small stub. Another possible placement, if the larval setal character is ignored, is next to the Amphipyrinae.

2. The Plusiinae are treated as "trifid" noctuids. The Plusiinae have paired hair pencils in the male eighth sternum. The hindwing venation is strongly "quadrifid," and a corona is absent. Basal hair pencils are absent. The condition of the SV group on the first larval abdominal segment can be two or three setae. A free pleurite is present.

3. The subfamily Amphipyrinae is restricted to a relatively small group of genera and has two tribes. The Amphipyrini contain *Amphipyra* Ochsenheimer and an undetermined number of Old World genera. The Phosphilini contain the North American genera *Phosphila* Hübner and *Helioscota* Grote and an undetermined number of Neotropical genera. The Amphipyrinae of the *Check List* represent a polyphyletic assemblage of tribes, most of which are placed in the Noctuinae of this phylogeny.

4. The Stiriinae are treated as a subfamily and not a tribe within the Amphipyrinae of the *Check List*.

5. The Eriopinae consist almost entirely of *Callopistria* Hübner, a large genus occurring primarily in the tropical and subtropical regions of the world and restricted, so far as known, to ferns as larval foodplants. A small number of related genera exists in the Old and New World tropics, and the exact limits of *Callopistria* itself are not clear. The males of *Callopistria* have a number of unique features including modifications to the antennae and the genitalia. Basal hair pencils are present, but distinctive. Each hair pencil is divided into two sub-pencils, and the apodeme attached to it is very long and articulated with the caudal margin of sternum 1+2. I was initially inclined to place *Callopistria* as a tribe of the Amphipyrinae, as defined in this fascicle, but strong basal hair pencils are present in some of the species. The larvae fall in Crumb's "larvae confluentae" group. The strong modifications found in *Callopistria* make it difficult to place this group with any certitude. The subfamily may represent only an extreme modification of some lineage within the Noctuinae. Some unplaced genera of the New World tropics may belong here, e.g., *Cropia* Walker, *Speocropia* Hampson, *Paratrachea* Hampson, and *Heterochroma* Guenée.

6. The old subfamily Acontiinae of Franclemont & Todd (1983) breaks up into at least three subfamilies, the Acontiinae, Bagisarinae, and Eustrotiinae. Some genera in the old definition of the subfamily belong to "quadrifid" lines of the Noctuidae. The Acontiinae, as here used, refers only to the tribe Acontiini of the *Check List* (Franclemont & Todd, 1983). The subfamily Bagisarinae appears to be a "quadrifid" group on the basis of other characters and is not discussed. The classical definition of the Acontiinae in the Hampsonian system (Hampson, 1903) was that the larvae were missing some prolegs (the number variable and the larva unknown for most groups) and that the adults were small and "quadrifid." The Acontiinae, as here defined, have the following: M_2 is well developed in the hindwing; a free pleurite is present; double, eversible hair-pencils are present in the male eighth abdominal sternum; a corona is present; and two SV setae are on abdominal segment one of the larva. The larvae fall within the "larvae liberae" grouping of Crumb.

7. The Acronictinae, Pantheinae, and Bryophilinae are united as a clade. In some classifications, for example the Hampsonian system, the subfamily name Acronictinae has been used for the groups included under the old subfamily name Amphipyrinae in the Hodges et al. (1983) Check List. The position of these three subfamilies and their relationships with each other are not clear. The three have been united because of the secondary hairs on the larvae and the "larvae liberae" condition. A few species of *Acronicta* have basal hair pencils as well. I have an unsupported feeling that the Pantheinae and Acronictinae are not closely related despite the similarity of their larvae. Of possible incidental interest, some species of *Acronicta* have spinules on the integumen of the larva of the same form and structure as those found in the Heliothinae. The significance of this observation, if any, is unknown.

8. The subfamily Psaphidinae is created to con-

tain the old tribe Psaphidini (of the old subfamily Cuculliinae) and a number of other genera scattered in different places and subfamilies in the *Check List,* but mostly in the old subfamily Amphipyrinae. Psaphidinae are based largely on a general similarity in the male genitalia; however, the male genitalia are greatly simplified, and the Psaphidinae possibly are not entirely a natural grouping.

9. The Cuculliinae have been greatly restricted and consist almost entirely of the genus *Cucullia* Schrank. The other elements of the old Cuculliinae have been moved elsewhere in the phylogeny (most to the Noctuinae as tribes Xylenini and Antitypini). The genus *Oncocnemis* Lederer and its relatives are now placed in the Oncocnemidinae. Some genera are merged with other genera in the Amphipyrinae of the *Check List* and placed in the Psaphidinae.

10. The largest change in the classification is the amalgamation of many lines in the Noctuinae. The Noctuinae as here presented include the Noctuinae, Hadeninae (less the Glotullini, e.g., *Xanthopastis* Hübner), and most of the old subfamily Amphipyrinae (the tribe Apameini, part of the old tribe Amphipyrini, and part of the "Unassociated genera") in the sense of the *Check List* (Franclemont & Todd, 1983). The tribal structure of the Noctuinae and all of the North American genera belonging to it remain to be determined.

11. The placement of the genus *Magusa* Walker is undetermined. The larva falls into the "larva liberae" group.

12. The placement of the tribe Glotullini, which includes the North American species *Xanthopastis timais* Cramer, is uncertain. This tribe is a small group of largely tropical genera primarily restricted to plant species of the families Amaryllidaceae and Liliaceae. The members of this tribe apparently differ from almost all other noctuid larvae in having two subventral setae on all three thoracic segments, whereas two subventral setae on the prothorax and one on the mesothorax and metathorax are standard for the rest of the Noctuidae. This character state is found in the Arctiidae, some species of Plusiinae, and the genus *Earias* Hübner of the Chloephorinae. The tribe traditionally has been included in the old subfamily Hadeninae because of hair on the eyes. For the moment I place it as a separate tribe within the Noctuinae.

13. The species *Diphthera festiva* (Fabricius) has a very problematical placement. It currently is placed in the "quadrifids" and most characters of the adult support that position. For example there is no free pleurite between the vinculum and tegumen of the male genitalia, and a corona is absent from the valve. The front of the head is completely scaled, whereas in most "quadrifids" (but not all) the ventral half of the frons is naked. A frontal tubercle is present, a structure normally associated with the "trifid" noctuids. The larva is more of a problem. There are only two SV setae on the first abdominal segment, and the larva falls into Crumb's "larva liberae" group. The placement of this genus, therefore, remains unknown.

SUBFAMILY AND TRIBAL NAMES. This section gives references to, or descriptions of, the subfamily and tribal names used in this introduction. The primary purpose of this section is to validate the new subfamily and tribal names.

SUBFAMILY
Eustrotiinae Grote.

Eustrotiinae Grote, 1882, *New Check List of North American Moths,* p. 37.
Type genus: *Eustrotia* Hübner, [1821].

SUBFAMILY
Condicinae Poole, NEW SUBFAMILY.

Condicinae Poole.
Type genus: *Condica* Walker, 1856.

The following description lists characters typical of *Condica* and related genera although the characters are not necessarily unique to the subfamily. The subfamily is divided into two tribes, the Condicini and the Leuconytini. The description applies specifically to the Condicini. The features characterizing the Leuconyctini are discussed under that tribal name. Head: male and female antennae simple in all known species of subfamily; all of front scaled, ventral half not bare; eye naked; eye not lashed from either antennal base or posterior margin of eye; palpus relatively large and long, usually closely scaled, particularly third segment, and usually nearly appressed to front of head. Forewing venation: accessory cell present; veins R_3 and R_4 stalked about ⅕ distance between apex of accessory cell and outer margin of wing; veins R_{3+4} and R_5 connate at apex of accessory cell; venation otherwise typical of family. Hindwing venation: M_2 strong, although slightly weaker than M_1 and M_3; cross veins mdc and ldc nearly perpendicular to axis of wing; inception of M_2 closer to M_3 than to M_1; M_3 and CuA_1 connate at position of cross vein ldc; hindwing venation otherwise as for family. External tympanic structure: alula present, moderate in development, with extensive long, white hairs, appearing larger than it actually is; tympanal membrane laterally elongate, arcuate, nodular sclerite well developed, unmodified; apicolateral corner of first tergum membranous; first abdominal tergum separated by depression into cephalic and caudal halves; hood post-spiracular, quadrate, well developed, projecting cephalad and partially covering tympanal cavity; a triangular, partially sclerotized bulla in dorsocaudal margin of intersegmental membrane of abdominal segment 1; triangular bulla separated by deep depression from hood, depression with elongate, sclerotized plate (plate visible only in cleared material); lower margin of intersegmental membrane of abdominal segment 1 slightly sclerotized, forming a weak lip over abdominal sternum 1+2; tympanal groove strong: second abdominal tergum with a slight hint of a lateral, linear depression following tympanal groove. Male abdomen: basal hair pencils and associated Stobbe's glands absent; abdominal sternum 1+2 quadrate, appearing as if hair pencils never existed; eighth abdominal sternum with two lateral sclerotized regions and two strong apodemes from anterolateral corners of sclerotized regions; central, double hair pencils sometimes present as two weak groups of elongate hairs (*Ogdoconta* Butler) or reduced to a weak line of hairs (*Condica*). Male genitalia: uncus elongate, usually hairy dorsally, and with an apical tooth; uncus with an ovate bulla on either side of base; uncus not freely articulated with tegumen, movement apparently highly restricted by lateral bullae; tegumen various, but without lateral shoulders (as in Noctuinae); free pleurite totally absent, tegumen directly articulating with vinculum; valve usually elongate, irregularly covered with long hairs; corona absent; digitus absent; valve not distinctly separated into cucullus and sacculus (although valve divided in *Ogdoconta*; clasper usually present, but normally small, oriented perpendicularly to long axis of valve; juxta a single plate, but lateral edges usually better developed than central region; saccal region of vinculum usually broad, not pointed; vesica of aedoeagus variable. Female genitalia: ovipositor lobes generally unmodified; ostium well sclerotized, variably shaped, but most commonly triangular; corpus bursae variable in shape; ductus bursae variable, but most commonly membranous; appendex bursae distinctive, weak, membranous, globular, connected to corpus bursae and ductus bursae by a thin, membranous stalk; corpus bursae without internal signa. Larva: usually feeding on herbaceous plants; posterior extremity of venter of abdominal segment 10 of "larvae confluentae" type; projecting portion of spinneret completely absent; crochets uniordinal; seta SV2 present on abdominal segment 1.

The following are the most diagnostic of these characters: 1) The direct articulation of the vinculum and tegumen of the male genitalia. 2) The bullae to either side of the base of the male uncus. 3) The absence of a corona and digitus and the perpendicular orientation of the clasper of the valve. 4) The presence of lateral apodemes on the eighth abdominal sternum. 5) The bulbous female appendix bursae connected to the rest of the bursae by a long, thin neck. 6) The presence of seta SV2 on the first larval abdominal segment and the absence of a projecting spinneret.

The three North American genera of the Condicini are *Condica, Ogdoconta,* and *Homophoberia* Morrison.

TRIBE
Leuconyctini Poole, NEW TRIBE.

Leuconyctini Poole.
Type genus: *Leuconycta* Hampson, 1908.

The tribe Leuconyctini is the sister group of the Condicini in the Condicinae. The most important synapomorphy of the tribe is the structure of the male valve: either one or both valves are massively swollen and inflated; a series of flattened scales usually occurs on the outer side of each valve; the clasper consists of a group of fused, socketed setae, the entire structure curled like a watch spring contained in a pocket of the valve and covered with a heavily sclerotized outer lip. Other differences between the Leuconyctini and the Condicini are the following: 1) Lateral bullae are absent at the base of the uncus. 2) Double hair pencils are usually strongly developed in the central region of the eighth abdominal sternum. 3) The bursae consists of equally important appendix and corpus bursae. The appendix bursae is better developed than in the Condicini, and the stalking is not as apparent.

The Leuconyctini are almost exclusively a New World group. The following North American genera are placed in the tribe; *Leuconycta, Bryocodia* Hampson, REVISED STATUS, *Crambodes,* Guenée *Diastema,* Guenée, *Micrathetis,* Hampson, and *Fotella* Grote. Several Neotropical species also belong to this tribe, but currently are misplaced in other genera. The single known Old World representative of the tribe is the genus *Kenrickodes* Viette from Madagascar.

SUBFAMILY
Plusiinae Boisduval.

Plusidi Boisduval, 1829, *Europaeorum Lepidopterorum Index Methodicus,* p. 91.
Type genus: *Plusia* Ochsenheimer, 1816.

SUBFAMILY
Acontiinae Guenée.

Acontidi Guenée, 1841, *Ann. Soc. Ent. France,* **10**: 219.
Type genus: *Acontia* Ochsenheimer, 1816.

SUBFAMILY
Amphipyrinae Guenée.

Amphipyridi Guenée, 1838, *Ann. Soc. Ent. France,* **7**: 107.
Type genus: *Amphipyra* Ochsenheimer 1816.

TRIBE
Phosphilini Poole, NEW TRIBE.

Phosphilini Poole.
Type genus: *Phosphila* Hübner, 1818.

The Phosphilini appear to be a New World group exclusively. The tribe is represented in North America by the two genera *Phosphila* and *Helioscota* Grote, REVISED STATUS. It is best developed in the New World tropics; however, the generic limits and the included species of the Neotropical fauna are undetermined. The most distinctive features of the Phosphilini are the use of Smilacaceae as larval foodplants and the often gaudy and conspicuous coloration of the larvae. The male genitalia are simplified, particularly the valves. The corpus bursae is elongate and membranous. The ostium is membranous except for a thin, lateral, sclerotized bar. A small white spot is present on the inner margin of the forewing following the basal line. The forewing has a superficially ovate appearance.

Head: antenna simple in most species, although pectinate in males of the Neotropical genus *Eucropia* Hampson; no lashes from base of antenna or posterior margin of eye; eye naked; ocelli present; front completely scaled; frontal processes absent; proboscis present. Wing venation: Forewing with accessory cell present although commonly irregular in shape and variable among specimens of same species; R_4 and R_5 stalked near apex of accessory cell; R_3 single, but almost adnate with R_4 and R_5 near junction of these two veins; hindwing with M_2 clearly present, although inner half considerably stronger than outer half; M_2 located about midway between M_1 and M_3, although slightly closer to base of M_3; M_3 and CuA_1 connate near their junction with cross vein ldc; venation otherwise as in Noctuidae as a whole. Legs: tibial spines absent; prothoracic tibial claw absent; tarsal segments scaled both dorsally and ventrally; otherwise variable. External tympanic region: alula present, but not strong; tympanal membrane elongate ovate with strong nodular sclerite; hood post-spiracular, weak, membranous with only outer edge distinctly sclerotized; a large, kidney-shaped sclerite and bulla in intersegmental membrane following hood and curving around hood; no distinct dorsocaudal bulla present; abdominal tergum 1 relatively narrow with membranous lateral and apical margins; no lobe at apex of the lateral margins of sclerite; tergum 1 without a definite division into cephalic and caudal halves. Male abdomen: basal abdominal hair pencils and Stobbe's glands absent from abdominal sternum 1+2; sternum 1+2 quadrate, appearing as if hair pencils never existed; eighth sternum with strong anterolateral apodemes; eighth sternum without double, eversible hair pencils, represented at most by a thin line of long hairs or entirely absent; eighth sternum U-shaped or delta-shaped; eighth tergum X-shaped with caudal part of "X" larger than anterior part. Male genitalia: uncus long, slightly hairy, generally unmodified; base of uncus an ovate plate resting in apical membrane of tegumen; tegumen without projections near base of uncus; free pleurite absent (orientation of the free pleurite and tegumen refer to the genitalia *in situ* in the moth, not on a slide of a dissected specimen); apex of vinculum bifurcate, one branch broadly articulated with dorsal-outer margin of tegumen, other branch expanded into a plate and fused with ventral surface of tegumen; saccus rounded; valve very simple, slightly membranous and expanded; corona absent, apex of valve irregularly covered with hairlike setae; valve not divided into distinct saccular and cucullar regions; digitus absent; clasper absent; basal coremata absent; juxta variable; aedoeagal vesica (in examined species) ovate, without diverticula or spines in *Phosphila* but with diverticula and spine patches in *Helioscota.* Female genitalia: ovipositor lobes unmodified; ostium and ostial region membranous except for a thin, lateral sclerotized bar; ductus bursae long, membranous; appendix bursae absent; corpus bursae very long, slender, without signum; ductus seminalis arising from caudal end of corpus bursae. Larva: feed on members of the plant family Smilacaceae, in North America on *Smilax*; posterior extremity of venter of abdominal segment 10 of "larvae confluentae" type; spinneret present, with an open, apical pore; all prolegs present, although slightly reduced on A3 and A4; crochets uniordinal; seta SV2 absent on Al.

SUBFAMILY
Stiriinae Grote.

Stiriinae Grote, 1882, *New Check List of North American Moths,* p. 35.
Type genus: *Stiria* Grote, 1874.

SUBFAMILY
Eriopinae Herrich-Schäffer.

Eriopodides Herrich-Schäffer, 1845, *Systematische Bearbeitung der Schmetterlinge von Europa,* **2**: 379.
Type genus: *Eriopus* Treitschke, 1825.

NOTE—*Eriopus* Treitschke, 1825 is a junior objective synonym of *Callopistria* Hübner [1821].

SUBFAMILY
Psaphidinae Grote, NEW STATUS.

Psaphidiini Grote, 1895, *Abhandl. Naturwissenschaftliche Vereins Bremen,* **14**: 70.
Type genus: *Psaphida* Walker, 1865.

NOTE—The tribal name Psaphidiini was based on the misspelling *Psaphidia* of *Psaphida* Walker and, therefore, is correctly emended to Psaphidini. Although the family-group name Dicopinae (Grote, 1883) has priority, the rules of zoological nomenclature (Article 40b) state that if a genus-group name is recognized as a junior synonym of another genus-group name before 1961 and the younger family-group name based on the senior synonym has gained general acceptance, then the younger family-group name is to be used. In this case Grote (1895) recognized *Dicopis* Grote, 1874 as a junior synonym of *Psaphida* Walker, 1865, and the tribal name Psaphidini has gained general acceptance; hence, Psaphidinae is the valid name.

The subfamily Psaphidinae is discussed and described in detail on page 162.

TRIBE
Nocloini Poole, NEW TRIBE.

Nocloini Poole.
Type genus: *Nocloa* Smith, 1906.

The Nocloini are a distinctive tribe of genera and species best identified by the structure of the prothoracic leg. The tarsal segments are very short and wide, and the tarsal claws almost always are greatly enlarged. Other characters of the tribe are very similar to those of the Psaphidini except the spining of the tarsal segments occurs in three rows (although very irregular rows) in contrast to the irregular spining found in the Psaphidini. The species of Nocloini occur almost exclusively in the southwestern United States and Mexico.

The technical description follows. General appearance and maculation extremely variable. Head: antenna of both male and female simple (except in *Petalumaria* Buckett and Bauer), scaled dorsally, finely ciliate ventrally; frons with well-developed frontal process usually present; frontal process normally a circular ring with a central prominence, but may be reduced or absent, a solidly filled ring, or a three-toothed structure; ocelli present; eye naked; no lashes from either base of antenna or posterior margin of eye; palpus weak, third segment barely reaching lower margin of frontal process; proboscis normal. Venation: forewing with R_3, R_4, and R_5 arising from a common stalk about ⅕ distance from apex of accessory cell to apex of wing; R_3 and R_4 stalked about ⅖ distance from end of accessory cell to apex of wing; forewing otherwise as in family; M_2 of the hindwing weak, not as strong as in Triocnemidini or Psaphidini, arising from near middle of cell. Prothoracic leg: tibial claw and spining absent; epiphysis large; tarsal segments very fore-shortened, each about as long as wide; tarsal spines in three irregular tows, except on fifth segment, large; tarsal claw and related structures all greatly enlarged; leg otherwise variable. Mesothoracic and metathoracic legs: tibial spining absent; tarsal segments not fore-shortened as in prothoracic tibia, each with three very irregular rows of spines; tarsal claws and associated structures not enlarged as in prothoracic leg. External tympanic structure: alula large, appearing swollen; tympanic membrane ovate-elongate, with small nodular sclerite dividing membrane into usual opaque cephalad and transparent caudal halves; hood moderate, located primarily dorsal to spiracle, but curving caudally around spiracle and enclosing it in deep depression; ventral margin of intersegmental membrane projecting forward, usually more exaggerated in males than females because of presence of basal hair pencils in some genera. Male abdomen: basal hair pencils present in most genera, but reduced or absent in others; Stobbe's gland and pocket for hair pencils present in genera with hair pencils; seventh tergum more strongly sclerotized caudally, but whole tergum not particularly enlarged; eighth tergum with sclerotization V-shaped, opening in the caudal direction, arms fused at cephalic margin of tergum and base with two cephalad pointing projections; eighth sternum U-shaped with weak anterior fringe of elongate scales. Male genitalia: uncus usually short and stubby with long hairs on outer surface; tegumen broad, short; free pleurite long, thin, closely joined to outer margin of tegumen; saccus broadly V-shaped; valve simple; no distinct corona, but apical ⅕ of valve with irregular mass of long, coarse spines; clasper oriented perpendicularly to long axis of valve, sometimes greatly reduced in size; digitus absent; valve not divided into distinct cucullur and saccular regions; vesica of aedoeagus variable. Female genitalia: ovipositor lobes variable; ostium a simple, membranous, circular opening, without sclerotized structures; ductus bursae with caudal half usually membranous, cephalic half with sclerotized internal ridges; corpus bursae ovate to trapezoidal, with internal plicae; ductus seminalis arising from cephalic end of corpus bursae. Larva and foodplants: unknown.

The following North American genera are included in the Nocloini: *Ruacodes* Hampson, *Paramiana* Barnes and Benjamin, *Euamiana* Barnes and Benjamin, *Petalumaria* Buckett and Bauer, *Oslaria* Dyar, *Nocloa* Smith, *Phaioecia* Dyar, *Lythrodes* Smith, *Redingtonia* Barnes and McDunnough, and *Sexserrata* Barnes and Benjamin.

TRIBE

Triocnemidini Poole, NEW TRIBE.

Triocnemidini Poole.

Type genus: *Triocnemis* Grote 1881.

NOTE—The genitive singular of *cnemis* is *cnemidis*. Therefore, the family-group name is correctly formed as Triocnemidini.

A number of disparate genera previously scattered throughout the old subfamilies Amphipyrinae and Cuculliinae are included in Triocnemidini. They possess a number of features that are synapomorphies for the tribe or intermediate between the Psaphidini and Nocloini. The genera included in the tribe are dissimilar superficially. All of them share two apparently derived characters to some extent. The structure of the abdominal spiracles, in particular, is unique: The distal half of the circle of the spiracle is greatly reduced or absent in most genera of the Triocnemidini, and the flange plate of the internal lever is greatly expanded, whereas the spiracle normally is circular or ovate with an internal lever and flange plate. The distinctness of this character varies among genera. The derived condition is strongest in *Poliocnemis* Benjamin and *Oxycnemis* Grote; the spiracle is almost normal in *Triocnemis* Grote and two related, but undescribed, genera. Secondly, M_2 of the hindwing is strong. The hindwing medial vein is present, at least at the base of the wing, and the mdc cross-vein is triangular with the apex of the triangle intercepting the medial vein or the fold representing the medial vein. This hindwing venation is found throughout the tribe, but to varying degrees. The medial vein is not as strong, and the cross-vein mdc is straight, not triangular, in the Psaphidini, Nocloini, and most other "trifid" noctuids. The triangular condition of mdc is strongest in *Poliocnemis,* and *Oxycnemis,* but *Triocnemis* and its two undescribed sister genera tend toward the more conventional, straight mdc. The interpretation of this venation character is difficult. The presence of a medial vein and an indentation in mdc also is found in the Pantheinae and the genus *Acronicta* Ochsenheimer. No other group has yet been found with this condition.

In addition to these putative synapomorphies for the tribe, the genera seem to show a combination of characters intermediate between the Nocloini and Psaphidini. The corona is absent; the clasper usually is small to absent; and the uncus often is swollen. The spines of the vesica of the aedoeagus often are situated in a rugose area, a feature characteristic of the Nocloini. The tibial claw is strong in all of the genera, arising from a strong plate consisting of the end of the tibia. A strong tibial claw also is found in the Psaphidini. The Triocnemidini may represent a relict group of genera of the ancestral group that gave rise to the Psaphidinae. The so-called synapomorphies defining the tribe are not always consistent, however, and the tribe may not be a monophyletic group.

The following combination of characters is used to define the tribe: 1) The antenna of males and females is simple or weakly bipectinate. 2) The front tends to be slightly bulging but is otherwise unmodified. 3) The dorsum of the metathorax usually has two lateral scale tufts and a distal, median scale tuft although the tufts may be degenerate. 4) The prothoracic tibia has a strong claw, and the claw is always associated with a knifelike ridge along the distal margin of the tibia. A secondary, distal claw is always absent but sometimes a small accessory claw exists just anterad of the main claw. 5) The cross-vein mdc in the hindwing usually is indented into a developed medial vein, more so in *Poliocnemis* and *Oxycnemis,* less so in *Triocnemis* and its relatives. M_2 of the hindwing generally is well developed and not degenerate as in most "trifid" noctuids. 6) The distal wall of the abdominal spiracles tends to degenerate. This tendency is strongest in *Poliocnemis* and *Oxycnemis,* less so in *Triocnemis* and its relatives. 7) The valve is simple and has a simple clasper. The corona is absent and is replaced by dense setae except in *Crimona* Smith. 8) The uncus usually is swollen and always has an apical tooth. 9) The vesica is elongate and generally has two roughly parallel rows of spines. An apical spine is always absent. 10) Basal hair pencils and accessory hair pencils are always absent except in *Crimona.* 11) The eighth abdominal tergum of the male is always enlarged, sclerotized, and its distal edge is weakly bilobed. 12) The ovipositor lobes usually are unmodified, but in a few genera there are various forms of modification. 13) The corpus bursae usually is ovate. The appendix bursae is always absent, and the ductus seminalis arises from a distal projection of the corpus bursae. 14) When signa are present, they occur in the form of an elongate ridge, shaped like a row of bricks. 15) The external tympanic region consists of a large hood followed, in the middle of the intersegmental membrane, by an ovate or rectangular, large bulla.

The following North American genera are placed in the Triocnemidini: *Poliocnemis* Benjamin, *Triocnemis* Grote, *Crimona* Smith, and *Oxycnemis* Grote. In addition species representing undescribed genera currently are placed in *Oncocnemis* and *Oxycnemis.*

TRIBE
Feraliini Poole, NEW TRIBE.

Feraliini Poole.
Type genus: *Feralia* Grote, 1874.

NOTE—The tribe Feraliini was first proposed by Franclemont and Todd (1983) but was not described. No earlier description has been found. Therefore, the tribal name Feraliini is treated as new.

The tribe Feraliini is described in detail later in this fascicle.

TRIBE
Azeniini Poole, NEW TRIBE.

Azeniini Poole.
Type genus: *Azenia* Grote 1882.

The tribe Azeniini consists primarily of the genus *Azenia* (=*Stiriodes* Hampson, NEW SYNONYMY), a group of small moths occurring mostly in the southwestern United States and Mexico. The species of *Azenia* can generally be recognized by their chisel-shaped or tridentate frontal process, the absence of a tibial claw, the absence of spines on the legs, and the extreme reduction or absence of the hood in the external tympanic region. The other genera included in this tribe are placed here mostly as a matter of convenience and because of their overall similarity to *Azenia* not because they have a group of shared, derived characters. These genera possess most of the characters of *Azenia,* but each included genus has its own unique features. Many of the similarities could be reductions. The genera in Azeniini, besides *Azenia,* could be treated as *incertae sedus.*

The formal description is based on *Azenia implora* Grote. Head: antenna of male and female with short, cylindrical segments, scaled above, heavily ciliate below; each segment with two parallel long setae, longer toward apex of antenna than at base; eyes large, naked; lashes absent; ocelli present; apex and front of head flattened, strongly sloping toward frontal prominence and covered with elongate, toothed scales; scale pattern of apex of head and front without defined ridges of scales; frontal prominence a flattened plate, outer edge either smooth and convex or tridentate; frontal prominence separated from clypeal plate by deep depression, depression filled with elongate scales; proboscis normal; palpus slightly elongate, paralleling frontal process, third segment approximately 5 × length of third. Venation: forewing; with accessory cell R_5 single, arising from posterior margin of accessory cell; R_3 and R_4 arising from apex of accessory cell, stalked very close to apex of accessory cell; venation otherwise as in family; hindwing with M_2 quite strong, discal cell cross vein triangularly indented at M_2; M_2 arising from cell closer to M_3 than to M_1; veins M_3 and CuA_1 joined at cross vein ldc. Thorax: dorsum covered with slightly elongate, dentate scales; no tufts present; wing scales rectangular, dentate; no specialized scaling present dorsally or ventrally. Prothoracic leg: femur and tibia closely scaled, without fringes of hair; tibia about ⅔ × length of femur; tibia without tarsal claw or spines; tarsal segments equal to length of tibia; first four tarsal segments with usual three rows of spines, but spines weak and deeply buried in scales; tarsal claw with small tooth. Mesothoracic and metathoracic legs: essentially as for prothoracic leg except for usual differences. External tympanic region: first abdominal tergum trapezoidal, much narrower anteriad than posteriad; anterior ⅔ lightly sclerotized, separated by cleft from distal ⅓; hood very small to nearly absent, located dorsal to spiracle; intersegmental membrane rectangular, without bullae. Male abdomen: basal abdominal hair pencils and Stobbe's glands absent; supporting arm for hair pencil apparently present, fused to body of sternum 1+2 to give basal apodemes of sternum a bifid appearance; eighth sternum U-shaped with weak, linear hair row; last tergum chalice shaped, proximal arms short, weakly developed. Male genitalia: uncus long, not swollen, with small apical tooth; hairs weak and few; free pleurite present, unmodified; valve elongate; clasper small, usually capitate; basal process of sacculus absent; corona present but sometimes weak; juxta a single quadrate plate; saccus large, rectangular with apical point; vesica elongate with a single mass of spines, which enlarge toward apex of vesica. Female genitalia: ovipositor lobes small, weakly sclerotized; a few elongate spines present, particularly as a basal row with each spine in a large, light circle; apophyses posteriores short; ovipositor neck short, unmodified; eighth sternum narrow, with large unsclerotized region distal to apophysis anterioris; ostium with lateral lobes, lamella antevaginalis cup shaped; ductus bursae with distal ¼ unsclerotized, anterior ¾ heavily rugose; corpus bursae elongate, blending imperceptibly with ductus bursae; appendix bursae a small distal lobe on corpus bursae. Larva and foodplants: unknown.

The following genera are included in the Azeniini: *Azenia* Grote, *Aleptinoides* Barnes and McDunnough, *Tristyla* Smith, and two undescribed genera for misplaced species.

TRIBE

Grotellini Poole, NEW TRIBE.

Grotellini Poole.

Type genus: *Grotella* Harvey, 1875.

The Grotellini are an anomalous group consisting primarily of the genus *Grotella*. The adults are commonly, although not always, small and white with rounded forewings and a pattern of black dots or lines. The most striking feature of the tribe is the presence of strong spines on the prothoracic and mesothoracic legs. Isolated spines are sometimes found on the metathoracic leg as well. The prothoracic tibial spines are arranged in rows along both outer margins. The apical spines are strongest, in particular the apical spine on the inner side of the tibia. This inner, apical spine may be homologous with a tibial claw. The mesothoracic tibial spines are arranged in a single row along the inner margin of the tibia. This tibial spining pattern is basically the same as found in the tribe Agrotini of the Noctuinae and some genera in the Heliothinae. The genera and species of the Grotellini are all found in the southwestern United States and the drier areas of Mexico. The tribe is placed in the Psaphidinae largely for convenience. There is no compelling evidence to place it in the Psaphidinae, but none to suggest any alternative placement. The larvae and foodplants are unknown.

The following description is based on the type genus *Grotella*. Head: male and female antennae simple, entirely naked, not scaled dorsally; lashes from base of antenna and rear margin of eye absent; eye naked; vertex and front of head closely scaled; a circular frontal ring with a central prominence usually present; frontal process may be reduced or modified into a triangular protuberance; clypeal shelf strong, projecting; pilifers strong; palpi weak, with third segment angled outward; proboscis present, strong. Venation: forewing accessory cell present; R_{3+4} and R_5 arising jointly from apex of accessory cell, in rare cases slightly stalked; R_2 arising from near middle of anterior margin of accessory cell; hindwing with M_2 moderate, arising closer to M_3 than to M_1; M_3 arising from base of cross vein ldc; vein CuA_1 arising just basal to ldc; venation otherwise as in family. Prothoracic leg: both femur and tibia closely scaled, without either fringes or tufts of scales or hairs; tibia with row of spines along each margin and a stronger spine at each distal apex; apical spine along inner margin very strong, possibly homologous to tibial claw; tarsal segments scaled dorsally and ventrally; tarsal spines strong, arranged in three rows. Mesothoracic leg: femur and tibia closely scaled, without fringes or tufts of scales or hairs; tibia with a single row of strong spines from the basal to apex. Metathoracic leg: tibia without spines, or at most, one or two small spines. External tympanic region: alula present but weak; tympanal membrane reduced, with strong nodular sclerite; anterior portion of tympanal membrane very small, triangular; distal ½ of tympanic membrane ovate; hood very reduced, almost absent, a thin, weak postspiracular flap; no distinct bulla in intersegmental membrane following hood, region quadrate, membranous; first abdominal tergum quadrate, not divided into distinct caudal and cephalic regions. Male abdomen: terga and sterna typically heavily melanized; basal hair pencils and Stobbe's glands absent; shape of sternum 1+2 not clearly indicative of potential for basal hair pencils; sclerotization of eighth sternum U-shaped, with a weak basal row of hairlike scales; eighth tergum weakly sclerotized with a broad base and two caudal arms. Male genitalia: uncus elongate, slightly swollen at apex, and with an apical tooth; tegumen unmodified, without lateral shoulders; free pleurite present, broadly articulated with dorsolateral margin of tegumen; free pleurite and vinculum partially fused; saccus elongate, triangular; valve elongate, simple; corona absent; digitus absent; valve not distinctly separated into saccular and cucullar regions; clasper small, sometimes absent, oriented perpendicularly to long axis of valve; juxta a single, simple plate, but otherwise variously modified; transtilla a well-sclerotized strap; vesica variable among species but typically elongate-globular with irregular groups of spines. Female genitalia: very variable among species; ovipositor lobes unmodified in some species, elongate and pointed in others; membrane of ovipositor neck some times with heavy spines; apophyses anteriores and posteriores variable in length, short in some, elongate in others; ostium variable, but most typically membranous and continuous with intersegmental membrane; bursa variable, but most typically with an elongate corpus bursae and an arcuate, globular appendix bursae; no signa in the corpus bursae. Larva: unknown.

The following genera are placed in the Grotellini; *Grotella* Harvey, *Neogrotella* Barnes and Benjamin, *Hemigrotella* Barnes and McDunnough, *Podagra* Smith, *Grotellaforma* Barnes and Benjamin, and *Homolagoa* Barnes and McDunnough.

SUBFAMILY

Heliothinae Boisduval.

Heliothidi Boisduval, 1829, *Europaeorum Lepidopterorum Index Methodicus*, p. 94.
Type genus: *Heliothis* Ochsenheimer, 1816.

SUBFAMILY

Agaristinae Boisduval.

Agaristides Boisduval, 1833, *Faune Entomologique de Madagascar, Bourbon et Maurice*, p. 70.
Type genus: *Agarista* Leach, 1814.

SUBFAMILY

Acronictinae Smith and Dyar.

Acronyctini Smith and Dyar, 1898, *Proc. U. S. Natl. Mus.*, **21**: 6.
Type genus: *Acronicta* Ochsenheimer, 1816.

NOTE—The original spelling Acronyctinae is based on the unjustified emendation *Acronycta* for *Acronicta*. Therefore, the family group name is correctly formed as Acronictinae. Apatelinae Grote, 1883 with type genus *Apatele* Hübner, 1822 is an older available family-group name for this subfamily. The generic name *Apatele* Hübner, 1822 is a junior objective synonym of *Acronicta* Ochsenheimer, 1816. Smith and Dyar treated *Apatele* as a junior synonym of *Acronicta*. This substitution of Aronictinae for Apatelinae occurred before 1961 and has gained general acceptance in the sense of the Code of Zoological Nomenclature (article 40b).

SUBFAMILY

Pantheinae Smith and Dyar.

Pantheini Smith and Dyar, 1898, *Proc. U. S. Natl. Mus.*, **21**: 6.
Type genus: *Panthea* Hübner, [1820].

SUBFAMILY

Bryophilinae Guenée.

Bryophilidae Guenée, 1852, *Histoire Naturelle des Insectes, Lépidoptères*, **5**: 21.
Type genus: *Bryophila* Treitschke, 1825.

SUBFAMILY

Cuculliinae Herrich-Schäffer.

Cuculidae Herrich-Schäffer, 1845, *Systematische Bearbeitung der Schmetterlinge von Europa*, **2**: 306.
Type genus: *Cucullia* Schrank, 1802.

SUBFAMILY

Oncocnemidinae Forbes and Franclemont.

Oncocnemidini Forbes and Franclemont, 1954, *in* Forbes, Lepidoptera of New York and neighboring states. Part 3, Noctuidae. Cornell Univ. Agric. Experiment Station, Mem., **329**: 122.
Type species: *Oncocnemis* Lederer, 1853.

NOTE—An earlier available name for this subfamily (Cleophanini Grote, 1890) must be considered. Cleophanini was proposed in a check list, and the only two included names in *Cleophana* Boisduval are considered to be species of *Cucullia* in the Cuculliinae. Grote's biological concept of the family-group name was incorrect. However, the name is available with *Cleophana* as the type genus. *Cleophana* is a junior subjective synonym of the Palearctic genus *Calophasia* Stephens. Careful reading of the code leads me to treat Cleophanini as having a misidentified type genus. Under the rules of zoological nomenclature, the case should be referred to the Commission. In my opinion Cleophanini should be suppressed and Oncocnemidinae recognized as the valid name for the subfamily.

SUBFAMILY

Noctuinae Latreille.

Noctuidae Latreille, 1809, *Genera Crustaceorum et Insectorum,* **4**: 191.
Type genus: *Noctua* Linnaeus, 1758.

SUBFAMILY

Cuculliinae Herrich-Schäffer.

Cuculidae Herrich-Schäffer, 1845, *Systematische Bearbeitung der Schmetterlinge von Europa,* **2**: 306.
Type genus: *Cucullia* Schrank, 1802.

The definition of the subfamily *Cuculliinae* employed in this work represents a radical departure from its use in the past. Traditionally, following the classification of the Noctuidae used by Hampson in his series *Catalogue of the Lepidoptera Phalaenae in the British Museum,* the subfamily Cuculliinae (Hampson, 1906) was used for any "trifid" noctuid with lashes from the base of the antenna and without hairy eyes or spines on any of the thoracic legs. His concept of the "subfamily" is polyphyletic, representing an assemblage of elements of the "trifid" noctuids now dispersed among the Cuculliinae, Psaphidinae, Oncocnemidinae, and the tribes Xylenini, Cleocerini, and Antitypini of the Noctuinae.

The subfamily Cuculliinae consists almost entirely of the genus *Cucullia.* A few other genera have been grouped with *Cucullia,* but not with much confidence. The sister group of the Cuculliinae may be the Oncocnemidinae although I make this statement without much justification. Other possible sister groups to the Cuculliinae occur outside of North America. One group is a collection of Palearctic genera without a formal name but including *Metopoceras* Guenée and *Calophasia* Stephens. A group of species and genera (*Desertullia* Köhler, *Neuquenioa* Rosas Costa, and *Metacullia* Köhler) in southern South America, primarily Patagonia, seems to bear some relationship with the Cuculliinae and the Oncocnemidinae. The generic limits in this Patagonian group are not well defined, and most of the species are incorrectly placed in Palearctic genera. Mario Gentili (Instituto Patagonico de Ciencias Naturales) is trying to determine the generic limits and species composition of the genera within this group.

The species of *Cucullia* have a distinctive superficial appearance and are readily recognized. Both the superficial and external features of the genus are discussed under the generic heading. The formal description of the subfamily follows. It serves for the generic description of *Cucullia* because only the genus *Cucullia* is undoubtedly placed in the subfamily. Comments on the larvae and their life histories are discussed under the generic heading.

Head: antenna of male simple, laminate, or serrate, but most commonly simple; antennal base with tuft of scales; frons smooth or roughened, sometimes with a conspicuous raised ridge or triangular region above the clypeus; clypeus usually slightly projecting; pilifer strong, sclerotized; proboscis well developed; gena trapezoidal, well developed; ocelli present; eye lashed below antenna and from posterior margin of eye. Prothoracic leg: tibia usually somewhat shortened, with or without a tibial claw; tibial spines absent; tarsal spines of all three legs arranged in three equal rows. Wing venation: forewing usually elongate; venation distorted by elongation of forewing; accessory cell present; R_1 arising from main radial vein, R_2 arising from ⅔ length of accessory cell; R_3, R_4, and R_5 from apex of accessory cell, R_3 and R_4 stalked about half way to apex. Thorax: with long, hairlike scales; patagia capable of elevation into a pointed hood. External tympanic region (text figure 2 *g*): hood not greatly developed and its edge flush with abdominal sternum 1+2; tergo-pleural groove continued by a sclerotized invagination of the second abdominal membrane separating the sternum and tergum. Internal tympanum (text figure 2 *f*) pocket IV large and triangular. Abdomen: weak tufts on first three or four abdominal segments; male abdomen with or without basal hair pencils and Stobbe's Glands; eighth sternum with sclerotization U-shaped and with or without a linear, internal setal patch. Male genitalia (text figure 2 *e, h*) valve usually elongate; clasper and basal process of the sacculus normally present; clasper oriented perpendicular to long axis of valve; rudimentary digitus present in scattered North American and Palearctic species; free pleurite present, rounded, articulation with tegumen simple; penicillus absent; corona usually present on apex of valve but sometimes reduced or absent; vesica of aedoeagus with 1–3 strong spines borne on diverticula, sometimes minutely spined and/or with a straplike sclerotized area near the base of the vesica. Female genitalia (text figure 3 *a*): ovipositor lobes usually unmodified; ductus bursae of variable length, usually with sclerotized ridges; corpus bursae apparently absent; ductus seminalis arising from bottom of appendix bursae.

GENUS
Cucullia Schrank.

Cucullia Schrank, 1802, *Fauna Boica,* **2**(2): 157.
Type species: *Phalaena Noctua umbratica* Linnaeus, 1758.
Subsequent designation by Curtis, 1824, *British Entomology,* **1**: 45.

Euderaea Hübner, [1821] 1816, *Verzeichniss Bekannter Schmettlinge* [sic], p. 245.
Type species: *Noctua asteris* [Denis and Schiffermüller], 1775. Subsequent designation by Stephens, 1835, *Illustrations of British Entomology,* **4**: 390.

Eucalimia Hübner, [1821] 1816, *Verzeichniss Bekannter Schmettlinge* [sic], p. 245.
Type species: *Noctua gnaphalii* Hübner, [1813]. Subsequent designation by Hampson, 1906, *Catalogue of the Lepidoptera Phalaenae in the British Museum,* **6**: 14.

Callaenia Hübner, [1821] 1816, *Verzeichniss Bekannter Schmettlinge* [sic], p. 246.
Type species: *Phalaena Noctua umbratica* Linnaeus, 1758. Subsequent designation by Hampson, 1906, *Catalogue of the Lepidoptera Phalaenae in the British Museum,* **6**: 14.

Empusa Hübner, [1821] 1816, *Verzeichniss Bekannter Schmettlinge* [sic], p. 247. NEW SYNONYMY.
Type species: *Noctua lactea* Fabricius, 1787. Monotypy.
NOTE—A junior homonym of *Empusa* Illiger, 1798, Orthoptera.

Argyritis Hübner, [1821] 1816, *Verzeichniss Bekannter Schmettlinge* [sic], p. 247.
Type species: *Noctua artemisiae* [Denis and Schiffermüller], 1775. Subsequent designation by Hampson, 1906, *Catalogue of the Lepidoptera Phalaenae in the British Museum,* **6**: 14.

Tribunophora Hübner, 1822, *Systematisch-alphabetischen Verzeichniss Europäischer Schmetterlinge,* pp. 20, 37.
Type species: *Phalaena Noctua umbratica* Linnaeus, 1758. Subsequent designation by Nye, 1964, *Bull. Zool. Nomenclature,* **21**: 75.

Lophia Sodoffsky, 1837, *Bull. Soc. Naturalistes Moscou,* **10**(6): 88.
Type species: *Phalaena Noctua umbratica* Linnaeus, 1758. Subsequent designation for *Cucullia* Schrank 1802.
NOTE—*Lophia* Sodoffsky, 1837 was proposed as an objective replacement name for *Cucullia* Schrank and has the same type species.

Lathosea Grote, 1881, *Bull. U. S. Geol. Geog. Surv. Terr.,* **6**: 270. NEW SYNONYMY.
Type species: *Lathosea pulla* Grote, 1881. Monotypy.

Nycterophaeta Smith, 1882 (February), *Bull. Brooklyn Ent. Soc.,* **4**: 45. NEW SYNONYMY.
Type species: *Nycterophaeta magdalena* Hulst, 1882. Monotypy.

Epinyctis Grote, 1882 (April), Can. Ent., **14**: 75. NEW SYNONYMY.
Type species: *Epinyctis notatella* Grote, 1882. Monotypy.

Rancora Smith, 1892, *Ent. News,* **3**: 253. REVISED SYNONYMY.
Type species: *Rancora strigata* Smith, 1892. Monotypy.

Copicucullia Smith, 1894, *Trans. Amer. Ent. Soc.,* **21**: 84. NEW SYNONYMY.
Type species: *Cleophana eulepis* Grote, 1876. Subsequent designation by Hampson, 1906, *Catalogue of the Lepidoptera Phalaenae in the British Museum,* **6**: 9.

Cheligalea Hampson, 1906, *Catalogue of the Lepidoptera Phalaenae in the British Museum,* **6**: 13.

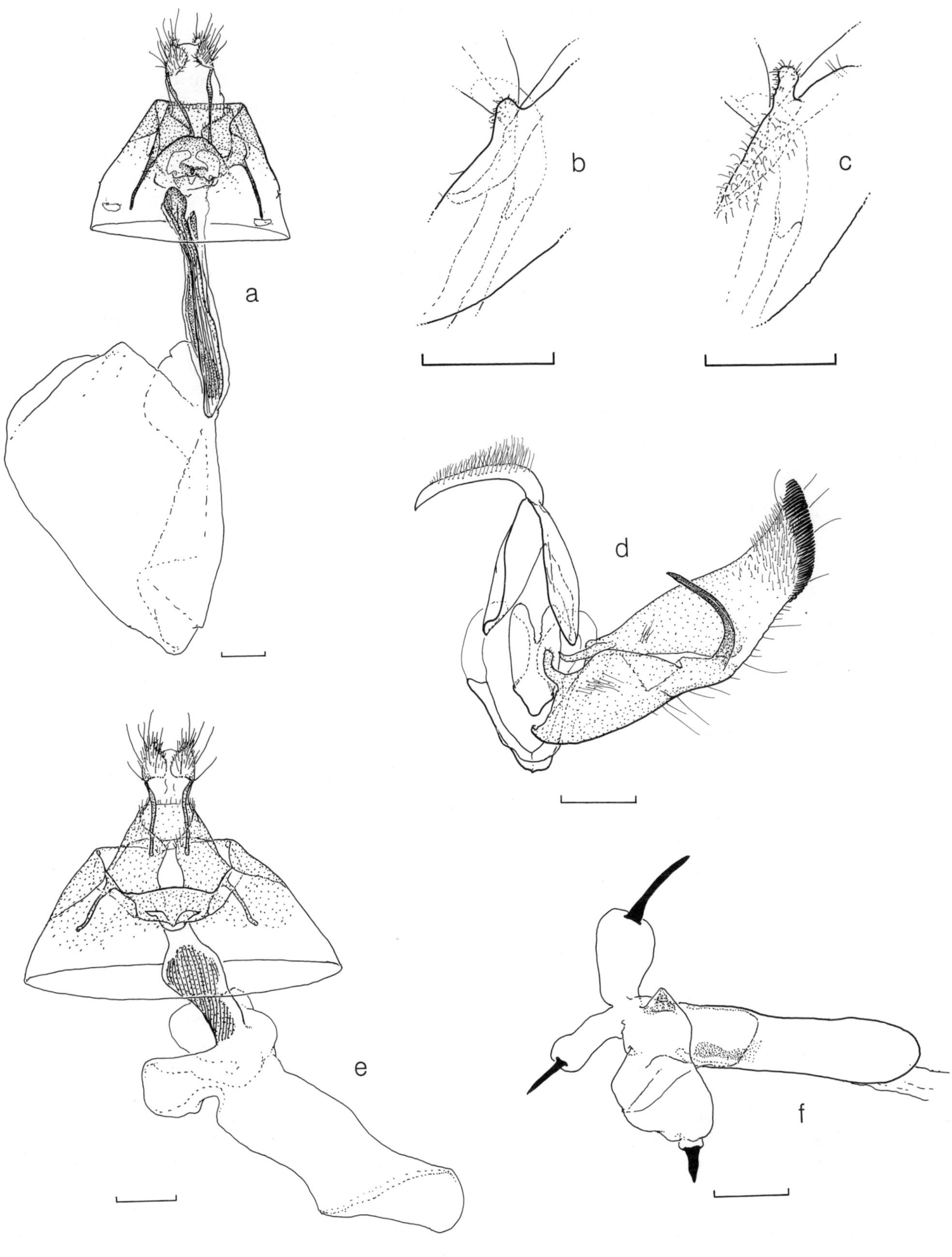

FIGURE 3: MALE AND FEMALE GENTALIA OF *CUCULLIA* SPECIES

a. Female genitalia of *Cucullia omissa*. *b.* Base of the male valve in *Cucullia florea*. *c.* Base of the male valve in *Cucullia postera*. *d.* Male genitalia of *Cucullia styx*. *e.* Female genitalia of *Cucullia styx*. *f.* Aedoeagus of *Cucullia styx*. Scale = 1 mm.

Type species: *Cucullia scopariae* Dorfmeister, 1853. Original designation.

Argyrogalea Hampson, 1906, *Catalogue of the Lepidoptera Phalaenae in the British Museum,* **6**: 81.
Type species: *Phalaena argentea* Hufnagel, 1766. Original designation.

Argyromata Hampson, 1906, *Catalogue of the Lepidoptera Phalaenae in the British Museum,* **6**: 82. NEW SYNONYMY.
Type species: *Phalaena Noctua splendida* Stoll, 1782. Original designation.

Empusada Hampson, 1906, *Catalogue of the Lepidoptera Phalaenae in the British Museum,* **6**: 85. NEW SYNONYMY.
Type species: *Noctua lactea* Fabricius, 1787. Monotypy for *Empusa* Hübner, [1821].
NOTE—proposed as an objective replacement name for *Empusa* Hübner [1821], a junior homonym of *Empusa* Illiger, 1798, Orthoptera.

Pseudocopicucullia Dumont, 1928, *Encyclopedie Entomologie, Series B,* **3**: 20. NEW SYNONYMY.
Type species: *Cucullia syrtana* Mabille, 1888. Original designation.

Pseudonycterophaeta Berio, 1934, *Boll. Soc. Ent. Italiana,* **66**: 124. NEW SYNONYMY.
Type species: *Pseudonycterophaeta melanoglossa* Berio, 1934. Original designation.

Calocucullia Ronkay and Ronkay, 1967, *Acta Zool. Hungarica,* **33**: 463.
Type species: *Cucullia celsiae* Herrich-Schäffer, 1850. Original designation.
NOTE—Described as a subgenus of *Cucullia* Schrank 1802.

Shargacucullia Ronkay and Ronkay, 1992, *Acta Zool. Hungarica,* **38**: 345. NEW SYNONYMY.
Type species: *Cucullia lychnitis* Rambur 1833. Original designation.

Cucullia is a large genus of approximately 250 species found almost exclusively in the temperate regions of the Northern Hemisphere with the exception of a few species found in Africa. Despite the diversity of the genus, its species are almost always characterized by a long forewing with a pointed apex. The patagia are capable of being raised into a pointed hat-shaped "hood;" the generic name means a hood or cowl. Not all specimens, however, have the hood raised when the moth dies. The male genitalia are characterized by the presence of one to three large spines in the vesica of the aedoeagus, each borne on a diverticulum (or in some case two of them borne on a bifid diverticulum) and the female genitalia by the emergence of the ductus seminalis from the anterior end (bottom in the figures) of the bursa except in a few Palearctic species. If my interpretation of the structure of the bursa normally found in the "trifid" Noctuidae is correct, the bursa consists solely of the appendix bursae, and the corpus bursae is absent or represented by a outpocketing of the bursae near its junction with the ductus bursae. To interpret the vesica of the aedoeagus, it usually is helpful to take notes on the vesica just after it has been inflated, but before the genitalia are mounted on a slide. The vesica often is highly three dimensional and commonly shaped like a caltrop. Necessarily, therefore, the vesica becomes distorted when put under a coverslip on a slide.

In North America several species-groups have been given generic status on the basis of a specific character such as the presence of a tibial claw or serrate antennae. In my opinion the similarities far outweigh the differences, and the differences are no greater than those found in the Palearctic species of the genus, although superficially they usually are more apparent. Given the evanescent nature of tibial claws (e.g., *cuculliioides* Barnes and Benjamin, formerly in *Copicucullia,* lacks a tibial claw, but its nearest relative has one), I have opted for the broader generic concept.

All of the larvae of *Cucullia* in North America feed on the flowers, seeds, and leaves of various composites, apparently preferring the flowers. However, several Palearctic species in one species-group (e.g., *Cucullia verbasci* (Linnaeus) and *Cucullia prenanthis* Boisduval) are found on species in the family Scrophulariaceae. The larvae are often brilliantly colored or colored to match the flowers of their foodplant. The larvae pupate in an earthen cell in the autumn; the adults emerge in early to late spring and apparently are long lived, flying throughout the summer. Many specimens have extensive deposits of pollen on the proboscis, and the adults may be important pollinators. The description of the subfamily is based on the genus *Cucullia*; thus, no formal description of the genus is necessary.

Initially, I had hoped to derive a phylogeny for the genus; however, it soon became clear that by far the greatest diversity of the genus occurs in the Old World with few, possibly only two, lineages occurring in the New World. A revision of the entire genus would have been necessary to construct a reasonable phylogeny, a task beyond the scope of this series. The North American species fall into well-defined groups, but the relationships between these groups or with Old World groups is not always clear. One of the North American groups, the *asteris* group, is well developed in the Palearctic. The species *Cucullia intermedia* is a close sister species of the Palearctic species *Cucullia lucifuga* [Denis and Schiffermüller]. The possible common ancestors of the rest of the North American species are not clear. Some speculations on relationships of the North American species with Old World groups are given under each group, but these are intended as only suggestions and should not be taken too seriously. I have a strong suspicion that most of the North American species of *Cucullia* except for the species in the *asteris* group and *Cucullia intermedia* may have arisen from a single, common ancestor despite the sometimes great superficial differences between the groups. One of the primary problems I have faced in trying to make meaningful conclusions about phylogenetic trends in *Cucullia* has been my failure to identify a sister group to the genus. I have included *Opsigalea* Hampson and *Emariannia* Benjamin, all genera of the southwestern United States or tropical America, in the subfamily Cuculliinae because of similarities in superficial appearance and tympanic structure. I have not found any Old World genus I would be comfortable associating with *Cucullia* even though the greatest diversity of species and groups occurs in the Old World. Logically, I might expect *Cucullia*'s nearest relative to be found in the Old World.

Two generic names have recently been proposed for Palearctic species-groups of *Cucullia*; *Calocucullia* Ronkay and Ronkay and *Shargacucullia* Ronkay and Ronkay. I agree that these two generic concepts contain monophyletic and distinctive species groups. The groups may even be basal to the genus. However I am strongly against the subdivision of already monophyletic genera. Subdivision of existing and well established genera should only be proposed if the generic concept is polyphyletic or paraphyletic. The ever finer splitting of genera is a serious problem in systematics causing instability in the nomenclature and leading to the ultimate absurdity of a multitude of closely related genera each with a single species.

KEY TO THE SPECIES OF *CUCULLIA*

1. Foretibia with a distinct claw 2
– Foretibia without a distinct claw 11

2. Forewing silvery white .... *luna* p. 64
– Forewing not silvery white 3

3. Foretibial claw projecting both anteriorly and posteriorly 4
– Foretibial claw with a single, anterior projection 6

4. Forewing orbicular and reniform marks obscure or absent; only prominent mark a black dash from lower part of outer margin to postmedial line and continued along inner margin to base of wing .... *eulepis* p. 61
– Forewing orbicular and reniform marks well defined; maculation more complex than described above 5

5. Forewing postmedial line moderately well defined, lower half marked with dark gray .... *basipuncta* p. 62
– Forewing postmedial line very weak or absent .... *mcdunnoughi* p. 63

6. Forewing orbicular and reniform marks absent or very obscure 7
– Forewing orbicular and reniform marks present although not always conspicuous 8

7. Forewing with a dark line along inner margin from base to postmedial line; dark dash present from lower third of outer margin to postmedial line; wing length from base to apex about 18 mm; Colorado, New Mexico, southern California .... *astigma* p. 60
– All forewing maculation obscure; no dark line along inner margin; no conspicuous dash from lower third of outer margin to postmedial line; wing length about 16 mm; Mohave Desert .. .. *heinrichi* p. 59

8. Hindwing with dark outer margin contrasting with light basal ⅔; reniform and orbicular marks defined by conspicuous narrow black lines; general coloration dark gray .... *antipoda* p. 55
– Hindwing without a color contrast between outer margin and light inner ⅔; reniform and orbicular marks sometimes defined by narrow black lines, but not as conspicuously; general coloration not dark gray 9

9. Front with a distinct triangular or heart-shaped process set off from remainder of front; hindwing of males usually white; apparently restricted to Mohave Desert .... *incresa* p. 58

— Front with a ridge but without a distinct triangular or heart-shaped frontal process; males usually with hindwing not white; primarily not Mohave Desert 10

10. Length of forewing from base to apex less than or equal to 16 mm; forewing coloration uniform dark gray; orbicular and reniform marks defined by thin black lines (plate 2, figure 18) .... *eurekae* p. 56

— Length of forewing from base to apex usually, but not always, greater than 16 mm; forewing coloration usually not uniform dark gray; orbicular and reniform marks normally defined by thick black lines .... *antipoda* p. 55

11. Palpus studded with thin brown hairs, appearing bristly; each antennal segment ventrally expanded into a laterally projecting plate of variable development, each plate with distinct hairs; adults early spring fliers 12

— Palpus not studded with thin brown hairs, not appearing bristly; antennal segments simple ... 17

12. Proboscis greatly reduced, covered and hidden by palpi 13

— Proboscis not reduced, not covered nor hidden by palpi 14

13. Rusty-brown patches in subterminal area of forewing; costa of forewing noticeably lighter than rest of wing .... *dammersi* p. 70

— Patches in subterminal area of forewing black; costa of forewing not noticeably lighter than rest of wing .... *pulla* p. 69

14. Forewing smooth dark gray; forewing with markings usually strong, thick, black lines, particularly basal streak, lower half of antemedial line, and streaks from outer margin in subterminal area (plate 3, figure 3); dark streaks in subterminal area of forewing usually haloed with brown; Pacific Northwest and Great Basin region ....*strigata* p. 67

— Forewing a lighter gray; forewing with markings not as strong as above (plate 3, figure 1); streaks from outer margin not haloed with brown; not from western British Columbia or Great Basin region 15

15. Middle California to southern California, Arizona 16

— Western British Columbia to Saskatchewan south in the Rockies through western Montana into Colorado .... *albida* p. 68

16. Hindwing of male pure white, but with patches of dark scaling projecting into the middle of the wing (plate 3, figure 1) .... *comstocki* p. 67

— Hindwing of male pure white, but without patches of dark scales projecting into middle of wing .... *serraticornis* p. 65

17. Length of forewing 15 mm or less; maculation as in plate 2, figure 31 .... *cucullioides* p. 62

— Length of forewing greater than 15 mm; maculation not as above 18

18. Posterior third of postmedial line a conspicuous light line edged with black on both sides (plate 1, figure 24); orbicular and reniform marks usually clearly present; forewing not gray, but if gray with a distinct violet or blue tint ... 19

— Posterior third of postmedial line not conspicuous although often vaguely present (plate 1, figure 26); orbicular and reniform marks usually obscure or absent; forewing some shade of gray without a blue or violet tint 28

19. A dark lateral streak through the middle of the forewing along the cubital vein (plate 1, figure 24) 20

— No dark lateral streak through the middle of the forewing along the cubital vein (plate 1, figure 18) 21

20. Light brown or brown; eastern North America .... *convexipennis* p. 42

— Violet brown or black brown; southwestern United States .... *oribac* p. 41

21. Forewing lilac or violet with a distinct fawn-brown patch between the reniform mark and outer margin .... *lilacina* p. 40

— Forewing not as above 22

22. Forewing light gray with costa and inner mar-

gin significantly darker than rest of wing; orbicular and reniform marks with dotlike patches of black; hindwing dirty brown, base neither white nor contrasting with outer margin; Great Basin area of United States .... *similaris* p. 36

– Forewing not light gray, or if gray, with brown tints on costa and inner margin not contrasting with rest of wing; orbicular and reniform marks without conspicuous black dots; hindwing various; locality various 23

23. Hindwing pure white or with a white base and dark-brown outer margin 24

– Hindwing dull brown and if base lighter than outer margin, not conspicuously so 26

24. Basal area of forewing with a large area of light brown; hindwing of male with broad band of dark brown (plate 1, figure 3); forewing with median and middle subterminal areas contrastingly lighter than rest of wing; western United States primarily Great Basin area ... . *montanae* p. 35

– Basal area of forewing without large patches of light brown (plate 1, figure 1); hindwing of male (but not of female) without broad band of dark brown; forewing with median and middle subterminal areas not significantly lighter than rest of forewing; eastern North America or extreme southwestern United States 25

25. Male and female forewings sexually dimorphic, female much darker than male; dark shade on posterobasal margin of postmedial line of male triangular (plate 1, figure 20); southern Arizona .... *eucaena* p. 40

– Male and female forewings not sexually dimorphic; females not significantly darker than males; dark shade on posterobasal margin of postmedial line of male ovate or square; eastern North America, apparently not in southern Arizona .... *asteroides* p. 34

26. Forewing a smooth dark violet gray with costa and inner margin significantly darker than rest of wing; forewing with brown shade near reniform mark blending with rest of dark costal margin (plate 1, figure 7) .... *omissa* p. 36

– Forewing not smooth violet gray, or if smooth, not violet gray nor with costa and inner margin contrasting with rest of wing; forewing generally without smooth-brown patch near reniform mark, or if patch present, not blending with dark costal margin 27

27. Rusty-brown streaks along costa of forewing from reniform mark to outer margin and in angle between outer and inner margins; median area and middle of subterminal area usually lighter than rest of wing, with an almost shiny appearance (plate 1, figure 9) .... *postera* p. 38

– Forewing without rusty-brown streaks along costa from reniform mark to outer margin and angle between outer and inner margins; median area and middle of subterminal area not significantly lighter than rest of wing, not appearing shiny (plate 1, figure 13); color dark violet gray or gray .... *florea* and *obscurior* pp. 37, 39

28. Hindwing uniformly brown, base not significantly lighter than rest of wing (plate 2, figure 6) .... *intermedia* p. 43

– Hindwing not uniformly brown, either almost all white or with base of wing significantly lighter than rest of wing .. *speyeri* species complex p. 44

asteris GROUP

The *asteris* group is based on the Palearctic species *Cucullia asteris* [Denis and Schiffermüller]. This species is closely related to the four North American species *Cucullia postera, florea, obscurior,* and *omissa. Cucullia asteris* seems most closely allied to *omissa* in both superficial appearance and male genitalia. A number of Old World species belong to the *asteris* group, e.g., *asteris, amota* Alpheraky, *elongata* Butler, and *ledereri* Staudinger; this list is not intended to be complete.

Species in the *asteris* group are most likely to be confused superficially with some species in the *convexipennis* group. Both groups are characterized by their pattern and coloration, simple antennae, smooth frons, lack of a prothoracic tibial claw, presence of basal hair pencils and Stobbe's glands in the male, and a long, linear valve in the male genitalia. The basal process and clasper of the male valve are relatively shorter and the clasper is farther toward the apex of the valve in the *asteris* group than in the *convexipennis* group. I consider the apical placement of the clasper to be a derived condition versus a primitive placement near the middle of the valve. The vesica of the aedoeagus almost always has a single spinelike cornutus. I consider one cornutus to be a derived state, because two or three cornuti is the condition in most other species of *Cucullia* and because rare individuals will sport a small, second cornutus. The denticulate patch of the vesica found in the species of the *convexipennis* group is absent in the *asteris* group. A strap-shaped process at the base of the vesica originates from the end of the aedoeagal shaft and terminates in a variably developed conical cornutus. A very strong, denticulate patch is in the membrane between the two halves of the tegumen. This patch is particularly well developed in the *asteris* group although spinose patches occur elsewhere in the genus. The lamella antevaginalis is produced caudally as a joined pair of fleshy, spiculate lobes covering the ostium. The sclerotized portion of the ductus bursae projects as a knob near its junction with the caudal, unsclerotized part. The ductus bursae is 8–12 × longer than wide. The seventh sternum of the female abdomen has a median lobe distally and has lateral patches of modified scales, which are difficult to remove during dissection. These scales possibly are pheremonal in function. The components of the forewing maculation generally contrast to some degree and are never entirely absent or totally gray as in the *speyeri* or *lucifuga* groups. All of the species of the group feed on species of the Asteraceae.

Cucullia asteroides Guenée (Cucullie de l'Aster (f.), Fr.)

PL. 1, FIGS. 1, 2; PL. B, FIGS. 9, 10 (♂ gen.); PL. G. FIG. 6 (♀ gen.); TEXT FIG. 4 (map) (RWH 10200).

Cucullia asteroides Guenée, 1852, *in* Boisduval and Guenée, *Histoire Naturelle des Insectes. Species General des Lépidoptères,* **6**: 133.
Type locality: North America. [USNM]

NOTE—Guenée described *asteroides* from six specimens in the collections of Doubleday, Boisduval, and Guenée. One of these specimens, a male, is in the USNM. It is designated and has been labeled the LECTOTYPE.

Cucullia asteroides is characterized by a white hindwing, narrowly margined with dirty brown in males and broadly margined in females. This white base of the hindwing separates *asteroides* from *florea, omissa,* and *postera,* the three species most likely to be confused with it. Some specimens of *postera* have a lighter hindwing base, but it is dirty white. The orbicular and reniform marks of *postera* are tinged with rusty brown and ringed with dull white but are more obscure in *asteroides* and not distinctly white

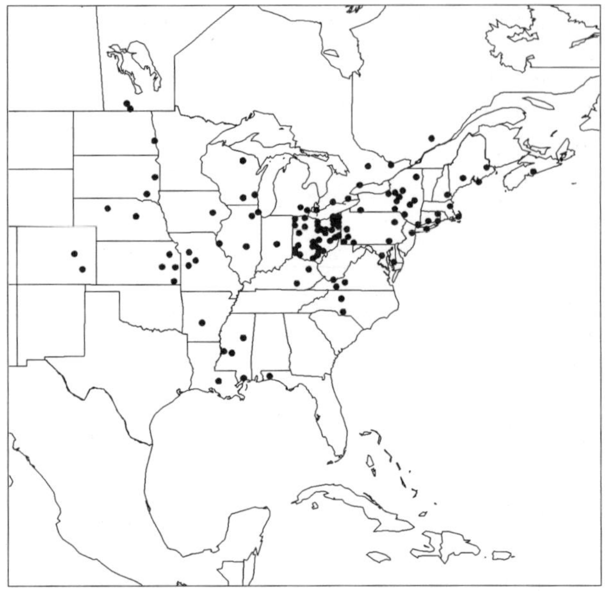

FIGURE 4: DISTRIBUTION OF SPECIMENS EXAMINED OF *CUCULLIA ASTEROIDES*

ringed. *Asteroides* is narrowly sympatric with *montanae* in the western Great Plains. The orbicular and reniform marks of the forewing of *montanae* are far more distinct than in *asteroides,* and the maculation is crisper and generally more distinct. The forewing of *montanae* lacks the violet hue of *asteroides.* There is a single large cornutus in the vesica of the male genitalia. This cornutus is borne on a globular, single diverticulum. The diverticulum is neither elongate, as in most other species in the group, nor double as it is in *similaris* and the *florea* complex of species. A second small diverticulum exists below the larger one. This small diverticulum corresponds to the small diverticulum found in *montanae* but does not bear a small, second cornutus as it does in *montanae.* Wing length: mean = 22.77 mm, standard deviation = 1.05 mm, n = 10.

The larva has been described by several persons including Crumb (1956) and Dethier (1944). The general larval pattern is green or yellow green with longitudinal dark-brown to black lines. The dorsal region is green with two wide, dark bands on either side. Two thin lines are dorsal to the spiracle and ventral to the subdorsal band. A pair of lines runs above and below the spiracle, and this double band is filled with black on the spiracles. A thin, subspiracular line is ventral to the spiracles. The spiracles have black rims with light centers. The prolegs are green and unmarked. The head is lightly reticulated with red-brown patches. The larvae feed on the flowers of various species of *Aster* and *Solidago,* pupating in the autumn in the ground and emerging next summer as adults.

Asteroides occurs throughout most of the eastern United States, west to the foothills of the Rockies, and northward into southern Ontario and Quebec. It is apparently more common in the north than in the south, although this may be a sampling artifact. I have seen specimens from Florida and Louisiana. *Cucullia asteroides* is most characteristic of disturbed, eastern deciduous forest and is replaced in the mixed deciduous-conifer forests of northern New England and Canada by *omissa, florea,* and *postera,* apparently without much precision. The range of *asteroides* narrowly overlaps that of *montanae* in the short grass prairies of the Dakotas and Montana. There is no significant geographical variation in this species. Adults have been collected from May to September and possibly have more than a single generation in the southern United States.

Cucullia montanae Grote

PL. 1, FIGS. 3, 4; PL. B, FIGS. 11, 12 (♂ gen.); PL. G, FIG. 7 (♀ gen.); TEXT FIG. 5 (map) (RWH 10201).

Cucullia montanae Grote, 1882, *Can. Ent.,* **14**: 175.
Type locality: [USA], Montana. [USNM]

Cucullia montana Hampson, 1906, *Catalogue of the Lepidoptera Phalaenae in the British Museum,* **6**: 63.

NOTE—An unjustified emendation of *montanae* Grote and therefore an available name with the same type and type locality as *montanae.*

Cucullia montanae is a distinctive species superficially closest to *similaris* and *asteroides.* The forewing of *similaris* is bluish gray and lacks the brown tints of *montanae.* The forewing of *asteroides* has a far more suffused coloration than *montanae,* and the costa above the reniform and orbicular marks is not distinctly black as in *montanae.* The clasper of the male genitalia is larger than in *similaris* and the band at the base of the vesica of the aedoeagus is diffuse, not dark and straplike as in *similaris.* The vesica is more highly modified than in other North American species of the group. There are always two cornuti instead of one, a small cornutus on a small diverticulum and a large cornutus corresponding to the single cornutus in the other species of the group. This large cornutus is borne on a single long diverticulum, and the diverticulum does not have another projection as is found in such species as *florea, postera,* and *similaris.* The diverticulum

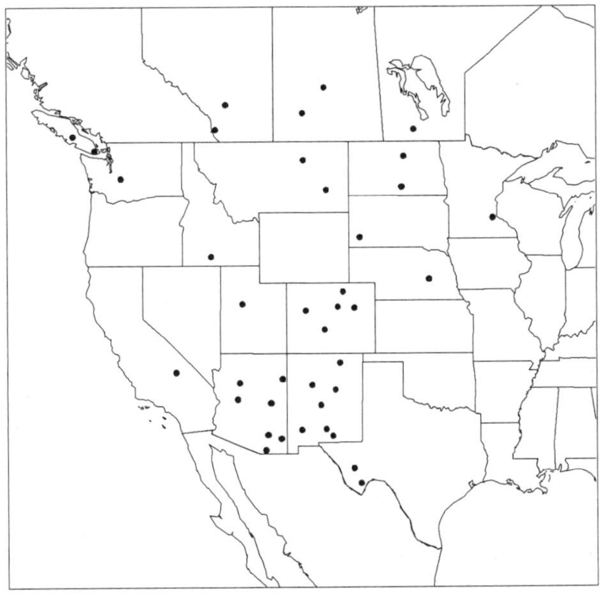

FIGURE 5: DISTRIBUTION OF SPECIMENS EXAMINED OF *CUCULLIA MONTANAE*

bearing the small cornutus is the same as the small diverticulum found in *asteroides,* but in *asteroides* the diverticulum does not bear a cornutus. The diverticulum bearing the large cornutus is long in *montanae,* but globular and not elongate in *asteroides.* Wing length: mean = 22.75 mm, standard deviation = 0.59 mm, n = 10.

Crumb (1956) has described the larva. It is similar to *asteroides*; however, the brown freckles of the posterior margin of the head are distinctly darker than those on the rest of the head. These freckles in *asteroides* are unicolorous over the entire head. The broad subdorsal band in *montanae* is dark only in the central one-half; in *asteroides* the band is completely filled with dark color except at the anterior and posterior margins. Finally, the spiracular lines are not filled with dark black in the region of the spiracle in *montanae,* but are in *asteroides.* Crumb collected his specimens from *Grindelia* (gum plant) (Asteraceae). Cook (1930) records the larva from *Chrysothamnus* (rabbitbush) (Asteraceae). A reared specimen in the U. S. National Museum of Natural History from Wellington, British Columbia was reared from *Grindelia* sp.

Cucullia montanae is widely distributed throughout the western United States, often sympatrically with *similaris.* The scattered distributional data suggest *montanae* is most often found in dry localities at moderate elevations, often as high as 7,000 to 8,000 feet. *Cucullia montanae* occurs in eastern Montana and western South Dakota, north into Al-

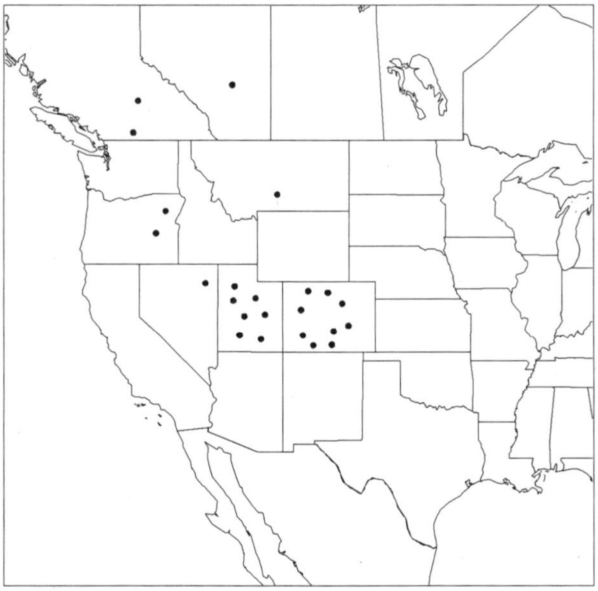

FIGURE 6: DISTRIBUTION OF SPECIMENS EXAMINED OF *CUCULLIA SIMILARIS*

berta and British Columbia, and south through the Rockies into Colorado and Utah. The species is also known from New Mexico, western Texas, and southern Arizona. Adults have been collected from June through August, with most having been collected in July and August.

Cucullia similaris Smith

PL. 1, FIGS. 5, 6; PL. B, FIGS. 13, 14 (♂ gen.); PL. G, FIG. 8 (♀ gen.); TEXT FIG. 6 (map) (RWH 10195).

Cucullia similaris Smith, 1892, *Proc. U. S. Natl. Mus.*, **15**: 41, pl. 2, fig. 3.
Type locality: [USA], Colorado, Platte Canyon. [USNM]
NOTE—The lectotype was designated by Todd (1982).

Cucullia similaris is a species of the Great Basin region occurring in drier locations at moderately high elevations. The forewing is characterized by a medium-gray coloration suffused with light gray, the absence of a brown tint as in *montanae,* and distinct light rings around the orbicular and reniform marks. This species is sympatric with *obscurior* and *montanae,* but its distribution is more restricted than either of them. Superficially, *similaris* strongly resembles *obscurior* where the two species are sympatric. The forewing of *similaris* is strongly marked with light gray, and the reniform and orbicular marks are each clearly marked with a light ring. Sympatric populations of *obscurior,* in contrast, are uniformly dark gray, and the reniform and orbicular marks are obscure. The diverticulum bearing the single large cornutus in the vesica is double, as it is in the *florea* complex of species. There is no small, second cornutus or associated small diverticulum as in *montanae.* A small, lateral diverticulum is near the junction of the ductus seminalis. Wing length: mean = 21.87 mm, standard deviation = 0.88 mm, n = 10.

The larva and foodplants are unknown.

Cucullia similaris is commonest in Utah, central to western Colorado, and eastern Nevada. Two specimens are known from southeastern Washington. Adults have been collected from late June through August.

Cucullia omissa Dod

PL. 1, FIGS. 7, 8; PL. B, FIG. 15; PL. C, FIG. 1 (♂ gen.); TEXT FIG. 2 *e*, *h* (♂ gen.); PL. G, FIG. 9 (♀ gen.); TEXT FIG. 3 *a* (♀ gen.), TEXT FIG. 7 (map) (RWH 10199).

Cucullia omissa Dod, 1916, *Can. Ent.*, **48**: 58, fig. 7.
Type locality: [Canada], Alberta, Calgary. [CNC]
NOTE—Dod states that the types are a male in his collection, now in the CNC, and a female in the Barnes collection, now at the USNM. The male in the CNC labeled "type male" is designated and has been labeled as the LECTOTYPE.

Cucullia omissa is primarily a species of the southern Canadian forests and the Rocky Mountains. Superficially, *omissa* is most similar to *florea.* The forewing's costa and inner margin are usually distinctly darker than the middle of the wing (unlike *florea*), and *omissa* has a diffuse brown shade running from the orbicular mark to the apex of the wing. This mark is weak or absent in *florea.* The overall appearance of *omissa* is smoother and more velvety than in either *florea* or *postera* although this quality is difficult to describe, and the forewing appears broader versus length than in either of them. These differences are best appreciated by examining the specimens on the plates. *Cucullia omissa* is easily separated from *Cucullia postera* by the smooth coloration of the forewing, particularly the absence of distinct, rusty-red patches. *Cucullia omissa* is easily separated from both *florea* and *postera* by the greatly reduced basal process of the sacculus. The base of the hindwing is dirty brown in *omissa,* but dirty white in *postera.* The vesica of the aedoeagus of *omissa* has a single cornutus on a diverticulum. The diverticulum is a single small, round, unmodified structure. There is another small round divertícu-

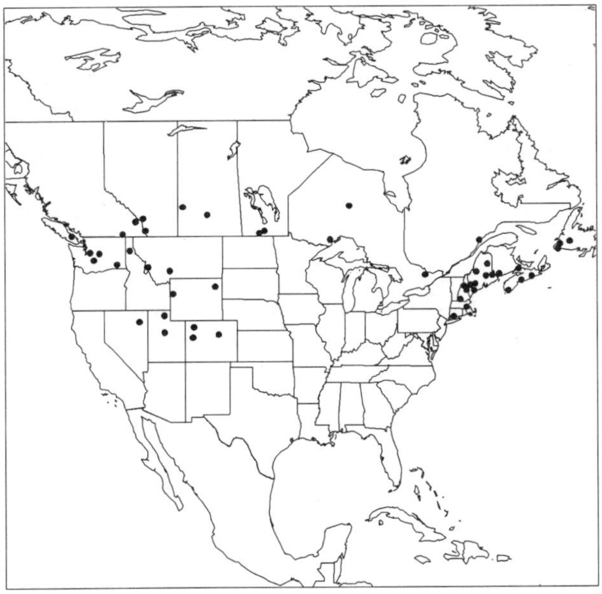

FIGURE 7: DISTRIBUTION OF SPECIMENS EXAMINED OF *CUCULLIA OMISSA*

lum near the origin of the tail of the vesica. In contrast the cornutus in *postera* and *florea* is on a long diverticulum, which is itself part of a double, large diverticulum. There is no small diverticulum from near the origin of the tail of the vesica in *postera,* but a small one exists in *florea.* Wing length: mean = 21.67 mm, standard deviation = 0.80 mm, n = 10.

The early stages have not been described, but there is an adult from Nova Scotia in the U. S. National Museum of Natural History reared from *Aster umbellatus* (Mill.) (Asteraceae) by D. C. Ferguson.

Cucullia omissa ranges from the mountains of New York and New England, throughout the Maritime Provinces of Canada as far north as Newfoundland, west through southern Quebec and Ontario, the southern plains provinces and Montana to British Columbia and Washington. There are specimens of *omissa* from Colorado and Utah, but the forewing is distinctly brighter with a stronger brown shade along the costa. The southern most extension of the range of the species seems to be in northern Arizona along the Mogollon Rim. Adults have been collected in June and July.

Cucullia florea Guenée

PL. 1, FIGS. 13, 14; PL. C, FIGS. 2, 3 (♂ gen.); PL. G, FIG. 10 (♀ gen.); TEXT FIG. 3 *b* (base of valve); TEXT FIG. 8 (map) (RWH 10197).

Cucullia florea Guenée, 1852, *in* Boisduval and Guenée, *Histoire Naturelle des Insectes. Species General des Lépidoptères,* **6**: 134, pl. 7, fig. 9.
Type locality: [USA], New York. [BMNH]

Cucullia indicta Smith, 1904, *Can. Ent.,* **36**: 154.
Type locality: [Canada] Alberta, Calgary, South Fork of Sheep Creek. [CNC]
NOTE—The lectotype was designated by Todd (1982).

The three species, *Cucullia florea, obscurior,* and *postera,* pose a real problem. Ignoring the taxon *obscurior* for the moment, the identification of *postera* and *florea* is reasonably direct. They are relatively easy to separate if series of well-prepared specimens are available. The forewing of *florea* is uniformly gray with only indistinct patches of purple brown along the costa. The forewing of *postera* has distinct rusty red patches along the costa from the orbicular mark toward the apex and in the inner angle between the outer margin and the lower end of the postmedial line. *Cucullia postera* also has a distinct light patch between the reniform mark and the anal dash, and the patch contrasts with the dark costa. This patch is absent or indistinct in *florea.* The forewing of *postera* has a much more striate or blotchy appearance overall than *florea. Cucullia postera* appears to have a slightly shorter forewing than *florea,* although available specimens have been collected too non-randomly to make statistical comparisons. The male genitalia of both species are variable in every aspect, and I have not found any completely consistent characters to separate *postera, florea,* or *obscurior.* The basal process of the sacculus is short and knoblike in *florea* and longer in *postera,* but only on average. The diverticulum bearing the single cornutus in the vesica is a large double structure in both *postera* and *florea,* with one lobe bearing the cornutus, and the other not. The cornutus in *omissa* is borne on a small, single diverticulum. A small diverticulum from near the origin of the tail of the vesica usually exists in *florea*; this diverticulum is absent in *postera* but present in *omissa.* A small diverticulum exists in *postera* just distal to the large double diverticulum bearing the single large cornutus. This diverticulum appears to be absent in *florea.* I have not found any differences in the female genitalia among the three species.

The separation of *florea* and *obscurior* is not so clear. *Cucullia obscurior* has been treated as a "subspecies" of *florea,* and it may yet prove to be. Superficially, *obscurior* is generally lighter gray than

FIGURE 8: DISTRIBUTIONS OF SPECIMENS EXAMINED OF *CUCULLIA FLOREA* (CIRCLES) AND *CUCULLIA OBSCURIOR* (TRIANGLES)

florea. The forewing costa of *obscurior* lacks any hint of the rusty or purple-brown tinge present in *florea*. The ranges of the two species are largely allopatric, suggesting that *obscurior* may be a subspecies of *florea*. However, there are two apparent areas of overlap between *florea* and *obscurior*. The first spot is in Washington in the general vicinity of Mt. Rainier based on reared material collected by Crumb. Two very distinct entities are in this region, one a perfect *florea* type (type-1) with a strong purple tinge and a second light-gray entity (type-2) lacking any hint of purple in the color of the forewing. The second locality is the Provo region of north-central Utah. The two types are distinct and easily separated at this locality. The specimens I have identified as *obscurior* at most other localities are darker than type-2 specimens from these two localities, and considerable variation exists. This situation is particularly true in Colorado and North Dakota. Both states have specimens that cannot be reliably assigned to either species. There are specimens from Bottineau County in northern North Dakota that should be *florea* on biogeographical grounds but superficially are most like *obscurior*. The North Dakota specimens have been marked as *florea* and the specimens from Colorado as *obscurior* on the range map (text figure 8). The diverticulum of the vesica from near the origin of the tail of the vesica seems to be absent in what I am treating as *obscurior*, but present in *florea*. However, I am very reluctant to list this as a firm character. I am treating the two names *florea* and *obscurior* as separate, largely allopatric, species, for the moment, but postulate that a case of incomplete speciation exists. This is only an hypothesis, and resolution of the status of *obscurior* versus *florea* remains incomplete. Wing length: mean = 22.56 mm, standard deviation = 0.73 mm, n = 10.

The larvae of *Cucullia florea* and *postera* were described by Crumb (1956) under the name *Cucullia postera* race *omissa*. These reared specimens are not *omissa* but a mixture of *florea* and *postera* from Washington. A general description of this mixed lot is similar to *asteroides*; however, there are some significant differences. In particular, the dorsal margin of each of the subdorsal bands is very dark and thin and sets off a distinctive dorsal region. The two thin lines between the subdorsal band and the spiracular region are almost absent, obscure at best. The thin, isolated subspiracular line found in *asteroides* is replaced by a distinct dark line margined ventrally by a wide reticulate brown band. The result is a distinctive yellow-green to green band between the spiracle and the subspiracular line followed by a brown band. Both of these features are absent in *asteroides* and *montanae*. The recorded foodplants given by Crumb are aster (*Aster* sp.) and goldenrod (*Solidago* sp.) (Asteraceae). Specimens from Nova Scotia in the U. S. National Museum of Natural History were reared on *Erigeron* by D. C. Ferguson.

Cucullia florea ranges from the mountains of North Carolina and West Virginia north through New York, New England, and into the Maritime provinces of Canada, Quebec, and Ontario. It extends across the plains provinces, dipping south as far as St. Paul, Minnesota and westward to British Columbia and Washington. These localities seem to indicate that *florea* is primarily a species of Canadian transitional or coniferous forest. Possibly, *postera* is more adapted to open grassland, but this has not been verified by personal observation. Adults have been collected from late June through August. The distribution of dates suggests a slightly later average flight period than that of *omissa*. There is no indication of differences in the flight periods of *florea, postera,* and *obscurior*.

Cucullia postera Guenée

PL. 1, FIGS. 9–12; PL. C, FIGS. 4, 5 (♂ gen.); PL. G, FIG. 11 (♀ gen.); TEXT FIG. 3 *c* (base of valve); TEXT FIG. 9 (map) (RWH 10198).

Cucullia postera Guenée, 1852, *in* Boisduval

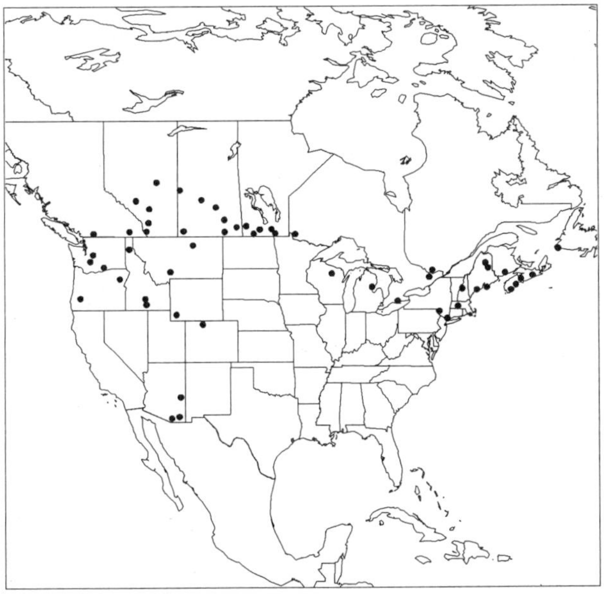

FIGURE 9: DISTRIBUTION OF SPECIMENS EXAMINED OF *CUCULLIA POSTERA*

and Guenée, *Histoire Naturalle des Insectes. Species General des Lépidoptères,* **6**: 133.
Type locality: North America. [BMNH]

The forewing of *postera* is violet brown with distinct rusty red patches along the costa from the orbicular mark toward the apex and in the inner angle between the outer margin and the lower end of the postmedial line. In particular the forewing between the reniform mark and the anal dash is distinctly lighter than the costal region. The forewing of *florea* lacks this lighter patch and does not have distinct rusty red patches, although the costa has a distinctly purplish-brown tinge to it. Wing length: mean = 20.65 mm, standard deviation = 0.78 mm, n = 10.

For a discussion of the larvae see *florea*.

Cucullia postera occurs in the mountains of New England, the Maritime provinces of Canada, Quebec, Ontario, Wisconsin, the plains provinces, Washington, Montana, Idaho, Colorado, and as far south as the White Mountains of Arizona. Adults have been collected from late June to August.

Cucullia obscurior Smith, REVISED STATUS

PL. 1, FIGS. 15–17; PL. C, FIGS. 6, 7 (♂ gen.); PL. G, FIG. 12 (♀ gen.); TEXT FIG. 8 (map) (RWH 10197, part).

Cucullia obscurior Smith, 1892, *Proc. U. S. Natl. Mus.,* **15**: 41.
Type locality: [USA], Colorado. [USNM]

NOTE—The lectotype was designated by Todd (1982).

Cucullia obscurior has a uniformly gray forewing without the purple-brown tinge found in *florea. Obscurior* has been treated as a subspecies of *florea,* but possible small areas of sympatry between the two have led me to treat them as distinct. The relationship between *obscurior* and *florea* is discussed more fully under *florea.* Wing length: mean = 21.49 mm, standard deviation = 1.54 mm, n = 10.

The larvae and its foodplants are unknown. Specimens reared by Crumb from the Mt. Rainier region of Washington were reared on *Aster* sp. However the larvae is unknown because Crumb confounded the larvae of the three species *florea, postera,* and *obscurior* under the name *omissa.*

Cucullia obscurior is a species of the Rocky Mountain and Great Basin regions of the United States. It ranges in the north from Washington and Idaho through Colorado to New Mexico and southern Arizona. In addition there are three specimens from north-central North Dakota that might be assigned to this species, but which I have determined as *florea* on biogeographical grounds. Adults have been collected in late June, July, and August.

convexipennis GROUP

The *convexipennis* group is a complex of five, sometimes superficially dissimilar, species closely united by the structure of the male and female genitalia. In addition to the four species treated here, a fifth species is in Mexico and Central America, *Cucullia costaricensis* Hampson, 1918 (= *Cucullia edificans* Dyar, 1919 NEW SYNONYMY and *Cucullia syggnomon* Dyar, 1919 NEW SYNONYMY). One species of the group, *convexipennis,* occurs in the northern and eastern regions of the United States and Canada while the other three (*lilacina* Schaus, *eucaena* Dyar, and *oribac* Barnes) occur in the southwestern United States and Mexico. The group is characterized by the medial placement and lack of modification of the clasper of the male valve. The corona is strong. There are two cornuti in the vesica, each borne on a separate diverticulum. The diverticulum bearing one of the cornuti is double with one of the subdiverticulae tipped with a scobinate patch. The diverticulum without the scobinate patch is itself double in all of the species except *convexipennis.* The female ductus bursae is approximately 6 × as long as wide. The sclerotized part of the ductus bursae is not produced into a knob as in the *asteris* group. Also the lamella antevaginalis (the membrane cephalad of the ostium) is not developed into fleshy lobes covering the ostium.

Only the life history of *convexipennis* has previously been published; the larvae of *lilacina* and *oribac* are described for the first time here.

I have been unable to identify any Old World species belonging to this group.

Cucullia lilacina Schaus

PL. 1, FIGS. 18, 19; PL. B, FIGS. 1, 2 (♂ gen.); PL. G, FIG. 2 (♀ gen.); TEXT FIG. 10 (map) (RWH 10196).

Cucullia lilacina Schaus, 1898, *Jour. New York Ent. Soc.*, **6**: 116.
Type locality: Mexico, Veracruz, Orizaba. [USNM]

Cucullia agua Barnes, 1905, *Can. Ent.*, **36**: 203.
Type locality: [USA], Arizona, Huachuca Mts. [USNM]

NOTE—This species was described from one male and one female. The male specimen bearing the genital label J. G. Franclemont 1905 is designated and has been labeled as the LECTOTYPE.

Cucullia lilacina is a pretty species from Mexico and the southwestern United States. It is easily recognized by the bright violet or lilac color of the forewing and the distinct brown patch along the costa from the reniform mark to the outer margin. The male clasper and basal process of the sacculus are long, and the vesica of the aedoeagus has one long and one short cornutus plus a scobinate patch. The male genitalia are indistinguishable from those of *eucaena* despite the superficial distinctness of the two species. There does not appear to be any significant individual or geographical variation in *lilacina* although the accuracy of this statement is limited by the scarcity of specimens. Males and females are not sexually dimorphic although females are slightly darker and larger than the males. Wing length: mean = 19.72 mm, standard deviation = 1.20 mm, n = 10.

The life history and foodplants have not been published; however, J. G. Franclemont has collected and reared the larva in the Huachuca Mountains of southeastern Arizona on *Erigeron divergens* Torr. (Gray) (Asteraceae). He has kindly furnished me with color slides of the larva from which this short description has been taken. The larva is strikingly different from the larvae of *convexipennis* and *oribac*, the other two species in the group whose larvae are known. It is light violet with a broad black annular band around each abdominal and thoracic segment. The anterior margin of each segment has small black bands. The prolegs are solid black. There is a hint of an obscure, orange dorsal longitudinal band. The eighth abdominal segment is humped, but not as conspicuously as in *convexipennis.*

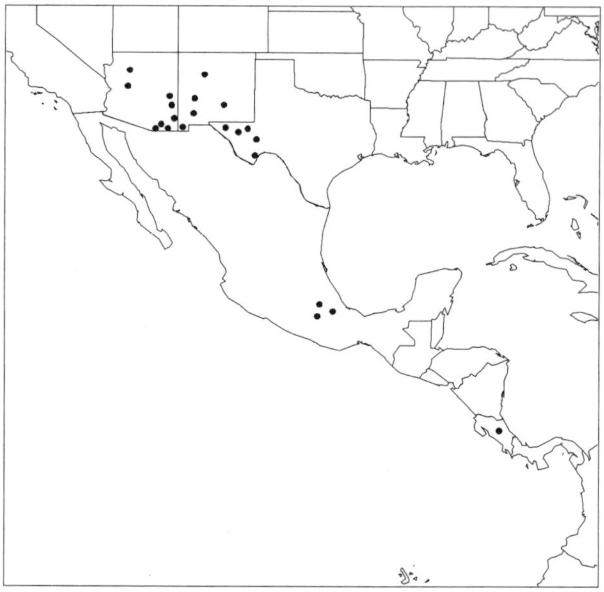

FIGURE 10: DISTRIBUTION OF SPECIMENS EXAMINED OF *CUCULLIA LILACINA*

Cucullia lilacina has the same basic geographical pattern as *eucaena* and *oribac,* but it may have a slightly greater range. In the United States it has been collected in the mountains of southern Arizona, reaching as far north as the White Mountains in the east and the Grand Canyon toward the west. In New Mexico it occurs throughout most of the state (based on scattered records), and it has been collected throughout western Texas. In Mexico it has been taken in the Federal District and Veracruz. A single female, which probably is this species, is known from Juan Vinas [state of Cartago] in central Costa Rica. Like its relatives, *eucaena* and *oribac, lilacina* is probably a denizen of open oak-pine habitats. Adults have been collected from June to October in the United States.

Cucullia eucaena Dyar

PL. 1, FIGS. 20, 21; PL. B, FIGS. 3, 4 (♂ gen.); PL. G, FIG. 3 (♀ gen.); TEXT FIG. 11 (map) (RWH 10201.1).

Cucullia eucaena Dyar, 1919, *Insecutor Inscitiae Menstruus,* 7: 76.
Type locality: Mexico, Mexico City. [USNM]

Cucullia eucaena is a Mexican species also occurring

FIGURE 11: DISTRIBUTION OF SPECIMENS EXAMINED OF *CUCULLIA EUCAENA*

in the mountains of southern Arizona. An interesting feature of the species is its marked sexual dimorphism, particularly in the Arizona populations. Sexual dimorphism also is found in *oribac.* The male forewing in the Arizona population is violet gray suffused with black and brown along the costa from the reniform mark to the apex. In contrast the female forewing is almost entirely suffused with black leaving only the basal area, outer margin, and a brown streak from the orbicular mark to the outer margin lighter in shade. The southern Mexican populations also exhibit this sexual dimorphism but to a lesser degree. The male in these Mexican populations is darker and more like the female than are Arizona males. Wing length: mean = 22.37 mm, standard deviation = 0.87 mm, n = 10.

The males of the Arizona population bear an uncanny resemblance to the eastern United States populations of *Cucullia asteroides* Guenée even though the two are not closely related. The two may be separated because 1) *asteroides* is not known to occur in southern Arizona, 2) the dark shade on the lower inside margin of the postmedial line is oval or square in *asteroides* and linear or triangular in *eucaena,* 3) the black dash from the lower end of the postmedial line to the outer margin tends to be slightly curved in *asteroides* and straight in *eucaena,* and 4) the male genitalia differ as illustrated in the figures.

The male genitalia of *eucaena* are indistinguishable from those of *lilacina* Schaus despite the superficial distinctness of the two species. The ribbing of the ductus bursae appears to be a bit stronger in *eucaena* than in *lilacina,* but not significantly so.

The life history and foodplants are unknown.

In the United States *eucaena* is known from the Huachuca, Baboquivari, and Santa Rita Mountains of southern Arizona. It is also known from the Big Bend region and Guadalupe Mountains of western Texas. I would be surprised if it were not eventually collected in southern New Mexico. In Mexico it has been collected in the states of Hidalgo, Federal District, Puebla, and Veracruz. It probably occurs throughout the mountains of Mexico in pine-oak habitats. Adults have been collected from July to October in the United States.

Cucullia oribac Barnes

PL. 1, FIGS. 22, 23; PL. B, FIGS. 5, 6 (♂ gen.); PL. G, FIG. 4 (♀ gen.); TEXT FIG. 12 (map) (RWH 10203).

Cucullia strigata Schaus, 1898, *Jour. New York Ent. Soc.,* **6**: 116.
Type locality: Mexico, Veracruz, Jalapa. [USNM]

NOTE—*Cucullia strigata* Schaus, 1898 is a junior primary homonym of *Cucullia strigata* Smith, 1892.

Cucullia oribac Barnes, 1904, *Can. Ent.,* **36**: 237.
Type locality: [USA], Arizona, Cochise Co., Huachuca Mts. [USNM]

Cuculia arizona Smith, 1905, *Jour. New York Ent. Soc.,* **13**: 206. NEW SYNONYMY
Type locality: [USA], Arizona, Cochise Co. [AMNH]

Cucullia perstrigata Hampson, 1906, *Catalogue of the Lepidoptera Phalaenae in the British Museum,* **6**: 72.
Type locality: Mexico, Veracruz, Jalapa. [USNM]

NOTE—*Cucullia perstrigata* Hampson, 1906 was proposed as an objective replacement name for *Cucullia strigata* Schaus 1898.

Cucullia oribac Barnes is a southwestern and Mexican species closely related to *convexipennis, lilacina,* and *eucaena.* Its two most notable characters are the violet-gray color of the forewing and the presence of a dark streak along the cubital vein from the bottom of the orbicular mark almost to the outer margin. *Cucullia convexipennis* has this streak, but it is a wood-brown color, not violet gray. The other two closely related species, *lilacina* and *eucaena,*

FIGURE 12: DISTRIBUTION OF SPECIMENS EXAMINED OF *CUCULLIA ORIBAC*

lack this streak. This streak gives both *convexipennis* and *oribac* a horizontally marked appearance. The male clasper is shorter and thicker than in *lilacina* and *eucaena* and the cornuti of the vesica are larger. Females are slightly larger and considerably darker brown than the males. However this sexual dimorphism is not as strong as in *eucaena.* Wing length: mean = 21.40 mm, standard deviation = 0.99 mm, n = 10.

Considerable variability exists within the Mexican populations of *oribac* in forewing color and the length of the male clasper. Two specimens from Zacualpan, Veracruz are noticeably lighter in color than specimens from the Federal District and Jalapa, Veracruz. The clasper in these two specimens from Zacualpan is also shorter and slightly thicker than in other Mexican specimens. The male genitalia of the Arizona population are intermediate between these two types, but the overall variability of the male genitalia in *Cucullia* must also be kept in mind. Possibly more than one species is involved.

No description of the larvae has been published. There are specimens from Morelos and Veracruz in the USNM reared from *Baccharis cordigera* DC. (Asteraceae). J. G. Franclemont (personal communication) collected and reared the larva in the Huachuca Mountains of southeastern Arizona on *Baccharis bigelovii* Gray, and has provided me with color slides from which this short description is taken. The larva is similar to that of *convexipennis*; however, the dorsal line is bright yellow and is not tinged with red or orange. The subdorsal black bands do not contain four thin, broken yellow lines as found in *convexipennis.* A conspicuous yellow patch is present on the hump on the eighth abdominal segment. This patch may or may not be present in *convexipennis,* but it is never as strong as in *oribac.*

Cucullia oribac has approximately the same distribution as *lilacina* and *eucaena,* occurring in the mountains of southern Arizona reaching as far north as Gila County. It occurs in southwest New Mexico and the Guadalupe Mountains of western Texas. The species is apparently primarily Mexican and is known from the Federal District and the states of Veracruz, Morelos, and Chiapas. Adults have been collected in July in the United States.

Cucullia convexipennis Grote and Robinson
PL. 1, FIGS. 24, 25; PL. B, FIGS. 7, 8 (♂ gen.); PL. G, FIG. 5 (♀ gen.); TEXT FIG. 13 (map) (RWH 10202).

Cucullia convexipennis Grote and Robinson, 1868, *Trans. Amer. Ent. Soc.* **2**: 201, pl. 3, fig. 76.
Type locality: "Atlantic District." [ANSP]
NOTE—*Cucullia convexipennis* was described from at least one male and one female. The female was figured. Therefore, the female is designated and has been labeled as LECTOTYPE.

Cucullia convexipennis is one of the most distinctive species of the genus. The wood-brown color of the forewing will immediately identify it. A distinct dark streak running from the orbicular mark to near the outer margin, is present on the forewing cubital vein. This streak is also possessed by *Cucullia oribac,* but the color of *oribac* is violet gray, not brown. This streak gives the forewing a horizontally streaked appearance. There does not appear to be any significant individual, sexual, or geographical variation in *convexipennis.* The closest relatives of *convexipennis* are *oribac, eucaena,* and *lilacina*; however, all three species are some shade of violet gray and superficially very different from *convexipennis.* Wing length: mean = 21.22 mm, standard deviation = 0.64 mm, n = 10.

The larva has been described by Crumb (1956) and figured by Holland (1903, plate 1, figure 3). A yellow to orange-red dorsal stripe is present margined on either side by a black to dark-brown subdorsal band. Four broken yellow lines occur within this band. Conspicuous, irregularly shaped yellow patches are present in the spiracular region. A broad, red band with yellow margining is ventral to the

FIGURE 13: DISTRIBUTION OF SPECIMENS EXAMINED OF *CUCULLIA CONVEXIPENNIS*

spiracular region. The ventral region is occupied by a series of yellow lines. The eighth abdominal segment is strongly humped. The head is solid black with yellow along the frontal sutures and in the ocellar region. The larvae mature in September and October, first feeding on the leaves and then the flowers of their foodplants as they come into bloom. The species passes the winter as a pupa in a silk and earth cocoon in the ground, the adult emerging in the early summer. The recorded host plants are *Aster* spp., *Solidago* spp., and *Callistephus chinensis* Nees, all composites.

Cucullia convexipennis occurs throughout the northern and eastern United States and Canada as far west as Wisconsin, as far south as North Carolina, and as far north as Nova Scotia. It is apparently rare west of Pennsylvania. Its primary habitat is undoubtedly eastern deciduous forest. This distribution is unusual because all four of its nearest relatives are Mexican species with northern range boundaries in the southwestern United States. Adults have been collected from May to September with most specimens caught in July and August.

lucifuga GROUP

The only North American species of this group, *intermedia,* is a close sister species of the Palearctic *Cucullia lucifuga* [Denis and Schiffermüller]. The *lucifuga* group is fairly large in the Palearctic, but the species-group limits are yet to be determined. The members of the group usually are moderately large, gray species, with little or no distinct maculation. The larvae feed on a number of composites. The male genitalia are not remarkable. The clasper is located about ⅓ the length of the valve from its base. The vesica has a single, small cornutus. Two diverticula are toward the base of the vesica, the most basal about 3 × longer than wide and without an apical cornutus. The second diverticulum is distal of the first and about as long as wide with a moderate to small cornutus (relative to other species in the genus). The uncus in *lucifuga* and *intermedia,* but not necessarily in other species of the group, is expanded and flattened into a hairy, round plate with a central ridge. A spiny, yoke-shaped structure is present in the female genitalia just ventral to the ostium (as in the *convexipennis* and *asteris* groups). The lamella antevaginalis and intersegmental membrane between the sixth and seventh sterna are enlarged, infolded, and strongly spiculate and are shaped somewhat like the fleshy folds found in the *asteris* group. The ductus bursae in *lucifuga* and *intermedia* is relatively short compared to *asteroides* and the corpus bursae somewhat longer and narrower.

Cucullia intermedia Speyer (Cucullie de la Verge d'Or (f.), Fr.)

PL. 2, FIGS. 6–9; PL. C, FIGS. 8, 9 (♂ gen.); PL, H, FIG. 1 (♀ gen.); TEXT FIG. 14 (map) (RWH 10194).

Cucullia intermedia Speyer, 1870, *Stettiner Ent. Zeit.,* **31**: 400.
Type locality: [USA], New York.

NOTE—Speyer based this name on "four males, six females" from New York. I have not been able to find any syntypes in North American museums. The Speyer collection eventually found its way to the Zoological Museum of Humboldt University in Berlin. Possibly the syntypes are there. The original description clearly refers to this species.

Cucullia cinderella Smith, 1892, *Proc. U. S. Natl. Mus.,* **15**: 49, pl. 2, fig. 1 (male valve).
Type locality: [USA], Colorado. [USNM]

Cucullia intermedia superficially resembles the various members of the *speyeri* complex. The forewing is uniform gray with little or no significant maculation. Two main differences separate *intermedia* from the species in the *speyeri* complex. First, the hindwing is a uniform, medium-brown color. The base of the hindwing (at least) is white in the *speyeri* complex and distinctly lighter than the outer margin. Indeed the male hindwing in the *speyeri* complex is almost always entirely white except for the outer margin. Second, the black streak in the forewing tornus of the *speyeri* complex is almost always absent in *intermedia.* The male and female genitalia are described under the species-group heading.

Cucullia intermedia is a close sister species of the Palearctic *Cucullia lucifuga.* In fact I suspect that the two are the same species. Two consistent differences exist between the two species or populations. First, *lucifuga* is slightly darker and the maculation slightly more distinct than specimens of the North American species *intermedia.* A pale dot in the medial area between the inwardly pointing teeth of the antemedial and postmedial line exists in *lucifuga,* but is usually absent in *intermedia.* Secondly the vesica diverticulum without a spine is about twice as long as wide in *intermedia,* but approximately equal in length and width in *lucifuga.* The narrow portion of the ductus bursae tends to be longer and narrower in *intermedia* than in *lucifuga.* However the difference is not consistent because of variation in relative length and width in populations of *intermedia.* The clasper and basal process of the sacculus are slightly thinner and longer in *intermedia* than in *lucifuga*, but again the difference is not consistent. *Cucullia intermedia* has no close relatives in North America, but it is one member of a large group of Old World species.

Cucullia intermedia and *lucifuga* may, or may not, be distinct populations of the same species. The key to delineating specific limits lies in the Old World, specifically Europe and Asia. For the moment I have made the decision to retain the North American entity as a separate species until the problem can be looked at in more detail. Wing length: mean = 21.65 mm, standard deviation = 0.71 mm, n = 10.

Lintner (1876) records the larva of *intermedia* from *Lactuca biennis* (Moench) Fernald (Asteraceae), commonly called tall blue lettuce. In a most perplexing foodplant switch, Prentice (1962) records *intermedia* on white birch. I have seen the reared specimens and they are this species. The species was collected 118 times from white birch so the record is unlikely to be accidental.

I have not seen any preserved larvae. Through the courtesy of J. D. Lafontaine I have examined color slides of both third and last instar larvae reared from *Lactuca canadensis* Linnaeus. The last instar larva is dull black. The most conspicuous feature is a series of salmon-orange patches on the side of the body near the spiracles. These patches are roughly rectangular on the thoracic segments, but become more ovate on the abdominal ones. There is a dorsal series of much smaller salmon-orange spots on the thoracic segments and the first two abdominal segments. The prolegs are black. The head is black with two small white spots in the adfrontal area. The third instar larva looks much different than the last instar larva. The third instar larva is also black, but a conspicuous cream-white dorsal line and a nearly white spiracular line are present. The head is black with conspicuous lateral white patches.

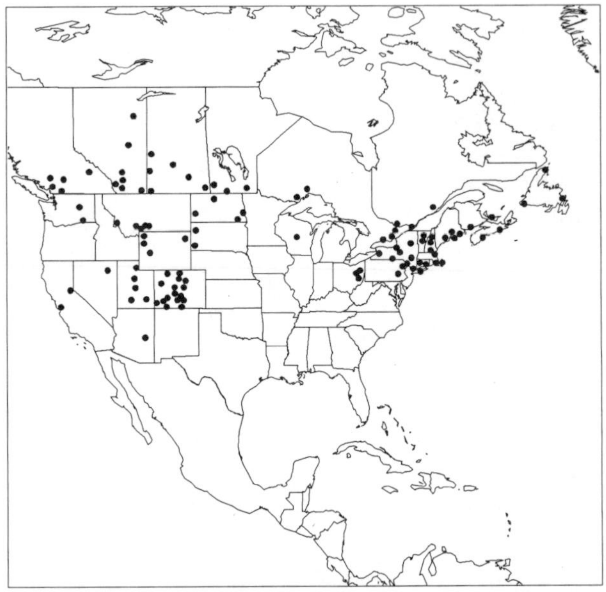

FIGURE 14: DISTRIBUTION OF SPECIMENS EXAMINED OF *CUCULLIA INTERMEDIA*

Cucullia intermedia occurs in most of Canada and the northern United States. It ranges as far south as Pennsylvania in the east. In the west it reaches northern California. In the Rocky Mountains it occurs as far south as the White Mountains in east-central Arizona and occurs commonly in Utah, Colorado, and northeastern Nevada. A single specimen from Nevada is much lighter than any other specimen I have examined except for two unusually pale individuals from Newfoundland collected with normally colored specimens. The species is absent from the Great Plains region of the United States, except for North Dakota and western South Dakota, and is absent from the southeastern quarter of the United States. Adults have been collected in May, June, and July.

speyeri GROUP

The *speyeri* complex is a large group of closely related species known only from North America. The species in the group, as treated here, are *Cucullia speyeri* Lintner; *C. styx,* new species; *C. laetifica* Lintner; *C. lethe*, new species; *C. charon,* new species; *C. dorsalis* Smith; *C. alfarata* Strecker; and *C. eccissica* Dyar. The species are all extraordinarily close superficially. It is usually possible to sort the species of the complex from the same locality if series of well prepared specimens are available. Accurate identification depends on an examination of the male genitalia, but even they are not always conclusive because extensive variation occurs, particularly in the shapes and sizes of the claspers.

The *speyeri* group is defined by an enlargement of the clasper of the right valve and often of the left valve as well. The clasper is a large hook (text figure 3 *d*) and, as a result, the middle of the right valve is expanded into a shallow pocket. The clasper of the left valve usually is enlarged to some degree: as large as for the right valve to no enlargement. Another characteristic of the group is the presence of an enlarged, spinose, and infolded female lamella antevaginalis. The female seventh abdominal sternum has the caudal margin enlarged with two lateral lobes. The membrane following the caudal margin is fleshy and has modified scales that are lighter in color than the scales on the rest of the abdomen. A twist is present in the ductus bursae, and the corpus bursae is constricted near its junction with the ductus. A major dichotomy in the *speyeri* group is the presence of two or three spines in the vesica of the aedoeagus. A T-shaped diverticulum is present at the base of the vesica. Each arm of the diverticulum may bear a cornutus depending on the species. Those species (*alfarata* and *eccissica*) with two cornuti in the vesica have the smaller branch of the T-shaped diverticulum greatly reduced in size and have lost the cornutus at the apex of the branch. A finely dentate band is present at the base of the T-shaped diverticulum. Another elongate diverticulum bearing a single apical cornutus is distal to the first diverticulum. It is not clear whether two or three cornuti in the vesica is the primitive condition or that the difference divides the group into two monophyletic entities. Occasional specimens of three-spined species will lose a cornutus and appear to be two-spined, but the branch from which it is lost will be unreduced in size. This possibility should be kept in mind when using two or three cornuti as an identification character. The correct polarization of the male clasper character is also not entirely clear. Unenlarged claspers is the primitive state by outgroup comparison; however, I cannot determine if the evolutionary sequence is an initial enlargement of both right and left claspers followed by a reduction in the left, or an enlargement in the right clasper followed by an enlargement in the left clasper. An examination of geographical variation in the size of the left clasper of *speyeri* has convinced me that the size of the left clasper is variable. I hypothesize that the putative ancestor of the group had equally enlarged right and left claspers and a three spined vesica. The loss of one cornutus from the vesica resulted in one branch with two-spined species. The species with a two-spined vesica are *eccissica* and *alfarata.* The left clasper was subsequently reduced in *eccissica* but remained large in *alfarata.*

The remaining three-spined species are *Cucullia speyeri, styx, laetifica, lethe, charon,* and *dorsalis.* The left clasper was subsequently reduced in size in *laetifica* and *dorsalis.* I am not proposing any particular phylogeny for the *speyeri* complex because of the difficulty of interpreting these differences in the male genitalia and the intra-specific variability found in some species. I have relied on small, but apparently consistent, differences in superficial appearance to define species because of the variability of the male genitalia. I have not found any reliable characters in the female genitalia for separating species. The larvae may have good characters, but very few larvae are positively known.

A combination of male genitalia, superficial appearance, and geographical location is helpful in identifying the species in this group. A compilation by region follows:

1. Eastern North America. *Cucullia speyeri* is the only species in eastern Canada and New England. Both *speyeri* and *alfarata* occur from New York to Virginia. The male vesica of *speyeri* has three cornuti, but only two cornuti occur in *alfarata. Cucullia alfarata* is a darker and slightly browner species. The forewing postmedial line of *alfarata* is generally marked by a black thickening at its interception with the anal dash. This dark thickening is followed by a light-gray blotch, which is almost always absent in *speyeri.* Apparently, only *alfarata* occurs in the southeastern United States although *speyeri* has been collected with *alfarata* in Missouri.

2. Central Canadian Provinces and the Great Plains. Throughout the region only *speyeri* occurs. *Cucullia laetifica* does occur in Kansas and Oklahoma, however. Males of the two species are readily separable because the left clasper of *laetifica* is markedly smaller than the right. The right and left claspers are approximately equally developed in *speyeri. Cucullia laetifica* is a superficially darker and generally slightly smaller animal, although the two species overlap in size and color. A light patch usually exists in the forewing median area before the intersection of the anal dash and the postmedial line in *laetifica.* This

light patch generally is absent in *speyeri*. The difference is best appreciated when series of the two species are available.

3. Western Canada and northwestern United States. Both *speyeri* and *dorsalis* occur in this area. Males are readily separable by the male genitalia. The claspers of the right and left valves of *speyeri* are equally enlarged, but the left clasper is smaller than the right in *dorsalis* (compare the photos of the male genitalia). A small, third projection of the T-shaped basal diverticulum of the vesica exists in *dorsalis* but is absent in *speyeri*. *Cucullia dorsalis* is slightly darker, and the forewing markings are more sharply defined with black than in *speyeri*. In particular the anal dash is much stronger in *dorsalis* than *speyeri*.

4. Texas. Three species of the *speyeri* group occur in Texas; *Cucullia laetifica*, *C. lethe*, and *C. charon*. *Cucullia lethe* is a species primarily of the eastern coastal area of Texas ranging as far west as central Texas on the eastern edge of the Edwards Plateau. There it co-occurs with *laetifica*. *Cucullia lethe* is lighter gray than *laetifica* (compare the color figures); *Cucullia laetifica* has a distinctly dark blue-gray cast. The differences in the male genitalia are minor and not consistent. The claspers of both the right and left valves, on average, are slightly larger and more contorted in *lethe* than in *laetifica*. The basal diverticulum of the vesica is Y-shaped in *lethe*, not T-shaped as in *laetifica*. A small outpocketing exists in *lethe* near the end of the denticulate band at the base of the basal diverticulum; this outpocketing is absent or extremely small in *laetifica*. The ranges of *Cucullia laetifica* and *charon* overlap in western Texas. *Cucullia charon* is uniformly dark gray while *laetifica* has a distinctly bluish cast to the forewing. Also *laetifica* normally has a distinctly lighter patch in the forewing median area; this patch is lacking in *charon*. There do not appear to be any completely consistent differences between the male genitalia of *laetifica* and *charon* although the left clasper of *charon* tends to be larger than in *laetifica*, often noticeably projecting past the costal margin of the valve.

5. Colorado. Both *speyeri* and *dorsalis* occur in eastern Colorado. They may be separated as indicated under # 3. Both *laetifica* and *dorsalis* occur in southern Colorado. *Cucullia dorsalis* is generally larger than *laetifica*. *Cucullia laetifica* tends to have a light patch before the postmedial line where it intercepts the anal dash. There is a slight brown tint to the forewing of *laetifica* that is absent in *dorsalis*. Western Colorado and possibly all of Utah is apparently the sole domain of *dorsalis*.

6. Arizona. Five species occur in southern Arizona; *Cucullia charon*, *C. styx*, *C. dorsalis*, *C. eccissica*, and *C. laetifica*. *Cucullia styx* is uniformly dark black gray, *C. charon* and *eccissica* a uniform medium gray, and *C. laetifica* a slightly bluish gray with a lighter blotch in the median area before the anal dash. *Cucullia charon* and *eccissica* are superficially almost identical. The definitive distinguishing characters are found in the male genitalia. *Cucullia eccissica* has two cornuti in the vesica; *charon*, *styx* and *laetifica* have three. The left and right claspers of *styx* are of approximately equal length, but the left clasper is significantly smaller than the right in *laetifica* and *charon*. Only *dorsalis* and *laetifica* are known to occur in northern Arizona.

7. California. In northern and central California only *dorsalis* is known to occur, in southern California only *eccissica*. The two species are virtually identical superficially. *Cucullia eccissica* has two spines in the vesica, *dorsalis* three. Male specimens from central California should be dissected because the boundary between the species has not been determined.

The characters separating species can generally be extrapolated for areas not covered above.

The sister group of the *speyeri* group is not definitely known. The enlarged claspers of the male valves seem to be an autapomorphy not found in any other group of *Cucullia* either here or in the Old World. The enlarged spinose and infolded lamella antevaginalis, the superficial similarity between this group and the *lucifuga* group, and the very strong overall similarities in larval pattern suggest the two are sister groups.

Cucullia speyeri Lintner

PL. 1, FIGS. 26, 27; PL. C, FIGS. 10, 11 (♂ gen.); pl. H, FIG. 2 (♀ gen.); TEXT FIG. 15 (map) (RWH 10190, part).

Cucullia speyeri Lintner, 1874, *Entomological Contributions*, no. **3**: 168, figs. 13, 14.

Type locality: [USA], New York, Albany. [USNM]

NOTE—The two specimens, a male and a female, cited in Lintner's description are in the USNM via the Otto Meske collection, but they do not bear type labels. The male specimen is labeled "Albany, 6.6.73" and the female "Sharon, 15.8.73," localities and dates opposite of those listed in the original description. Apparently the labels have been switched or Lintner made a mistake in the original description. The male's abdomen is glued on. Consequently the female specimen bearing the labels 1) "Sharon, 15.8.73," 2) "Collection O. Meske," 3) "Riley," 4) "USNM 76542 Type no." is designated and has been labeled as LECTOTYPE. The type locality given above is based on the assumption that the labels have been switched.

Cucullia speyeri is a light-gray species with little or no definite forewing maculation. The male hindwing is white with only a thin dark-brown border. The brown border is much wider in the female occupying ⅓–½ of the hindwing. In both sexes at least the base of the hindwing is white separating this species from

Cucullia intermedia with its uniformly brown hindwing.

Cucullia speyeri is a widespread species occurring in the eastern United States and Canada, southern Canada, the midwestern United States, and the state of Washington. It overlaps the ranges of three other species; *alfarata, laetifica,* and *dorsalis.* Both *speyeri* and *alfarata* occur in the eastern United States, from New York to Virginia. *Cucullia alfarata* from this region appears to be more of a coastal species than *speyeri.* Males of *speyeri* and *alfarata* are readily separated by differences in the genitalia; there are two cornuti in the vesica of *alfarata* but three in *speyeri.* Both *speyeri* and *dorsalis* occur in the eastern mountains of Colorado. The left clasper of *speyeri* is approximately the same size as the right, but the left is significantly smaller than the right in *dorsalis* (compare the photos of the male genitalia). These two species in this area of overlap are very similar. Mixed series of the two species, however, can usually be sorted fairly readily. *Cucullia dorsalis* is slightly darker than *speyeri,* and the markings are more distinctly accented with black. In particular the anal dash is stronger in *dorsalis* than in *speyeri.*

The ranges of *speyeri* and *laetifica* overlap in southern Kansas and Oklahoma. Males of the two species can be separated by differences in the male genitalia. The left clasper of *laetifica* is much smaller than the right, but the claspers are of approximately equal size in *speyeri. Cucullia laetifica* usually has a dark, slightly bluish cast, but *speyeri* is uniformly light gray. *Cucullia laetifica* tends to have a light patch in the median area of the forewing where the postmedial line intercepts the anal dash. This light area is absent or poorly defined in *speyeri.* This light area is not always well developed in *laetifica,* and the two species sometimes cannot be distinguished on superficial appearance alone. The closest relative of *speyeri* is probably *Cucullia styx* from southern Arizona and Mexico. *Cucullia styx* is a uniformly dark, black-gray species. The male genitalia are slightly more massive and the claspers more strongly developed than in *speyeri.* The differences, however, are slight and apparently partly dependent on the size of the individual. I am treating *styx* and *speyeri* as distinct species because of the relatively great superficial differences in coloration and the large disjunctions in range and ecological province. Wing length: mean = 21.95 mm, standard deviation = 1.07 mm, n = 10.

There is little superficial variation in *speyeri* other than the hindwing sexual dimorphism. The male genitalia exhibit some geographical variation. The

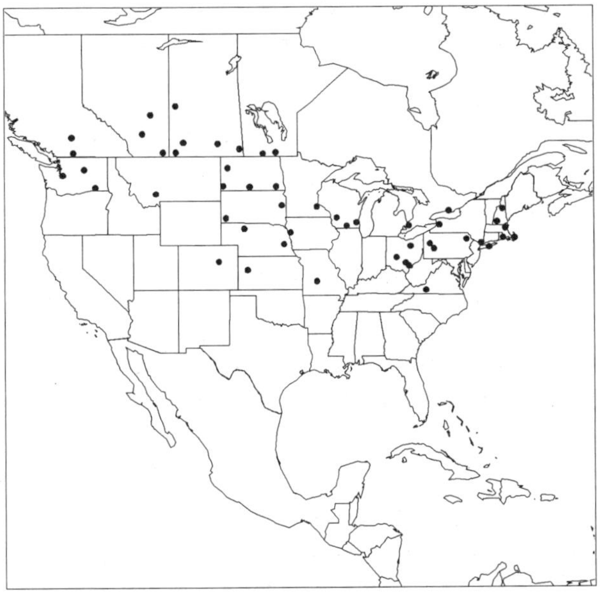

FIGURE 15: DISTRIBUTION OF SPECIMENS EXAMINED OF *CUCULLIA SPEYERI*

left valve tends to be narrower and the clasper of the left valve stronger in the western part of its range (from the Dakotas to Washington) than in the more eastern parts of its distribution.

The larva has not been unequivocally described for *speyeri.* In the USNM there are larvae from Illinois that have been reared from *Conyza canadensis* (L.) Cronq. (Asteraceae). The general larval pattern is a complicated series of yellow markings on a black background. Dorsally each abdominal segment has two large, yellow triangles with extensions from the apices running forward and backward to the next segment or triangle and ventrally to the yellow subdorsal line. The dorsal triangles are extensively connected and the black occurs as isolated patches running in a dorsoventral direction. The subdorsal line is yellow and broken with the yellow running through the spiracular region to the subspiracular area cutting off irregular, isolated islands of black. Each proleg is black with a large yellow patch on the apex of the leg. The head is black with broad yellow markings on the frons, the frontal sutures, and a broad yellow stripe running from near the occiput to the ocellar region. This ocellar stripe is connected to the yellow stripe on the frontal suture. In contrast the ocellar stripe does not reach the frontal suture in *laetifica.* Crumb (1956) gives a description of this species as *"Cucullia speyeri* race *dorsalis"* from Washington. The single adult from this rearing is a female and, therefore, cannot be identified absolutely as *speyeri* or *dorsalis.* Superficially, it most resembles *speyeri.* In Crumb's description the prolegs are described as all black, but in the larva of *speyeri* the legs have broad yellow apices. Crumb reared this female from larvae found on *Conyza canadensis* (L.) Cronq. (Asteraceae).

Cucullia speyeri is primarily a species of the eastern and central United States and Canada. It occurs in the East from New England and southern Ontario south to southwestern Virginia, throughout most of the Midwest, and south to Nebraska and southern Missouri. *Cucullia speyeri* occurs throughout southern Canada west to British Columbia and south into Washington. It has not yet been reported from Montana or Idaho, but undoubtedly lives there as well. Adults have been collected in May, June, July, and early August.

Cucullia styx Poole, NEW SPECIES

PL. 1, FIGS. 28, 29; PL. C, FIGS. 12, 13 (♂ gen.); PL. H, FIG. 3 (♀ gen.); TEXT FIG. 3 *d, f* (♂ gen.); TEXT FIG. 3 *e* (♀ gen.); TEXT FIG. 16 (map).

Cucullia styx Poole.
Type locality: USA, Arizona, Cochise Co., Chiricahua Mtns., Paradise. [USNM]

Cucullia styx is a primarily Mexican species occurring as far north as the Chiricahua and Huachuca Mountains of southern Arizona. *Cucullia styx* looks like a dark, smoky gray variation of *speyeri,* and it could be a much darker, highly disjunct race of *speyeri.* There are no significant differences in either the male or female genitalia between the two species. I have concluded that *styx* is a distinct species because of the highly disjunct ranges, the significant superficial differences in appearance, and the different ecological provinces. The formal description follows.

Upper surface as in plate 1, figures 28, 29. Head: palpus dark smoky gray, lighter anteroventrally, darker posterodorsally; front with small central tuft of smoky gray; two longitudinal tufts, one above and one below antennal bases, both tufts darker above than below giving appearance of two dark lines to the naked eye; lashes cream white, distinctly lighter than surrounding scaling; posterior and ventral margin of eye surrounded by dark smoky brown hair. Thorax: anteroventral edge of patagia black; median area with a light-gray line edged with dark smoky gray. Prothoracic leg: smoky gray; anterior surface of tibia with a black line. Mesothoracic and metathoracic legs: midleg smoky gray; tibia with dark line on exterior surface, spurs cream white with black bands; hindleg smoky gray, tibia with dark line, the line continued on the first two tarsal segments; both pairs of spurs cream white with dark bands. Ventral surface of wings: forewing dark smoky gray; hindwing dull white except for dark scaling along the costa, veins, and outer margin. Male genitalia: text figure 2 *d, f*; characterized by strong development of clasper in left valve and three cornuti in the vesica. Female genitalia: text figure 2 *e*; characterized by strong development of appendix bursae. Wing length: mean = 21.55 mm, standard deviation = 0.90 mm, n = 10.

FIGURE 16: DISTRIBUTION OF SPECIMENS EXAMINED OF *CUCULLIA STYX*

The larva and its foodplants are unknown.

TYPES. Holotype: ♂ [USA], Arizona, Cochise Co., [Chiricahua Mountains], Paradise, USNM genitalia slide 41,362. USNM. Paratypes: 4 ♀, same locality as the holotype. USNM.

Because of the difficulty of defining specific limits in this species complex, I am restricting the type series to the type locality. I have seen other specimens from the Huachuca Mountains in southeastern Arizona, and the Federal District of Mexico. *Cucullia styx* is probably primarily Mexican reaching northward to the mountains of the southwestern United States. Adults have been collected in June, July, and August in the United States.

Cucullia dorsalis Smith, REVISED STATUS

PL. 1, FIGS. 30, 31; PL. C, FIGS. 14, 15

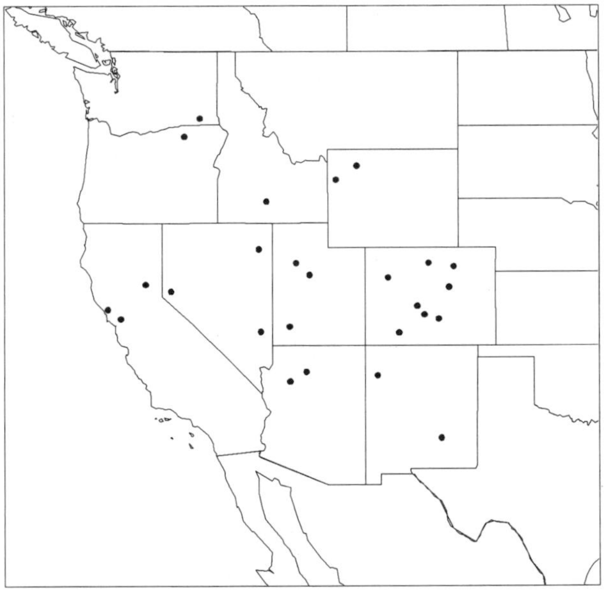

FIGURE 17: DISTRIBUTION OF SPECIMENS EXAMINED OF *CUCULLIA DORSALIS*

(♂ gen.); PL. H, FIG. 4 (♀ gen.); TEXT FIG. 17 (map) (RWH 10190, part).

Cucullia dorsalis Smith, 1892, *Proc. U. S. Natl. Mus.*, **15**: 47.
Type locality: [USA], Colorado. [USNM]; lectotype designated by Todd (1982).

Cucullia dorsalis has been considered a race or subspecies of *speyeri*. However there are small, but consistent, differences in superficial appearance and the male genitalia. The ranges of the two species appear to overlap marginally on the eastern edge of the Rocky Mountains, although the locality of the overlapping specimens, "Denver," is too vague to guarantee sympatry. A second area of possible sympatry is in southern Washington. *Cucullia dorsalis* is darker and the markings are more strongly etched with black than in *speyeri*, and the anal dash of the forewing is stronger and more likely to reach the postmedial line. The right clasper of the male genitalia is much smaller than it is in *speyeri*. The right and left valves are fairly symmetric in appearance, but the right clasper is still slightly larger than the left.

The ranges of *dorsalis* and *laetifica* overlap in southwestern Colorado. The two species in this region can be very difficult to separate superficially. In general *dorsalis* is larger and a more uniformly gray color than *laetifica*. *Cucullia laetifica* often has a slightly bluish cast, and the median area of the forewing in *laetifica* often has a lighter area preceding the postmedial line where it intercepts the anal dash. The claspers of the right and left valves of the male genitalia are almost equal in size in *dorsalis*, but the right clasper is markedly larger than the left in *laetifica* (compare the photos of the male genitalia). Wing length: mean = 22.55 mm, standard deviation = 1.48 mm, n = 10.

The larva of *dorsalis* has not been described. Two adult specimens from California in the USNM were reared from *Machaerantha shastensis* Gray, an herbaceous composite, but there are no associated larvae. There is also a single female from Weld County, Colorado reared from *Helianthus* sp. (Asteraceae). Crumb's (1956) "Cucullia #8" may be this species. It belongs to the group and is most similar to *speyeri* but has an isolated anterior dorsal yellow marking as in *laetifica*.

Cucullia dorsalis is a species of the western Rocky Mountains and the Great Basin area. It has so far been found in most of Colorado except the far eastern part of the state, north to southern Washington, western Wyoming, northern New Mexico and Nevada, northern Arizona, Utah, and central California. There does not appear to be any significant variation either geographical or individual in *dorsalis* other than the sexual dimorphism in hindwing wing coloration typical of the group. Adults have been collected from May to August.

Cucullia laetifica Lintner

PL. 1, FIGS. 32, 33; PL. D, FIGS. 1, 2 (♂ gen.); PL. H, FIG. 5 (♀ gen.); TEXT FIG. 18 (map) (RWH 10191).

Cucullia laetifica Lintner, 1875, *in* Grote, *Checklist of the Noctuidae of America North of Mexico. I. Bombyciae and Noctuelitae (Nonfasciatae)*, p. 24.
Type locality: [USA], Texas, Bastrop. [USNM]

Cucullia cita Grote, 1883, *Papilio*, **3**: 75.
Type locality: [USA], Arizona. [USNM]
NOTE—Grote did not indicate the sex or number of specimens in the original description. There is a single male specimen without an abdomen marked "Arizona," "*Cucullia cita* Grote type" in the USNM. This specimen has been labeled and is here designated as LECTOTYPE.

Cucullia hartmanni French, 1888, *Can. Ent.*, **20**: 69.
Type locality: [USA], Texas, Harris Co., Hockley.
NOTE—The type or types of this species are apparently lost. The locality, the Houston vicinity, is more

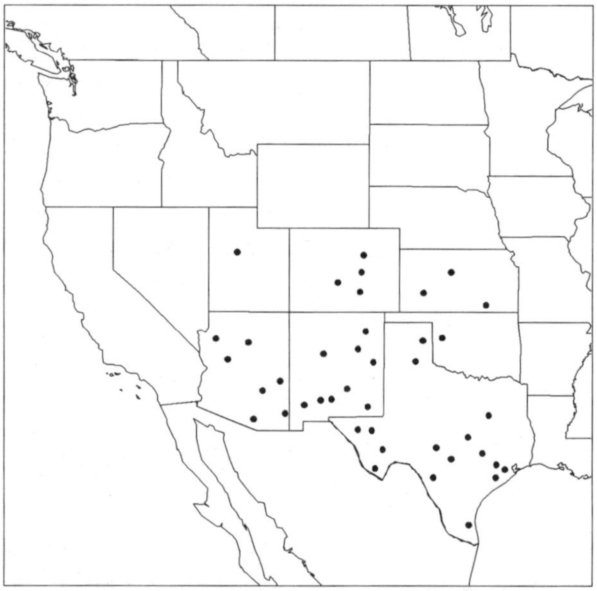

FIGURE 18: DISTRIBUTION OF SPECIMENS EXAMINED OF *CUCULLIA LAETIFICA*

typical of *lethe* than of *laetifica,* but both species do occur in the area. The original description is not precise enough to distinguish between the two species. I am retaining *hartmanni* as a synonym of *laetifica.* I could have used this name for *lethe* and designated a neotype. However, I believe stability is better served by retaining the ambiguous name *hartmanni* as a synonym of *laetifica* and treating the previously unrecognized species as undescribed.

Cucullia laetifica is a species of the south-central and southwestern United States and has been collected most commonly in shrubby desert and shortgrass prairie. The distribution of this species overlaps the ranges of several other members of the *speyeri* group. It is the only species of the group with a light blotch usually present in the median area of the forewing preceding the interception of the postmedial line and anal dash. The forewing also has a distinct steel-blue tint. The methods for separating *laetifica* from *dorsalis* and *speyeri* in their areas of overlap have been discussed. *Cucullia laetifica* occurs with *lethe* in eastern and central Texas. *Cucullia lethe* is lighter gray. The basal diverticulum of the male vesica is T-shaped in *laetifica* and both arms are similar. The two arms are approximately Y-shaped in *lethe*, and the lower arm is distinctly larger and broader than the upper. In addition a small outpocketing exists near the end of the denticulate band of the basal diverticulum in *lethe.* This outpocketing is absent in *laetifica* or greatly reduced. *Cucullia laetifica* overlaps the ranges of *eccissica, charon,* and *styx* in Arizona, New Mexico, and western Texas. *Cucullia eccissica* and *charon* are lighter and more uniformly colored than *laetifica.* Accurate identification depends on the male genitalia. There are three cornuti in the vesica of *laetifica* but only two spines in *eccissica. Cucullia styx* is a uniform dark black gray. Both the right and left claspers of the male genitalia are more strongly developed in *styx* than in *laetifica* (compare the photos of the male genitalia). There are no apparent, consistent genital differences between *charon* and *laetifica.* Wing length: mean = 20.25 mm, standard deviation = 0.59 mm, n = 10.

The larvae was described by Crumb (1956) who based his description on a series of larvae reared by Dyar from *Chrysothamnus* (rabbitbush) (Asteraceae). The larvae are easily distinguished from *speyeri.* The yellow band in the ocellar region of the head is not connected to the yellow stripe along the frontal suture. The yellow markings in *speyeri* usually form annular yellow bands. In contrast the anterior yellow dorsal marking on each abdominal segment in *laetifica* is surrounded by black and does not extend toward the yellow of the subdorsal region. A distinctive V-shaped black mark is present within this anterior yellow patch. Three adults in USNM from Maverick County, Texas were reared from *Baccharis neglecta* Britton (Asteraceae). A single specimen in the USNM supposedly was reared from "English Ivy."

Cucullia laetifica occurs throughout Texas, westward through New Mexico and Arizona, and north to Colorado. The northern limit of its range in the east is in southern Kansas. There is little superficial variation in *laetifica* although the color of the forewing fades badly in older specimens. There is the usual variation in the development of the claspers in the male genitalia. Adults have been collected from May to September.

Cucullia lethe Poole, NEW SPECIES
PL. 1, FIGS. 34, 35; PL. D, FIGS. 3, 4 (♂ gen.); PL. H, FIG. 6 (♀ gen.); TEXT FIG. 19 (map).

Cucullia lethe Poole.
Type locality: USA, Texas, Padre Island National Seashore. [USNM]

Cucullia lethe is a species known primarily from the eastern shores of Texas but ranging as far west as the central portion of the Edwards Plateau and as far north as the Oklahoma border and southern Arkansas. Superficially, *lethe* is most similar to *laetifica*

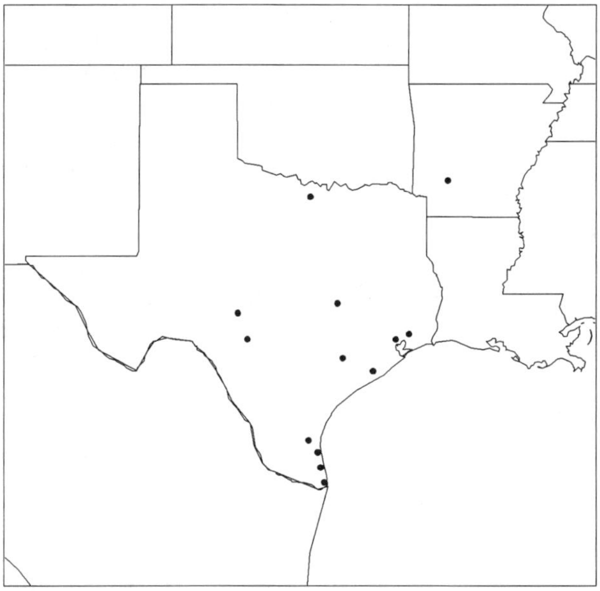

FIGURE 19: DISTRIBUTION OF SPECIMENS EXAMINED OF *CUCULLIA LETHE*

and undoubtedly is its sister species. The two species are readily separated, given good series of well-prepared material, because the forewing of *lethe* is distinctly lighter in color than *laetifica.* The male genitalia are very similar although the claspers of both right and left valves tend to be larger and more contorted in *lethe* than in *laetifica.* The basal diverticulum of the vesica is Y-shaped in *lethe,* not T-shaped as in *laetifica.* A small outpocketing exists in *lethe* near the end of the denticulate band at the base of the basal diverticulum; this outpocketing is absent or extremely small in *laetifica.*

Dorsal surface coloration and maculation as in plate 1, figures 34, 35. Forewing coloration light gray, darker toward costa and with brown shadings in cell between orbicular and reniform marks and on inner margin near base. Head: palpus smoky gray, lighter anteroventrally, darker posterodorsally; front with small central tuft of smoky gray; two longitudinal tufts, one above and one below antennal base, both tufts darker above than below giving the appearance of two dark lines to the naked eye; lashes light gray, lighter than surrounding scaling; posterior and ventral margin of eye surrounded by dark smoky brown hair. Thorax: anteroventral edge of patagia black; median area with light-gray line edged with dark smoky gray. Prothoracic leg: light gray; anterior surface of tibia with a black line. Mesothoracic leg: gray; tibia with dark line on exterior surface, spurs cream white with black bands. Metathoracic leg: light gray, tibia with dark line, line continued on first two tarsal segments; both pairs of spurs cream white with dark bands. Under side of forewing smoky gray. Under surface of hindwing white except for dark scaling along costa, veins, and outer margin. Male genitalia: plate D, figures 3, 4; little different from *laetifica,* claspers slightly larger and more contorted; vesica with basal diverticulum Y-shaped, lower arm fatter and larger than upper; small outpocketing of basal diverticulum present near base of basal diverticulum. Female genitalia: plate H, figure 6; nearly identical to *laetifica* but two membranous pockets at junction of ductus and corpus bursae equal in size in *lethe* but distal larger than proximal in *laetifica.* Wing length from base to apex: mean = 20.87 mm, standard deviation = 0.80 mm, n = 10.

TYPES. Holotype: ♂, [USA], Texas, Padre Island National Seashore, 24-ix-1979, A. & M. E. Blanchard. USNM. Paratypes (22 ♂, 3 ♀): Lake Brownwood State Park, Texas, A. & M. E. Blanchard, April and September, 14 ♂, 1 ♀; El Rancho Cima, Hays, Comal Cos, Texas, A. & M. E. Blanchard, August, 1, ♂; Sinton Welder Wildlife Refuge, Texas, A. & M. E. Blanchard, February and October, 5♂, 2♀; Houston, Harris County, Texas, A. & M. E. Blanchard, April 1♂; Hondo, Medina County, Texas, E. Knudson, September, 1 ♂. USNM, CNC, LACM, and AMNH. There are other, less well preserved specimens in collections; these specimens are excluded from the type series.

A specimen in the USNM was reared from *Amphiachyris dracunculoides* (DC.) Nutt., an annual composite, from Refugio County, Texas, but there are no associated larvae. Adults have been collected in April and May and again in September and October. The species is apparently two brooded.

Cucullia lethe has been most commonly collected along the eastern shore of Texas, but ranges as far west as the central portions of the Edwards Plateau and as far north as southern Oklahoma and Arkansas. There is no significant superficial variation, but the sizes and shapes of the claspers of the male genitalia vary greatly.

Cucullia charon Poole, NEW SPECIES

PL. 1, FIG. 36, PL. 2, FIG. 1; PL. D, FIGS. 5, 6 (♂ gen.); PL. H, FIG. 7 (♀ gen.); TEXT FIG. 20 (map).

Cucullia charon Poole.
Type locality: USA, Texas, Brewster County, Chisos Mountains, Panther Pass. [USNM]

Cucullia charon is one of the darker and smaller species of the *speyeri* group. It is known from southwestern Texas and southern Arizona. *Cucullia char-*

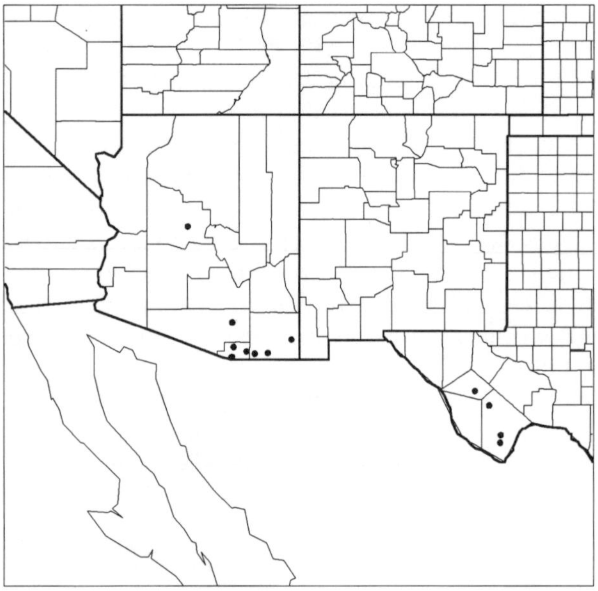

FIGURE 20: DISTRIBUTION OF SPECIMENS EXAMINED OF *CUCULLIA CHARON*

on is most likely to be confused with *laetifica,* but the forewing is soft, dark gray and lacks the steel-blue tint and the light patch in the forewing median area so often characteristic of *laetifica.* There do not appear to be any consistent differences between the male genitalia of *laetifica* and *charon,* although the left clasper of *charon* tends to be larger than that of *laetifica,* often noticeably projecting past the costal margin of the valve. *Cucullia charon* is a smaller animal than *styx.* The male hindwing is white with a small dark-brown outer margin, but the male hindwing is distinctly suffused with brown in *styx.* This species appears to overlap the range of *eccissica* in southern Arizona. The two species are superficially nearly identical. *Cucullia charon* tends to be slightly smaller and slightly darker than *eccissica.* The size difference is easily seen in side by side comparison of series of the two species, but is less obvious in isolated individuals of the two species. The size difference is most apparent in the male and female genitalia of the two species; *eccissica* is considerably larger than *charon.* The male genitalia must be examined for positive identification. There are two cornuti in the vesica of *eccissica,* three in *charon.* One specimen of *charon* from the type locality in western Texas has only two cornuti in the vesica; the upper arm of the basal diverticulum is slightly reduced in size and the spine absent. However, the arm is present, not absent as in *eccissica.* In addition a small outpocketing of the basal diverticulum exists near the end of the dentate patch in *charon* (opposite the two arms) that is absent in *eccissica.* The vesica must be inflated to see these critical differences between *charon* and *eccissica* because determining the number of cornuti in the uninflated vesica is not reliable. There do not appear to be any consistent differences in the valves or their processes between these two species.

Dorsal surface coloration and maculation as in plate 1, figure 36, and plate 2, figure 1. Forewing coloration gray to dark gray, almost uniformly colored across the forewing except for a slightly lighter streak along the anal dash; no light area in median area. Head: palpus smoky gray, lighter anteroventrally, darker posterodorsally; front with small central tuft of smoky gray; two longitudinal tufts, one above and one below antennal base, both tufts darker above than below giving the appearance of two dark lines to the naked eye; lashes light gray at tips, darker at their bases, but not much lighter than surrounding scaling; posterior and ventral margin of the eye surrounded by dark smoky brown hair. Thorax: anteroventral edge of patagium black; median area with a light-gray line edged with dark smoky gray. Prothoracic leg: gray; anterior surface of tibia with a black line. Mesothoracic leg: gray; tibia with dark line on exterior surface, spurs cream white with black bands. Metathoracic leg: gray, tibia with dark line, line continued on first two tarsal segments; both pairs of spurs cream white with dark bands. Underside of forewing smoky gray. Under surface of hindwing white except for dark scaling along costa, veins, and outer margin. Male genitalia: plate D, figures 5, 6, little different than *laetifica* although clasper of left valve usually longer than that of *laetifica,* projecting past costal margin. Female genitalia: plate H, figure 7, apparently identical with female genitalia of *laetifica.* Wing length: mean = 18.89 mm, standard deviation = 1.10 mm, n = 10.

TYPES. Holotype: ♂, [USA], Texas, Brewster Co., Chisos Mountains, Panther Pass, June 2, 1973, Douglas C. Ferguson, USNM 38,373. USNM. Paratypes (6 ♂, 2 ♀): from the type locality and collector. CNC and USNM. Other specimens from southwest Texas and southern Arizona are excluded from the type series.

The larva and its foodplants are unknown.

Cucullia charon is known only from the mountains of southwestern Texas and scattered localities in Arizona. There appears to be little if any superficial variation in *charon* except for the usual sexual dimorphism in the male and female hindwings. Specimens have been collected from March through August.

Cucullia alfarata Strecker
PL. 2, FIGS. 2, 3; PL. D, FIGS. 7, 8 (♂ gen.); PL. H, FIG. 8 (♀ gen.); TEXT FIG. 21 (map) (RWH 10192).

Cucullia alfarata Strecker, 1898 *Lepidoptera, Rhopaloceres and Heteroceres, Indigenous and Exotic, Supplement,* **1**: 9.
Type locality: [USA], Florida, St. Augustine. [FMNH]

Cucullia phila Smith, 1908, *Ann. New York Acad. Sci.,* **18**: 117.
Type locality: [USA], Pennsylvania, Philadelphia. [AMNH]; lectotype designated by Todd (1982).

Cucullia alfarata is one of two species with two cornuti in the vesica of the aedoeagus rather than three. This species is a denizen of the eastern and southeastern United States and has the superficial appearance of a dark, stubby winged *speyeri.* The forewing of *alfarata* usually has a blotch of dark gray brown at the interception of the anal dash and postmedial line. The ranges of *speyeri* and *alfarata* overlap in the central-eastern United States. The differences in forewing coloration will usually separate them. Accurate identification, however, depends upon the difference in the number of cornuti of the vesica of the male genitalia, two in *alfarata,* three in *speyeri.* Wing length: mean = 19.84 mm, standard deviation = 1.41 mm, n = 10.

The larva was described by Crumb (1956) and was reared from *Aster ericoides* L. (Asteraceae). I have not seen preserved material of this species, but Crumb describes it as having a continuous orange dorsal line and three white spots in the supra-spiracular region of each abdominal segment. These two characters should readily separate it from the larvae of *speyeri* and *laetifica,* the other two species of the group for which larvae are definitely known.

Cucullia alfarata appears to be primarily an inhabitant of the southeastern United States; however, it has been taken as far north as Long Island. The species is apparently confined to the coastal region in the northern part of its range. *Cucullia alfarata* is not common in collections, and specimens occur as single individuals from scattered localities. Females considerably outnumber males in collections. I have seen specimens from New York, Pennsylvania, Virginia, North Carolina, Tennessee, Missouri, and Florida. Adults have been collected in June and August.

FIGURE 21: DISTRIBUTION OF SPECIMENS EXAMINED OF *CUCULLIA ALFARATA*

Cucullia eccissica Dyar
PL. 2, FIGS. 4, 5; PL. D, FIGS. 9, 10 (♂ gen.); PL. H, FIG. 9 (♀ gen.); TEXT FIG. 22 (map).

Cucullia eccissica Dyar, 1919, *Insecutor, Inscitiae, Menstruus,* **7**: 77.
Type locality: Mexico, Puebla, Tehuacan. [USNM]

Cucullia eccissica is a light-gray or dark-gray species from the southwestern United States and Mexico. It is medium to light gray in Mexico and southern Arizona and superficially almost identical with *speyeri* although the ranges of the two species do not overlap. *Cucullia styx* is uniformly darker and more nearly black than *eccissica. Cucullia laetifica* is also darker than *eccissica* and often has a conspicuous light blotch in the forewing preceding the interception of the postmedial line and anal dash. *Cucullia eccissica* overlaps the range of the more eastern *charon* in Arizona. The two species are nearly identical, but *eccissica* is slightly larger and lighter in coloration than *charon.* The difference in size is easily seen in series of specimens, but not as apparent in isolated individuals. The difference in size is very evident in the male and female genitalia; the genitalia of *eccissica* are considerably larger than those of *charon.* The most important difference between the two species is the presence of two cornuti in the vesica of *eccissica* and three in *charon.* In rare individuals of *charon* the upper arm of the basal di-

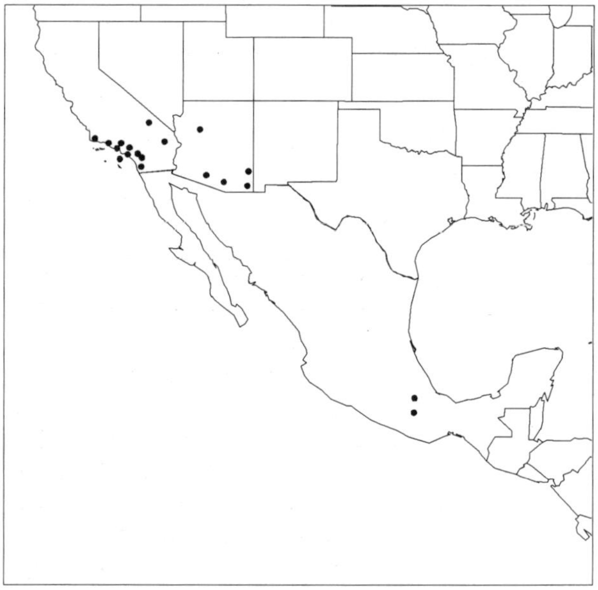

FIGURE 22: DISTRIBUTION OF SPECIMENS EXAMINED OF *CUCULLIA ECCISSICA*

verticulum will lack its cornutus (i.e., the vesica will be two spined), but the diverticulum is still present. Also on the basal diverticulum there is a small outpocketing near the base opposite the two arms in *charon* that is absent in *eccissica.* Wing length: mean = 20.81 mm, standard deviation = 1.11 mm, n = 10.

Adults in the USNM were reared from *Corethrogyne filaginifolia* (Hook. & Arn.) Nutt. (Asteraceae), but the larvae has not been described and I have not seen any preserved larvae I could definitely associate with this species.

Cucullia eccissica has been collected in the states of Mexico and Puebla in Mexico, southern and western Arizona, and southern California. It is seasonally dimorphic in southern California; early season specimens are darker than specimens from the late summer and autumn. All Mexican specimens so far seen are light gray. Specimens from the Los Angeles Basin region in southern California are smoother and slightly duller than specimens from elsewhere. This same geographic change in coloration is also found in *incresa.* The left clasper is, on average (but not consistently), approximately equal in size to the right clasper in the Los Angeles Basin specimens, but markedly smaller than the right in *eccissica* outside of this region.

Adults have been collected in April–May and August–September in southern California suggesting the possibility of two broods.

antipoda GROUP

The members of the *antipoda* group comprise a series of species traditionally placed under the generic name *Copicucullia* because of the presence of a tibial claw on the prothoracic leg. The species-group members are primarily found in the arid regions of the American southwest. The species are characterized, superficially, by the presence of the tibial claw, the absence of abdominal hair pencils, and their generally smaller size relative to the species, for example, in the *asteris* or *speyeri* groups. The front tends to be rough, sometimes developed into a distinct ridge or even triangular protuberance. None of these characters is constant within the group. The tibial claw is almost totally absent and abdominal hair pencils present in *Cucullia cuculloides* despite the fact that its closest relative, *eulepis,* has a well-developed claw and no abdominal hair pencils. There appears to be a strong evolutionary trend toward modification of the valves of the male genitalia, particularly a heavy sclerotization of the valve, the loss of the corona, and a reduction in clasper size. Some species in the *antipoda* group have a digitus-like structure, a feature also found in some unrelated Palearctic species. The eighth abdominal sternum of the male abdomen lacks the apophyses found in most of the rest of the genus.

The female genitalia of most species have a weak, spiculate lamella antevaginalis formed from the membrane between the seventh sternum and ostium. This lamella antevaginalis is sometimes two lobed, as in the *asteris* group. The apex of the seventh sternum is the same as in the *asteris* group with a distal projecting ledge, which may be bilobed, and lateral areas of specialized scales that are very difficult to remove during dissection. The membrane between the seventh and eighth sterna is typical for the genus, i.e., tough and leathery. The apophyses anteriores are very short relative to those of other species groups in *Cucullia.* The corpus bursae of *antipoda* has a perpendicular orientation relative to the ductus bursae unlike the proximodistal orientation found in the previous groups. This orientation is more pronounced in some species than others but generally is true of all of them. The vesica of the male genitalia is two spined, although adventitious cornuti do occur. The two cornuti are borne on separate diverticula located in a V-shaped configuration opposite the long tail of the vesica. A small ventral diverticulum exists basal to the two major diverticula. Minute cornuti are located in the basal area of the vesica near the two diverticula.

The minute spines are continued along the manica of the aedoeagal shaft. The sclerotized band running from the aedoeagal shaft onto the vesica is extremely weak.

The larva of only one species in this group is known, *Cucullia mcdunnoughi.* It has a simple pattern of stripes with a yellow dorsal stripe and a subspiracular yellow line. This pattern most closely resembles that of the species in the *asteris* group. However, a striped pattern may be primitive for the genus and may not help much in determining the group's phylogenetic affinities.

It is not completely certain that the species in this group comprise a monophyletic unit. Much morphological diversity exists among the species. A knowledge of the larvae of more than one species may help clarify the situation.

Species attributed to *Copicucullia* (i.e., species with a tibial claw) have also been described from central Asia and Africa, but an examination of some of the species, particularly from southern Africa, shows no relation between them and the American species. Undoubtedly the characters "small size" and "tibial claw" used to define the genus are convergent characters evolved in response to a desert environment.

Cucullia antipoda Strecker

PL. 2, FIGS. 10–17; PL. D, FIGS. 11, 12 (♂ gen.) PL. H, FIG. 10 (♀ gen.); TEXT FIG. 23 *b* (ductus bursae); TEXT FIG. 25 (map) (RWH 10206).

Cucullia antipoda Strecker, 1877, *Lepidoptera, Rhopaloceres and Heteroceres,* **14**: 129.
Type locality: [USA], Arizona. [FMNH]

NOTE—In the original description Strecker states "several examples from Arizona." However in a later list of the types in the Strecker collection he says "one example, Arizona." There is currently a single specimen in the FMNH and I suspect Strecker's original citation of many specimens is inaccurate. Therefore I consider this single specimen to be the holotype. I have a color photograph of the type through the courtesy of Eric Metzler.

Cucullia propinqua Smith, 1894, *Trans. Amer. Ent. Soc.,* **21**: 85.
Type locality: [USA], Colorado. [USNM]; lectotype designated by Todd (1982).

NOTE—*Cucullia propinqua* Smith, 1984 is a junior primary homonym of *Cucullia propinqua* Eversmann 1842.

Copicullia [sic] *luteodisca* Smith, 1909, *Jour. New York Ent. Soc.,* **17**: 63. NEW SYNONYMY.
Type locality: [USA], New Mexico, [Luna County], Deming. [USNM]; lectotype designated by Todd (1982).

Copicucullia antipoda race *adopitna* Barnes and Benjamin, 1927, *Can. Ent.,* **59**: 6.
Type locality: [USA], California, Inyo Co., Olancha. [USNM]

Cucullia antipoda is a confusing species. It exhibits significant inter- and intrapopulation variability. There is a bewildering polymorphism within populations involving coloration, wing shape, and the development of a ridge on the front of the head. Two basic morphs can be defined. The forewing of one morph is uniform light, dull gray; and the hindwing is usually suffused with light brown, the base not contrasting with the outer margin. This morph also tends to be smaller and to have broader wings than the second morph. The second morph has a darker forewing with a slight blue tint, and the middle of the median area is distinctly lighter than the rest of the wing. The base of the hindwing is often significantly lighter than the outer margin. This second morph is, on average, larger and more narrow winged than the first morph. This division into two morphs is not always clear. Although many specimens within a locality tend to fall neatly into one category or the other, other individuals exhibit characteristics of both. Wing length: mean = 18.44 mm, standard deviation = 0.89 mm, n = 10.

The early stages of this species are unknown, but the foodplant(s) is likely to be a shrubby composite.

Cucullia antipoda is widespread in the desert regions of western North America, particularly in the Great Basin. It is known from southern British Columbia, eastern Washington, Oregon, throughout Utah, Nevada, central and western Colorado, northwest Wyoming, western North Dakota, northern and central Arizona, New Mexico (except in the east), and California (except along the central and northern western coast). Two populations of *antipoda* are more lightly colored than the typical population; a population at Olancha in east-central California and an isolated population at Callao, Utah on the edge of the Great Salt Lake Desert. The name *adopitna* is available for the eastern-central California population if a subspecies concept is employed. Both of these light populations are probably local adaptations to very arid conditions possibly with light-colored or salty soil. Other pale populations, although not as distinctive as the two mentioned above, occur in southern California. Populations in New Mexico are slightly larger than normal with a

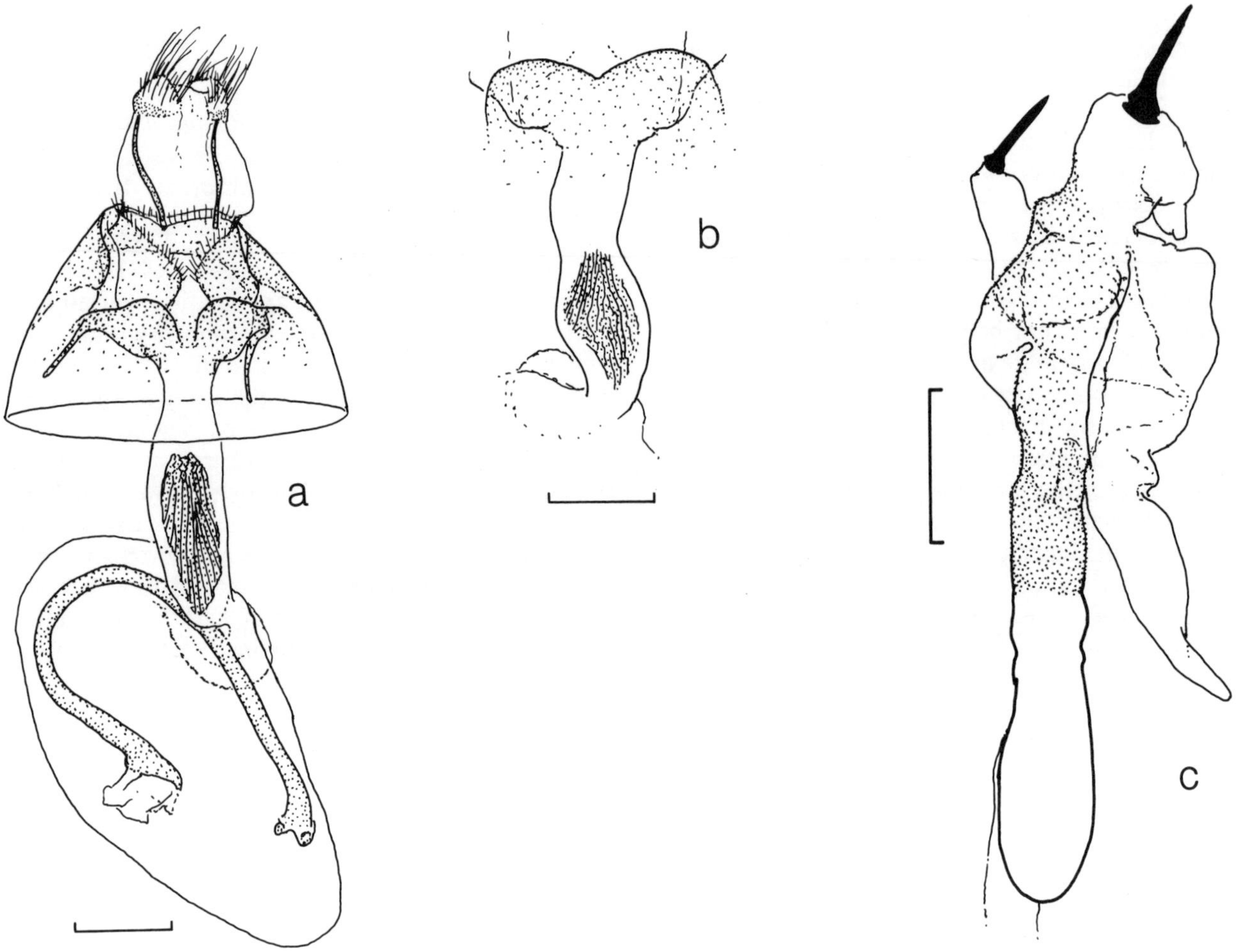

FIGURE 23: FEMALE AND MALE GENITALIA OF *CUCULLIA ANTIPODA* AND *CUCULLIA EUREKAE*
a. Female genitalia of *Cucullia eurekae. b.* Ductus bursae of *Cucullia antipoda. c.* Aedoeagus of *Cucullia eurekae.* Scale = 1 mm.

greater predominance of the dark form. Adults fly from June to August in most of the range, but fly earlier in southern California where specimens have been taken as early as late April and are commonest in May and June.

Cucullia eurekae Poole, NEW SPECIES
PL. 2, FIGS. 18, 19; PL. D, FIGS. 13, 14 (♂ gen.); PL. H, FIG. 11 (♀ gen.); TEXT FIG. 23 *a* (♀ gen.); TEXT FIGS. 23 *c,* 24 *a* (♂ gen.); TEXT FIG. 26 (map).

Cucullia eurekae Poole.
Type locality: USA, Utah, Juab Co., Eureka. [USNM]

I have segregated a small group of specimens from Eureka, Utah from the extensive series of *antipoda* collected by Thomas Spalding in the early part of this century. I believe these specimens represent a distinct species. On the other hand this entity may be only one more manifestation of the variability of *antipoda.* Four characteristics of these small specimens have led me to propose a specific name. 1) *Cucullia eurekae* is consistently smaller than *antipoda* from the same locality. The samples are not random or large enough to make statistical inferences, but a comparison of the figures in the colored plates should make the point. 2) the overall coloration of *eurekae* is duller and generally less contrasting than that of *antipoda.* 3) the valve of the male genitalia is narrower, on average, than in *antipoda,* and the corona tends to be reduced, reaching only half way across the apex of the valve. The larger of the cornutus-bearing diverticula in the vesica of *eurekae* is clearly double (i.e., with a second outpocketing of the diverticula), but clearly single in *antipoda.* A large bulbous outpocketing exists at the base of the long tail of the vesica in *eurekae,* but in

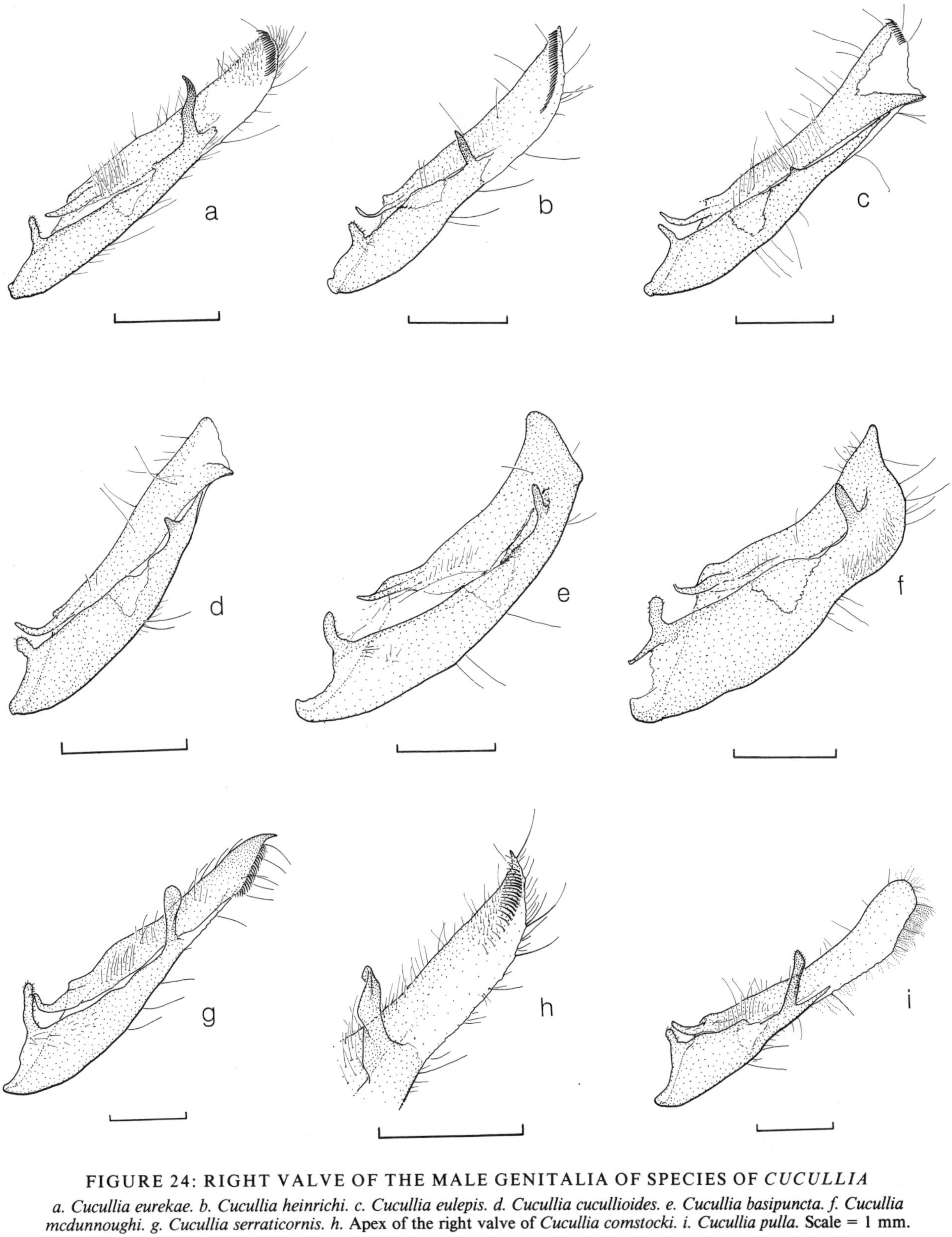

FIGURE 24: RIGHT VALVE OF THE MALE GENITALIA OF SPECIES OF *CUCULLIA*

a. Cucullia eurekae. b. Cucullia heinrichi. c. Cucullia eulepis. d. Cucullia cucullioides. e. Cucullia basipuncta. f. Cucullia mcdunnoughi. g. Cucullia serraticornis. h. Apex of the right valve of *Cucullia comstocki. i. Cucullia pulla.* Scale = 1 mm.

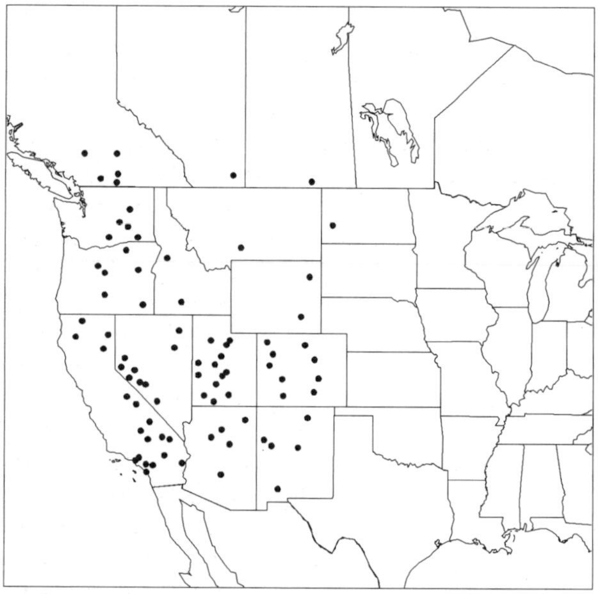

FIGURE 25: DISTRIBUTION OF SPECIMENS EXAMINED OF *CUCULLIA ANTIPODA*

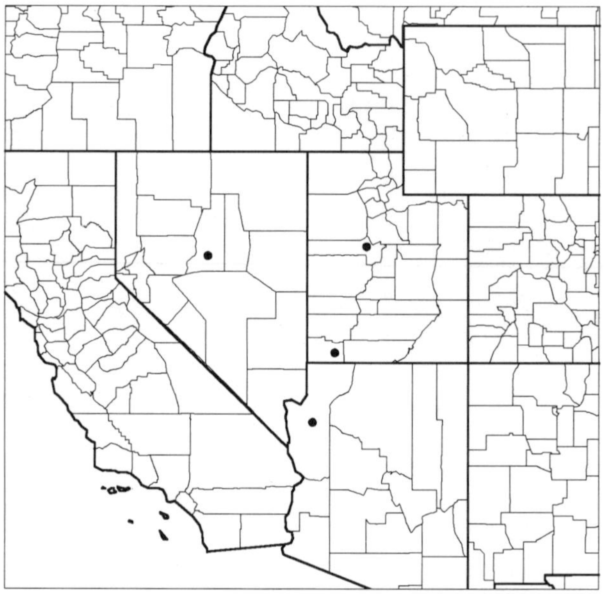

FIGURE 26: DISTRIBUTION OF SPECIMENS EXAMINED OF *CUCULLIA EUREKAE*

antipoda there are two or three small outpockets. 4) the lobes of the sternite covering the ostium of the female genitalia are deeply and distinctly bilobed in *eurekae,* but slightly invaginated in *antipoda.* This character of the female genitalia must be used with caution because of the tremendous variability of the genitalia in these two species. Only *eurekae* has been collected in Mohave County, Arizona.

Upper surface as in plate 2, figures 18, 19; coloration of forewing usually uniformly gray, with veins marked with black, or with a distinct, black inner margin. Head: frons with a variably developed ridge, scaled with intermixed black and white scales; raised row of black and white scales below antenna and another row between antennae; tuft below base of antenna well developed; antennae naked except for white scales dorsally; palpus with fringe of long white scales and even longer hairlike brown scales. Thorax: patagium with a least three strong, black horizontal lines on a gray-white background; tuft of long white hairs present below base of forewing and hindwing; metathorax with dorsal tuft; abdomen with tufts on abdominal terga two and three. Prothoracic leg: tibial claw well developed, tibia and femur strongly fringed posteriorly, scales of fringe banded with black; anterior margin of tibia with brown line; tarsus covered with mixed white and gray-brown scales. Mesothoracic and metathoracic legs: covered with mixed white and brown-gray scales. Male abdomen: basal hair pencils absent. Male genitalia: text figures 23 *c,* 24 *a*; variable; valve usually narrower and the corona less developed than in *antipoda.* Female genitalia: text figure 23 *a*; sternite covering ostium bursae invaginated, otherwise as in *antipoda.* Wing length: mean = 16.04 mm, standard deviation = 0.85 mm, n = 10.

The larva and its foodplants are unknown.

TYPES. Holotype: ♂, [USA], Utah, [Juab Co.], Eureka, VI-8-20, Tom Spalding. USNM. Paratypes, 9 ♂, 1 ♀ from the type locality. CNC, LACM, and USNM.

I have excluded from the type series a male from Bellevue, Washington Co., Utah, specimens from Nevada, and specimens from Mohave County, Arizona to restrict the name to the type locality. Adults have been collected in May and June in Utah and Nevada, and March in Arizona.

Cucullia incresa (Smith), NEW COMBINATION

PL. 2, FIGS. 20–23; PL. D, FIG. 15; PL. E, FIG. 1 (♂ gen.); PL. H, FIG. 12 (♀ gen.); TEXT FIG. 27 (map) (RWH 10212).

Copicucullia incresa Smith, 1910, *Jour. New York Ent. Soc.,* **18**: 97.
Type locality: [USA], California, Los Angeles Co., Claremont. [AMNH]; lectotype designated by Todd (1982).

Cucullia incresa is one of the perplexing problems that seem to pervade the genus *Cucullia.* Neither the male or female genitalia are consistently differ-

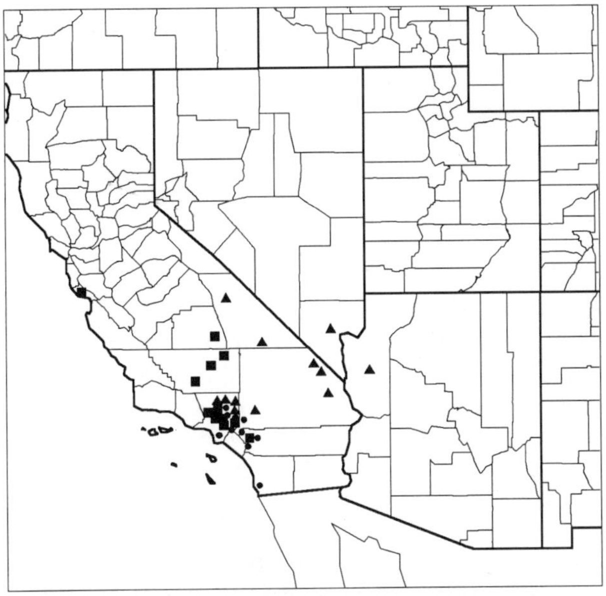

FIGURE 27: DISTRIBUTION OF SPECIMENS EXAMINED OF *CUCULLIA INCRESA*

The squares, circles, and triangles represent the three distinct populations of *incresa* discussed in the text.

ent from those of *antipoda*; the range in variability in each species overlaps that of the other. *Cucullia incresa,* however, has two distinguishing superficial characteristics: 1) the male hindwing is either pure white or white with light-brown accenting on the veins (varying among populations), and 2) the frontal ridge of the head is usually developed into a large, roughly triangular, in cross section, protuberance that is normally much more pronounced than in *antipoda.* The melanization of the abdominal sclerites in females is greater than in *antipoda.* This melanization is reflected, in particular, in a sclerotization of the seventh sternum into two bars that are not present in *antipoda.* Wing length: mean = 17.51 mm, standard deviation = 0.82 mm, n = 10.

The larva and its foodplants are unknown.

Cucullia incresa has three geographical races with relatively sharp boundaries, although many intergrades exist. One race occurs in the coastal and west-central mountains of California from Santa Cruz south to San Diego. If a subspecific concept is used, the nominal subspecies will refer to this race. These populations have a smooth-gray to dark-gray appearance, and the area of the forewing below the cell and toward the outer margin is usually strongly tinted with white. Specimens from the Los Angeles Basin are similar, but the white shading in the cell of the forewing is usually not nearly as pronounced. In both races the veins of the forewing are not distinctly accented, and the wing does not appear striate. The male hindwing in these two races is more nearly white, on average, than is normal in specimens from the Mohave region, and some specimens, notably specimens from Loma Linda in San Bernardino County, have a pure white male hindwing without any brown accenting on the veins. The third race of *incresa* is apparently a denizen of the Mohave Desert, from the western edge of the San Bernardino and San Gabriel Mountains, running north to the Owens Valley and Death Valley regions of Inyo County and east through southern Nevada and northwestern Arizona. This race is characterized by a black coloration of the forewings and a striate appearance caused by dark, black accenting on the veins. The middle of the forewing may be distinctly lighter than the costal or inner margins, but none of these contrasting specimens have the smooth, almost glossy, look of the coastal and central California populations. Although multiple intergrades exist between the first two races, the line of demarcation between them and the Mohave populations is very sharp at the eastern edge of the San Gabriel Mountains. The Mohave population is conceivably a distinct species. Adults have been collected from the middle of March until the middle of July. Most specimens were taken from April to early June.

Cucullia heinrichi Barnes and Benjamin, REVISED COMBINATION

PL. 2, FIGS. 24, 25; PL. E, FIGS. 2, 3 (♂ gen.); PL. I, FIG. 1 (♀ gen.); TEXT FIG. 24 *b* (♂ right valve); TEXT FIG. 28 (map) (RWH 10213).

Cucullia heinrichi Barnes and Benjamin, 1924, *Contrib. Nat. Hist. Lep. N. Am.,* **5**(3): 132.
Type locality: [USA], Arizona, Mohave Co. [USNM]

Cucullia heinrichi is a distinctive species in an obscure way. Its closest relative appears to be *incresa* and like *incresa* the color of the hindwing is sexually dimorphic; pure white in males and dirty brown in females. The characteristric features of the species are 1) the forewing maculation is obscure and in particular the orbicular and reniform marks are almost entirely absent, 2) sexual dimorphism of hindwing color exists, 3) a large, triangular frontal protuberance is present (as in *incresa*), and 4) the prothoracic tibial claw is shorter and more nearly

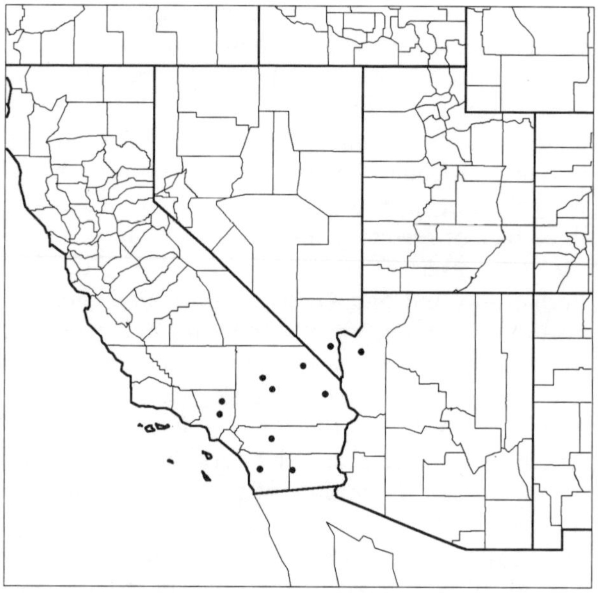

FIGURE 28: DISTRIBUTION OF SPECIMENS EXAMINED OF *CUCULLIA HEINRICHI*

triangular than in the preceding three species. The male genitalia are distinctive. A strong dentate band is present on the basal diverticulum of the male vesica. This band is present in *eulepis* and *cuculloides* but absent in *incresa.* A strong dentate patch exists next to one of the cornuti. The apex of the valve is pointed, not rounded, and the clasper is short. The uncus is greatly swollen about ¼ the distance from its pointed apex. The female genitalia are equally distinctive with a well-developed V-shaped sclerotization of the ostium. The specialized scales in the membrane of the seventh abdominal sternum are very weak and the membrane between the seventh and eighth sterna is not nearly as thick or as leathery as it is in *antipoda* and *incresa.* Wing length: mean = 16.75 mm, standard deviation = 0.42 mm, n = 6.

The larva and its foodplants are unknown.

This species has been collected in Mohave County, Arizona (the type locality) and San Diego, Los Angeles, Riverside, and San Bernardino Counties in California. All of the known localities are in the Mohave Desert. The few dated specimens are from February and April.

Cucullia astigma (Smith), NEW COMBINATION

PL. 2, FIGS. 26–28; PL. E, FIGS. 4, 5 (♂ gen.); PL. I, FIG. 2 (♀ gen.); TEXT FIG. 29 (map) (RWH 10211, 10214)

Copicucullia astigma Smith, 1894, *Trans. Amer. Ent. Soc.,* **21**: 85, pl. 4, fig. 8.
Type locality: [USA], Colorado. [USNM]; lectotype designated by Todd (1982).

Copicucullia jemezensis Dyar, 1922, *Insecutor Inscitiae Menstruus,* **10**: 167. NEW SYNONYMY.
Type locality: [USA], New Mexico, [Sandoval County], Jemez Springs. [USNM].

Cucullia astigma is an easily recognized species although the male and female genitalia indicate a close relationship with *antipoda* and its relatives. The species is larger than others in the group and is characterized by a smooth, gray forewing coloration. The only distinct forewing markings are the black inner margin, the usual black streak from the postmedial line to the outer margin, and a thin black basal dash running almost all the way to the postmedial line. There are two disjunct and distinct populations of *astigma,* one on southern California and the other in Colorado and New Mexico. The forewing usually is shaded with white in specimens from southern California. Both the reniform and orbicular marks are present, if not particularly distinct, in southern California. In contrast the forewing is uniform medium gray, and both marks are absent in Colorado and New Mexico specimens. The two populations may represent different species, but there is insufficient information at present to resolve the problem. The male genitalia are like those of *antipoda,* but the apex of the valve is narrower and the corona reduced. The clasper is long and straight lacking the slight kink found, to varying degrees, in *antipoda.* The female ductus bursae is longer and more heavily sclerotized than that of *antipoda.* Wing length (California): mean = 19.69 mm, standard deviation = 0.94 mm, n = 10.

The larvae of this species has been reported to feed on *Corethrogyne filaginifolia* (Hook. & Arn.) Nutt. and *Haplopappus squarrosus* Hook. & Arn. by McFarland (1975) under the name *jemezensis.* Both plants are shrubby composites. McFarland (1975) states that the larvae tend to predominate on one foodplant species or the other in successive years. The larva has not been described.

The typical population is known from only a few localities, notably Garfield County in Colorado and Jemez Springs in north-central New Mexico. The second population occurs in southern California. The valve of the male genitalia is slightly shorter and narrower than that of the Colorado-New Mexico population. The southern California population appears to live in the mountains. It is known from the San Bernardino and San Gabriel Mountains in the Los Angeles region, northward in the Tehachapi Mountains through Kern County to southern Tulare County. Adults have been captured from late May to early July.

Cucullia eulepis (Grote), NEW COMBINATION

PL. 2, FIGS. 29, 30; PL. E, FIGS. 6, 7 (♂ gen.); PL. I, FIG. 3 (♀ gen.); TEXT FIG. 24 *c* (♂ right valve); TEXT FIG. 30 (map) (RWH 10209).

Cleophana eulepis Grote, 1875, *Bull. Buffalo Soc. Nat. Hist.*, **3**: 86.
Type locality: [USA], Oregon. [AMNH]

Cucullia bistriga Smith, 1892, *Proc. U. S. Natl. Mus.*, **15**: 47.
Type locality: [USA], Colorado. [USNM]; lectotype designated by Todd (1982).

Copicucullia mala Smith, 1908, *Ann. New York Acad. Sci.*, **18**: 118.
Type locality: [USA], California, Witch Creek. [AMNH]

Cucullia eulepis is midway in size and appearance between *astigma* and *cucullioides*. Forewing length is approximately 16.5 mm compared with an average length of 19.5 mm in *astigma* and 13.5 mm in *cucullioides*. The forewing is smooth, light gray to gray without the striate effect found in *cucullioides*. The middle of the forewing is usually noticeably darker than the rest of the wing in *eulepis*, but the forewing is uniformly gray in *astigma*. The foretibia of *eulepis* bears a visible claw; the claw is absent in *cucullioides*. The first sternum of the male abdomen lacks hair pencils, but hair pencils are present in *cucullioides*.

The male genitalia (text figure 24 *c*) are more similar to those of *cucullioides* than to *astigma*. In particular the clasper is greatly reduced and the apex of the valve is produced into a ventral projection. The female ovipositer lobes are narrow and divided into a lower portion with a few spines and an upper portion bearing several strong spines. This condition is apparently intermediate between the normal ovipositer lobes of the genus and the extreme modification found in *cucullioides*.

Cucullia eulepis and *cucullioides* might not belong to the *antipoda* group. Although both species have a two-spined vesica, a strong dentate band exists at the base of the most basal cornutus-bearing diverticulum, an atypical condition for most of the species in the species group. Secondly, an obvious digitus exists in both species. A digitus is absent in the majority of species in *Cucullia* but is found in these two species and some species from the Palearctic, e.g., *Cucullia tanaceti* [Denis and Schiffermüller]. Third, the membrane between the seventh and eighth sterna of the female abdomen is not leathery and tough as in most other North American species of the genus. The seventh sternum is unmodified and lacks a distal ledge of specialized patches of scales in the lateral membranes. The eighth sternum is strongly sclerotized in both species. Two characters support the inclusion of *eulepis* and *cucullioides* in the *antipoda* group. The corpus bursae in both species is strongly perpendicular to the ductus bursae, even more so than in *antipoda*, and the apophyses anteriores are very short, a character shared with *antipoda* and its relatives. For the moment *eulepis* and *cucullioides* appear best placed in this group. Wing length: mean = 16.49 mm, standard deviation = 0.80 mm, n = 10.

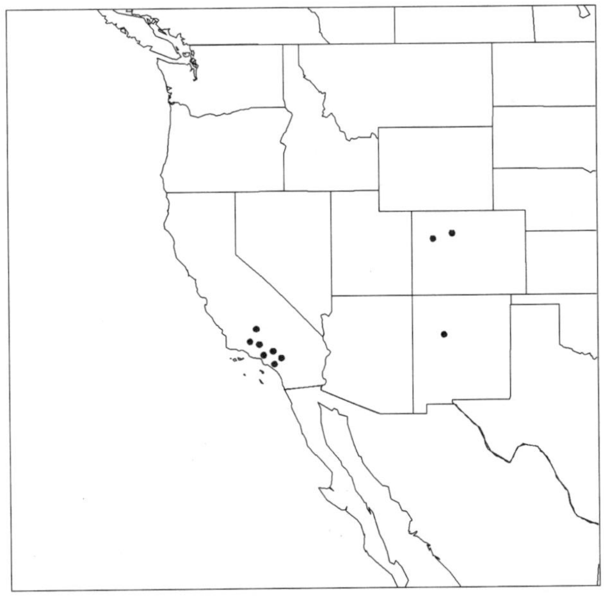

FIGURE 29: DISTRIBUTION OF SPECIMENS EXAMINED OF *CUCULLIA ASTIGMA*

The larva and its foodplants have not been recorded. A specimen in the LACM from Los Angeles County, California was reared from *Stephanomeria exigua* Nutt., an annual composite.

The distribution of *eulepis* is spotty. It occurs on the eastern, dry edge of the Cascades in Washington and Oregon. In California it has been collected throughout the state but perhaps more commonly in southern California, but this may be an artifact of collecting intensity. Its eastern limit is on the

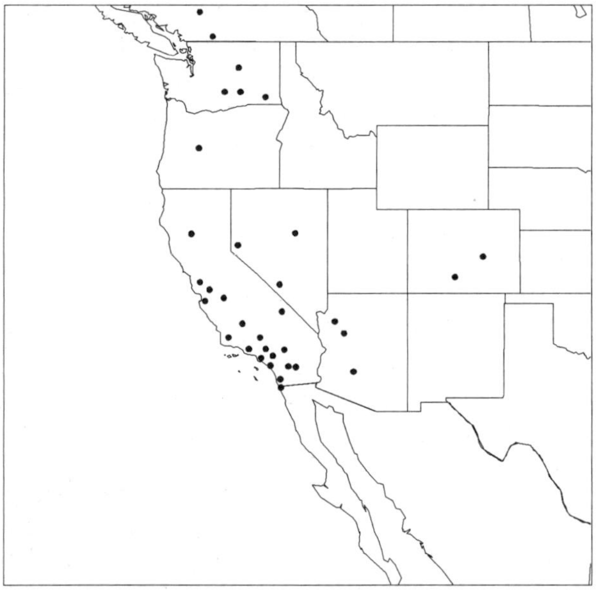

FIGURE 30: DISTRIBUTION OF SPECIMENS EXAMINED OF *CUCULLIA EULEPIS*

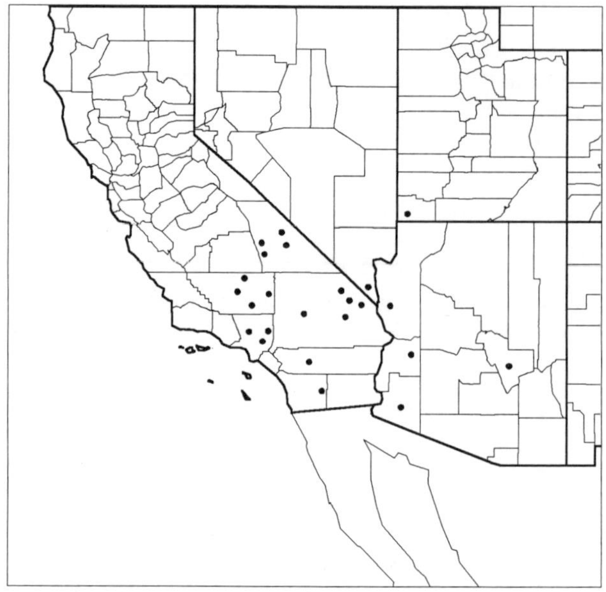

FIGURE 31: DISTRIBUTION OF SPECIMENS EXAMINED OF *CUCULLIA CUCULLIOIDES*

eastern slope of the Rockies in Colorado. It is also known from most of Nevada, and Mohave and Maricopa Counties in Arizona. Adults have been collected from March to October. There may be two peaks of abundance in southern California; one in late spring, and another in autumn.

Little geographical variability is apparent in *eulepis.* A small amount of within population variability is present in the distinctness of the markings and the degree of contrast between the costa and middle of the forewing. None of this variability is particularly significant. Females, on the whole, are slightly darker than males, particularly the hindwing.

Cucullia cucullioides (Barnes and Benjamin), NEW COMBINATION
PL. 2, FIGS. 31, 32; PL. E, FIGS. 8, 9 (♂ gen.); PL. I, FIG. 4 (♀ gen.); TEXT FIG. 24 *d* (♂ right valve); TEXT FIG. 31 (map) (RWH 10210).

Copicucullia cucullioides Barnes and Benjamin, 1923, *Contrib. Nat. Hist. Lep. N. Am.,* **5**(2): 75. Type locality: [USA], Arizona, Mohave Co. [USNM]

This species is remarkable. The tibial claw is almost entirely vestigial and the first sternum of the male abdomen bears both basal hair pencils and Stobbe's glands. These structures are absent in all other members of the *antipoda* group, despite the obvious close relationship between *cucullioides* and *eulepis.* Additionally the ovipositor lobes of the female are highly modified. They are large and armed with a few large, strong spines. These spines are also enlarged in *eulepis,* but in *cucullioides* the spines are larger and fewer in number. The apophyses posteriores of *cucullioides* are reduced in size and their apices rounded, but are unmodified in *eulepis. Cucullia cucullioides* resembles *eulepis* superficially. However, *cucullioides* is smaller (an average wing length of 14 mm versus 16.5 mm) than *eulepis.* The forewing of *cucullioides* has a slightly striate appearance, *eulepis* a smooth one. The hindwing is whiter at the base than in *eulepis.* Wing length: mean = 13.58 mm, standard deviation = 0.90 mm, n = 10.

The larva and its foodplants are unknown.

Cucullia cucullioides is primarily a species of the Mohave desert region of California and Nevada, but extends into western and central Arizona and southwestern Utah. There is no discernible geographical variation. A bit of within population variability exists in the distinctness of the markings. Adults have been collected in May through October.

Cucullia basipuncta (Barnes and McDunnough), NEW COMBINATION
PL. 2, FIGS. 33, 34; PL. E, FIGS. 10, 11 (♂ gen.); PL. I, FIG. 5 (♀ gen.); TEXT FIG. 24

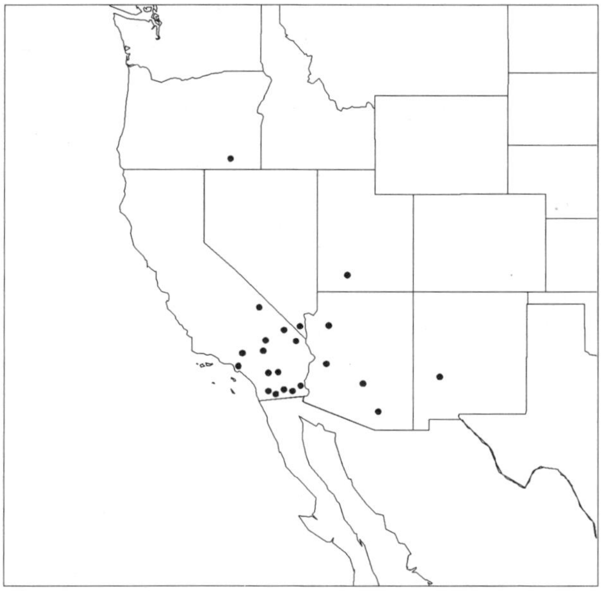

FIGURE 32: DISTRIBUTION OF SPECIMENS EXAMINED OF *CUCULLIA BASIPUNCTA*

e (♂ right valve); TEXT FIG. 32 (map) (RWH 10207).

Copicucullia basipuncta Barnes and McDunnough, 1918, *Contrib. Nat. Hist. Lep. N. Am.*, **4**(2): 100, pl. 16, fig. 5.
Type locality: [USA], California, Riverside Co., Palm Springs. [USNM]

Cucullia basipuncta is the first of two closely related species from the western United States; the other is *Cucullia mcdunnoughi.* The apophyses anteriores in both species are double and Y-shaped. The ductus bursae in both species is wide relative to its length, and the sclerotized area lacks the ridges found in other species in the genus. The male valve of both species is strongly sclerotized, and the clasper is located near the apex of the valve. The corona is absent. Both species are light gray with blotches of black, particularly on the costa of the forewing between the reniform and orbicular marks. *Cucullia basipuncta* and *mcdunnoughi* are superficially similar; the main difference is in the strength of the forewing postmedial line. The postmedial line of *basipuncta* is apparent all the way from the costa to the inner margin but is distinct only in the lower half of the wing. The postmedial line of *mcdunnoughi* is completely absent. The best differences between the two species are in the male genitalia. The apex of the valve is produced into a blunt spine in *mcdunnoughi* (text figure 24 *f*) but is not in *basipuncta* (text figure 24 *e*). The clasper is also larger in *mcdunnoughi,* and its tip touches the costa of the valve. The clasper is much smaller in *basipuncta* and does not approach the costal margin of the valve.

The affinities of these two species are problematical. They share the tibial claw and general superficial appearance of *antipoda* and *incresa,* but the male and female genitalia are highly modified. Wing length: mean = 15.75 mm, standard deviation = 1.03 mm, n = 10.

The larva has been described and figured by Comstock and Henne (1943). The larvae were reared on *Stephanomeria runcinata* Nutt., an herbaceous composite. The larval color is variable from light green to dark gray. There is a dorsal yellow stripe and a lateral wide, white band.

Cucullia basipuncta has been most commonly collected in the Mohave Desert region. It has been collected widely in southern California from the Mexican border throughout the Mohave region, as far north as Inyo County. It also appears to be fairly common in southern Nevada and western Arizona. There are also single specimens from more scattered localities. It has been collected in Harney County in southeastern Oregon, Garfield County in south-central Utah, Pima County in south-central Arizona, and as far east as Lincoln County in central New Mexico.

Adults have been collected from March to late May.

Some within population variation occurs in the male genitalia. In specimens from Palm Springs, the type locality, the clasper varies from absent to moderately well developed. Males and females are sexually dimorphic for hindwing color: the male hindwing is pure white, but the female hindwing is suffused with brown.

Cucullia mcdunnoughi (Henne), NEW COMBINATION

PL. 2, FIGS. 35, 36; PL. E, FIGS. 12, 13 (♂ gen.); PL. I, FIG. 6 (♀ gen.); TEXT FIG. 24 *f* (♂ right valve); TEXT FIG. 33 (map) (RWH 10208).

Copicucullia mcdunnoughi Henne, 1940, *Bull. So. California Acad. Sci.*, **39**: 71.
Type locality: [USA], California, San Bernardino County, Providence Mountains. [LACM]

The differences separating *mcdunnoughi* from its close relative *basipuncta* are discussed under *basipuncta.* Wing length: mean = 16.27 mm, standard deviation = 0.67 mm, n = 10.

The larva has not been described, but there is a

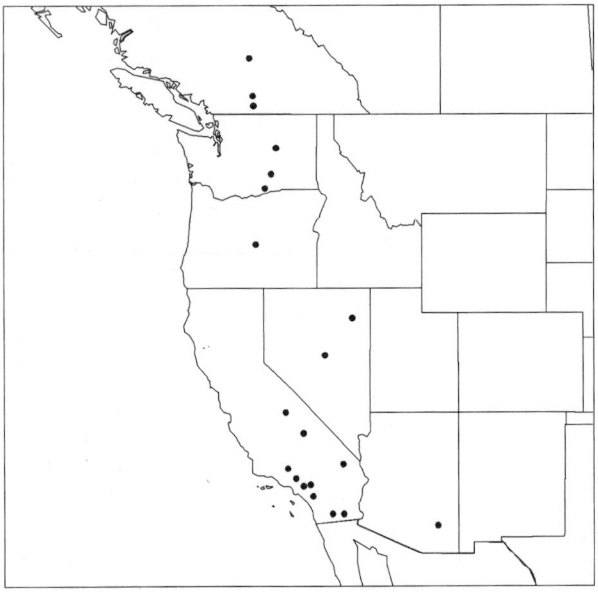

FIGURE 33: DISTRIBUTION OF SPECIMENS EXAMINED OF *CUCULLIA MCDUNNOUGHI*

specimen in the LACM reared by Christopher Henne from *Stephanomeria* sp. (Asteraceae) and apparently collected along with larvae of *basipuncta* from the same foodplant.

Although *mcdunnoughi* and *basipuncta* have both been collected at Palm Springs in southern California, *mcdunnoughi* has a wider and more northerly distribution. It has been collected in southern and east-central California, southern Arizona, northern and central Nevada, and eastern Washington. *Cucullia mcdunnough* is undoubtedly a desert species as are the other species in this group. Adults have been collected from April through July.

luna GROUP

Only a single species is involved and it is easily identifiable with no indecision, equivocation, or difficulties. The only species in this species group, *Cucullia luna,* is characterized by the pure white, pearly forewings and a strong foretibial claw with a small backwardly projecting plate. *Cucullia luna* has no obvious North American relatives. At first glance the pearly white scaling of the forewing would make the Palearctic group of species centered about *Cucullia argentea* (Hufnagel) (a beautiful silvery white and metallic-green species from Europe and Asia) the most obvious sister group candidate. In fact *argentea* has an extremely reduced foretibial claw; however, the two species have very different male and female genitalia.

I strongly suspect that *Cucullia luna* is a relative of or highly divergent member of the *antipoda* group. Some features they share 1) the tibial claw on the prothoracic leg, 2) the absence of lateral apodemes on the eighth abdominal sternum of the male abdomen, and 3) the presence of two cornuti in the vesica. The most basal of the two diverticula in the vesica of *luna* has a dentate band as in *eulepis* and *cucullioides* of the *antipoda* group. This band also occurs in the *convexipennis* group. The female apophyses anteriores are relatively short, a feature found in the *antipoda* group, but not limited to it. The ductus bursae is wide and lacks distinct internal ridges, a condition reminiscent of *basipuncta* and *mcdunnoughi,* if possibly not synapomorphic. Larval morphology might help to determine the relationships of *luna* with the other species in the genus, but the larva is unknown.

Cucullia luna Morrison

PL. 2, FIGS. 37, 38; PL. E, FIGS. 14, 15 (♂ gen.); PL. I, FIG. 7 (♀ gen.); TEXT FIG. 34 (map) (RWH 10205).

Cucullia luna Morrison, 1875, *Proc. Boston Soc. Nat. Hist.,* **18**: 122.
Type locality: [USA], Wyoming, Banks of the Yellowstone River. [FMNH]

Nycterophaeta magdalena Hulst, 1882 [February], *Bull. Brooklyn Ent. Soc,* **4**: 45.
Type locality: [USA], South Dakota, Black Hills. [AMNH]

NOTE—Hulst based this name on several specimens from several localities. However, the only locality listed was for the specimen from the Black Hills. Consequently, this specimen in the AMNH is designated LECTOTYPE.

Epinyctis notatella Grote, 1882 [April], *Can. Ent.,* **14**: 75.
Type locality: [USA], Montana.

NOTE—Grote gave neither the number or sex of specimens in the type series. The type or types are apparently not in the BMNH. In manuscript notes that came with the Barnes collection to the USNM, the type is listed as in the Neumoegen collection. This collection came to the USNM, but I have not been able to find such a specimen in the USNM.

How such a pretty and distinctive species could have been described three times in three different genera is a mystery. The silvery white color and elongate forewings will immediately distinguish *luna* from any other North American noctuid. It has served as the type of two generic names, both based on the

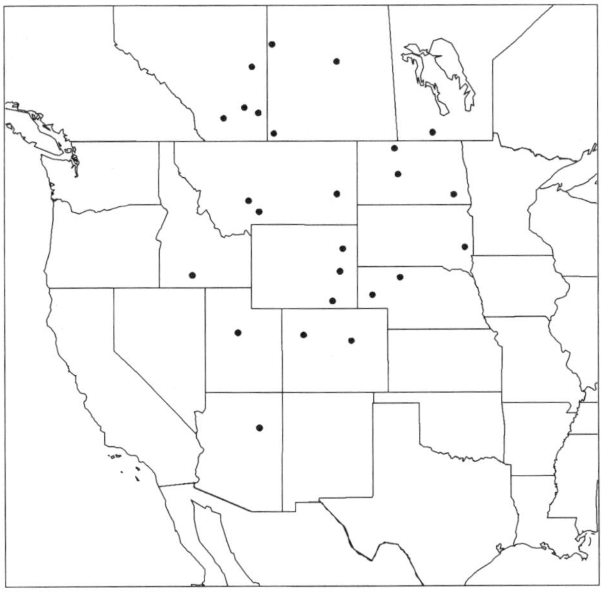

FIGURE 34: DISTRIBUTION OF SPECIMENS EXAMINED OF *CUCULLIA LUNA*

presence of a foretibial claw projecting into anterior and posterior points. Wing length: mean = 20.40 mm, standard deviation = 0.75 mm, n = 10.

The larva and its foodplants are unknown.

Cucullia luna is found primarily in the northern plains states of the United States and the Prairie Provinces of Canada. It is moderately common in North and South Dakota, Montana, Wyoming, south into Nebraska and Colorado, and as far west as central Utah. It occurs as far south as Navajo County in northeastern Arizona. The collecting localities with which I am familiar are all located in grassland of one sort or another. Adults have been collected in June and July.

strigata GROUP

This western North American species group was the basis of the generic concept *Rancora.* The adults fly in the spring, and they possess many adaptations often found in early flying moths. In particular they are "fuzzier" than other members of the genus, and the palpi are reduced in size. The palpi are covered with erect, fine brown hairs giving them a bristley look. A characteristic feature of the group is the presence on the ventral side of the hindwing of a black line running from the discal dot to the wing base. Each male antennal segment is ventrally expanded into a laterally projecting plate covered with fine hair. This antennal character, the discal dot line, the palpal characters, and the spring flight period are shared with the *pulla* group. However, the proboscis is greatly reduced in the *pulla* group but normally developed in the *strigata* group.

The apex of the valve is produced into a point and the clasper is located near the apex. The clasper is usually flattened into a spoon-shaped process. There are two cornuti in the vesica, each on a separate diverticulum. A broad scobinate band is present at the base of the more basal diverticulum, and this band constricts the diverticulum's base. Basal abdominal hair pencils are absent. The lamella antevaginalis is fleshy and bilobed as in the *asteris* group. The lobes are particularly strong in *serraticornis* but much less well developed in other species of the group. A small signum is in the bursa. The fleshy, bilobed lamella antevaginalis argues for a relationship to the *asteris* group, but in habits and superficial appearance this group seems to have more in common with *intermedia* and the *speyeri* group. Indeed *Cucullia intermedia* has a vaguely fleshy, bilobed lamella antevaginalis, and in all other characters of the female genitalia the *strigata* group is apparently most closely related to *intermedia* and its Old World relative *lucifuga.*

The four species of the *strigata* group are 1) *Cucullia serraticornis* Lintner from southern Arizona and the coastal areas of California from the Mexican border north to northern California, 2) *Cucullia comstocki* McDunnough from the Mohave desert, 3) *Cucullia strigata* Smith from Oregon north through British Columbia to Alaska, east to the border of Alberta, and south into Utah and western Colorado, and 4) *Cucullia albida* Smith from the Alberta and British Columbia border east to Saskatchewan and south to central Colorado.

Cucullia serraticornis Lintner

PL. 2, FIGS. 39–41; PL. F, FIGS. 1, 2 (♂ gen.); PL. I, FIG. 8 (♀ gen.); TEXT FIG. 24 *g* (♂ right valve); TEXT FIG. 35 (map) (RWH 10184).

Cucullia serraticornis Lintner, May 1874, *Entomological Contributions,* **3**: 174.
Type locality: [USA], California. [NYSM]

NOTE—This name and Strecker's two names *matricariae* and *solidaginis* are listed as published in May of 1874. However, Strecker was notorious for falsifying his dates of publication, and the earliest reliable date for his two names is September 1874 (Brown, 1964). *Cucullia serraticornis* was described from two males. These syntypes should be in the New York State Museum. However they were not listed by McCabe and Johnson (1980). It is clear from the original description, however, that they represent this species.

Cucullia solidaginis Strecker, September 1874, *Lepidoptera, Rhopaloceres and Heteroceres, Indigenous and Exotic,* **10**: 94. NEW SYNONYMY.
Type locality: [USA], California. [FMNH]

NOTE—*Cucullia solidaginis* is the female of *serraticornis.* The name has been widely misapplied to specimens of *strigata.* Strecker had two specimens. I have seen color photographs of both syntypes through the courtesy of Eric Metzler and both are clearly females of *serraticornis.* Both specimens are identically labeled. I have not seen the actual specimens; therefore, I am not designating a lectotype. They probably came from the San Francisco region.

Cucullia matricariae Strecker, September 1874, *Lepidoptera, Rhopaloceres and Heteroceres, Indigenous and Exotic,* **10**: 94.
Type locality: [USA], California. [FMNH]

NOTE—*Cucullia matricariae* is the male of *serraticornis.* Strecker had two specimens. I have seen photographs of both syntypes, again through the efforts of Eric Metzler. As before, I am not designating a lectotype. The specimens probably came from the San Francisco region.

Cucullia serraticornis Lintner is a sexually dimorphic species known from southern Arizona and California from the Mexican border as far north as the Oregon border. The males are uniformly gray with little significant maculation except for various fine black streaks. The male hindwing is pure white dorsally except for darker veins and a thin brown outer margin. The female is darker than the male and suffused with smoky brown. The female hindwing is dark smoky brown, not white.

Cucullia serraticornis is superficially similar to *Cucullia comstocki.* Males of *comstocki* are more contrastingly marked than those of *serraticornis* and have distinct patches of dark scales between the veins of the hindwing. Males of *comstocki* have a distinctly striate appearance normally lacking in *serraticornis,* and the veins of the hindwing are more strongly streaked with black. Females of *serraticornis* and *comstocki* are very similar although females of *comstocki* tend to have more sharply defined light and dark areas in the forewing. The reniform mark in particular is distinct in *comstocki* but normally diffuse to absent in *serraticornis.* There are distinctive differences between the species in the shape of the male valve. The valve is longer and narrower in *serraticornis* than in *comstocki.* The ratio of the length of the valve from the base of the sacculus to

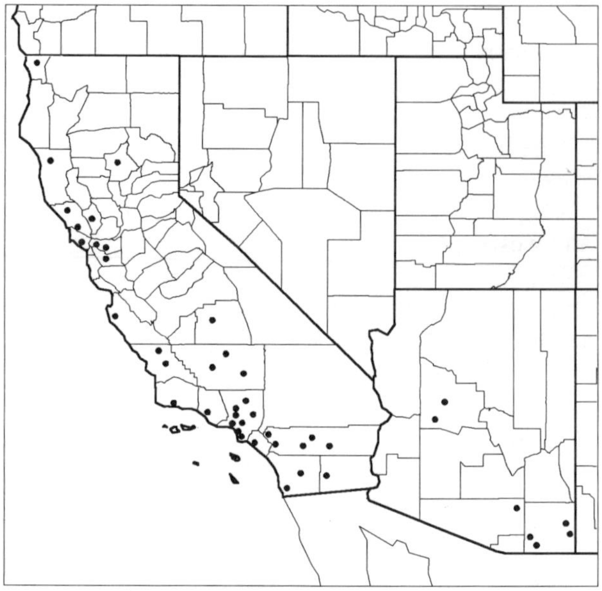

FIGURE 35: DISTRIBUTION OF SPECIMENS EXAMINED OF *CUCULLIA SERRATICORNIS*

the apex is approximately 5.5 in *serraticornis* versus 4.7 in *comstocki.* The apex is elongate and pointed outward in *serraticornis* (text figure 24 *g*). The apex is pointed in *comstocki* as well, but the apex is bent upward (in the orientation of the figure) (text figure 24 *h*). The clasper is shorter and thicker in *comstocki* than in *serraticornis.* The spines of the corona run along the outer edge of the valve in *serraticornis,* but in *comstocki* the coronal spines turn inward on the valve at about half the width of the valve. The basal processes of the sacculus are slightly longer in *serraticornis* than in *comstocki.* No differences have been found between vesicas of the two species. I have been unable to find any consistent differences between the female genitalia of *serraticornis* and *comstocki.*

Cucullia serraticornis may be distinguished from *strigata* by the lack of sexual dimorphism and the more distinct maculation of *strigata.* The clasper of the male valve is much thicker and broader in *strigata* (about 3 × as long as wide) than in *serraticornis* (about 5–6 × as long as wide). Females of the two species are easily separated by the absence of sexual dimorphism in *strigata* (i.e., the female looks like the male). The female genitalia are easily separable as well. The ovipositor neck is twice as long as wide in *strigata,* but equally wide as long in *serraticornis.* In addition the ductus bursae is about 4–5 × longer than wide in *strigata,* but only 3 × as long in *serraticornis. Cucullia albida* occurs well outside the range of *serraticornis*. The male genitalia of *albida* are virtually identical to those of *strigata*. Wing length: mean = 21.61 mm, standard deviation = 1.08 mm, n = 10.

The life history and foodplants appear to be unknown.

Cucullia serraticornis lives in southern Arizona and western California. In California it occurs from San Diego in the far south, throughout the western and central portions of the state at least as far north as Del Norte County on the Oregon border. It appears to be most common in the coastal regions of California and absent from the eastern portions of the state. The apparent lacuna between the Arizona and California populations may be the result of inadequate collections of these early spring flying moths. Southern California specimens have been taken in January and February, San Francisco ones in March, and Arizona specimens in March and April.

Cucullia comstocki (McDunnough), NEW COMBINATION

PL. 3, FIGS. 1, 2; PL. F, FIGS. 3, 4 (♂ gen.); PL. I, FIG. 9 (♀ gen.); TEXT FIG. 24 *h* (♂ right valve); TEXT FIG. 36 (map) (RWH 10185).

Rancora comstocki McDunnough, 1937, *Can. Ent.*, **69**: 77.
Type locality: [USA], California, San Bernardino Co., Mohave Desert, Yermo. [CNC]

Cucullia comstocki closely resembles *serraticornis*. The characters used to separate the two are given under *serraticornis*. Wing length: mean = 20.82 mm, standard deviation = 0.93 mm, n = 10.

The life history and foodplants are unknown.

Cucullia comstocki occurs in the California Mohave Desert region. It is found in a restricted area including Los Angeles County, north into Kern and Santa Barbara Counties, and southward into San Diego and Riverside Counties. It does not seem to occur in the southeastern regions of California. Adults have been collected from December to February.

Cucullia strigata (Smith), NEW COMBINATION

PL. 3, FIGS. 3, 4; PL. F, FIGS. 5, 6 (♂ gen.); PL. I, FIG. 10 (♀ gen.); TEXT FIG. 37 (map) (RWH 10183, 10186, 10187.3).

Rancora strigata Smith, 1892, *Ent. News*, **3**: 253, pl. 10, upper left corner.

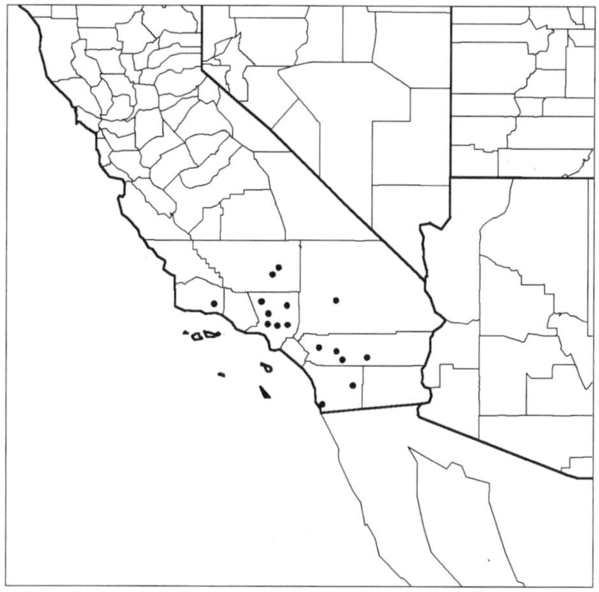

FIGURE 36: DISTRIBUTION OF SPECIMENS EXAMINED OF *CUCULLIA COMSTOCKI*

Type locality: [Canada], British Columbia, Victoria. [USNM]

NOTE—Smith repeated the description of *strigata* in 1894, *Trans. Amer. Ent. Soc.*, **21**: 83.

Rancora brucei Smith, 1903, *Can. Ent.*, **35**: 136. NEW SYNONYMY.
Type locality: [USA], Colorado, Garfield Co. [AMNH]

Rancora ketchikana Barnes and Benjamin, 1922, *Contrib. Nat. Hist. Lep. N. Am.*, **5**(1): 30. NEW SYNONYMY.
Type locality: [USA], Alaska, Ketchikan. [USNM]

Cucullia strigata is characterized by 1) the fairly uniform dark-gray color of the forewing, 2) the presence of dark, prominent black streaks in the terminal and subterminal areas of the forewing, 3) a prominent black basal dash in the forewing, and 4) the dark coloration of the hindwing in both males and females. All of these characters are subject to pronounced within population variability. In particular the black streaks on the forewing vary from pronounced with black accenting to almost entirely absent. The color of the male hindwing varies from brown to white with patches of dark-brown scales.

Cucullia strigata may be separated from *serraticornis* by the darker, more uniform gray forewing color, the brown coloration of the male hindwing, and the presence of black streaks on the forewing.

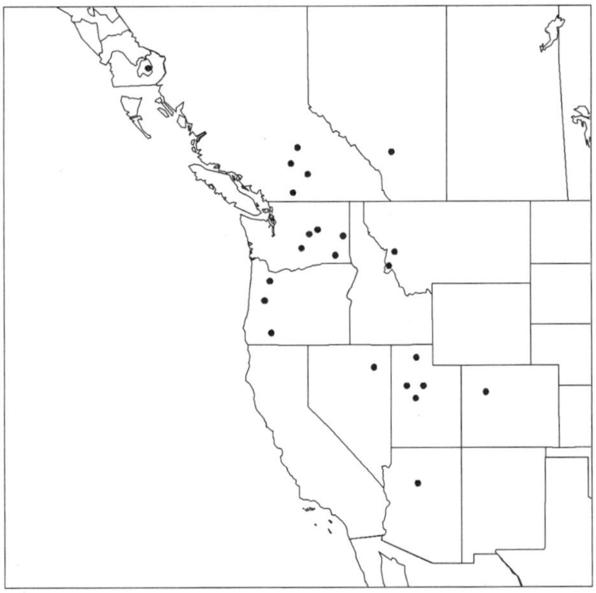

FIGURE 37: DISTRIBUTION OF SPECIMENS EXAMINED OF *CUCULLIA STRIGATA*

The differences in the male and the female genitalia between *strigata* and *serraticornis* are discussed under *serraticornis* but in summary involve a much longer neck of the ovipositor and ductus bursae and the wider clasper in *strigata*. *Cucullia strigata* is usually much darker than *albida*. There are small differences in the male genitalia between *albida* and *strigata*. In particular the apex of the valve is slightly more strongly produced in *strigata*. The variability within a species and between species is so broad that I am not sure this difference will be consistent. The major genital difference between the two species is found in the female. The ovipositor neck is about as long as wide in *albida*. In contrast the ovipositor neck is approximately 3 × longer than wide in *strigata*. The two species appear to be mostly allopatric, but there are indications of some overlap in their ranges. Specimens of both species are labeled Calgary, Alberta. The types of *albida* and of one of the synonyms of *strigata* (*brucei*) are both labeled Garfield County, Colorado. In all the material collected by Barnes at Glenwood Springs, the presumed collecting locality in Garfield County, there are no specimens of *albida*. Calgary, Alberta as a locality was used in a very broad sense in earlier days. Therefore neither case of range overlap can be considered proven. Wing length: mean = 21.74 mm, standard deviation = 0.53 mm, n = 10.

The life history and foodplants are unknown.

Cucullia strigata is primarily a species of the Pacific Northwest and the Great Basin. It occurs in Utah and western Colorado, Oregon, north through Washington, British Columbia, and as far north as the southern most extension of Alaska. These Alaskan specimens were the basis of the name *ketchikana* and are darker than typical *strigata*. The southernmost extension of the species is along the Mogollon Rim at Sedona in north-central Arizona. Adults have been collected in April and May.

Cucullia albida Smith

PL. 3, FIGS. 5, 6; PL. F, FIGS. 7, 8 (♂ gen.); PL. I, FIG. 11 (♀ gen.); TEXT FIG. 38 (MAP) (RWH 10187.1, 10187.2).

Cucullia albida Smith, 1894, *Trans. Amer. Ent. Soc.*, **21**: 84, pl. 4, fig. 9. Type locality: [USA], Colorado. [USNM]

Rancora albicinerea Smith, 1903, *Can. Ent.*, **35**: 137. NEW SYNONYMY. Type locality: [Canada], Alberta, Calgary, head of Pine Creek. [AMNH]; lectotype designated by Todd (1982).

Cucullia albida is a much lighter species than *strigata*, and the forewing of fresh specimens has a distinctly bluish tinge. The name *albida* is based on extremely light, almost white, Colorado specimens collected in the latter part of the nineteenth century. The extreme lightness of these Colorado specimens is probably due to their poor condition and fading. However, no new, fresh specimens are known from Colorado, and a comparison cannot be made. If a subspecific concept is employed, the name *albida* applies to the Colorado population and *albicinerea* to populations in Alberta and Saskatchewan. Other than the color differences, these two populations are very similar and the male genitalia do not differ significantly. The separation of *albida* and *strigata* and areas of possible overlap in their ranges are discussed under *strigata*. Wing length: mean = 20.50 mm, standard deviation = 0.74 mm, n = 10.

The life history and foodplants are unknown.

Cucullia albida occurs in Saskatchewan, Alberta, and extreme eastern British Columbia. It has also been collected in central Colorado. Adults with collection dates were all taken in May.

pulla GROUP

The *pulla* group contains two species comprising the former generic grouping *Lathosea*. The species group is characterized by a short, stubby forewing, heavily

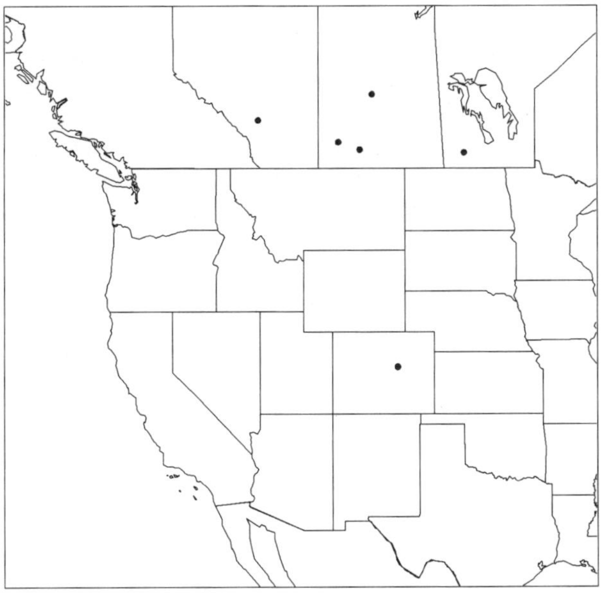

FIGURE 38: DISTRIBUTION OF SPECIMENS EXAMINED OF *CUCULLIA ALBIDA*

serrate male antennae, a marked reduction in the development of the proboscis, bristly, hairy palpi, and the stocky, woolly appearance of the thorax and abdomen. Basal hair pencils are absent from the male abdomen. The valves lack a corona and are very long. There are two cornuti in the vesica, a denticulate band at the base of the vesica, and a scobinate band around the base of the basal cornutus-bearing diverticulum (as in the *serraticornis* group). The female genitalia are virtually identical to those found in the *serraticornis* group. The superficial characters of serrate antennae and hairy vestiture are apparently adaptations to the early emergence of the adults, December and January in southern California, March through May in Utah, Colorado, Oregon, and Washington. The *pulla* group is clearly closely related to the *serraticornis* group despite the autapomorphic characters. The male genitalia are variable in shape, both within and between localities. Apparently the larvae are as well because Crumb (1956) divides the single species *pulla* into four species on the basis of larval coloration. I have found no consistent genital characters separating *pulla* and *dammersi*, despite the superficial distinctness of the two. Crumb separated the genus "*Lathosea*" from *Cucullia* on the basis of the absence of a basal process on the oral face of the mandible in *Lathosea*. A basal process is present in the other North American species of *Cucullia* I have seen.

Cucullia pulla (Grote), NEW COMBINATION
PL. 3, FIGS. 7–9; PL. F, FIGS. 9, 10 (♂ gen.); PL. I, FIG. 12 (♀ gen.); TEXT FIG. 24 *i* (♂ right valve); TEXT FIG. 39 (map) (RWH 10180, 10182).

Lathosea pulla Grote, 1881, *Bull. U. S. Geol. Geog. Surv. Terr.*, **6**(2): 270.
Type locality: [USA], Oregon. [AMNH]

Lathosea pullata Grote, 1890, *Revised Check-list of the North American Noctuidae. Part I. Thyatirinae–Noctuinae,* p. 14.
NOTE—*Lathosea pullata* is an unjustified emendation of *Lathosea pulla,* and is, therefore, a junior objective synonym of *pulla* with the same type and type locality.

Lathosea spauldingi Barnes and Benjamin, 1922, *Contrib. Nat. Hist. Lep. N. Am.,* **5**(1): 29. NEW SYNONYMY.
Type locality: [USA], Utah, Eureka. [USNM]

NOTE—*Lathosea spauldingi* is often emended to *spaldingi* to correspond to the correct spelling of the collector's name.

Cucullia pulla is a distinctive moth with a robust, hairy, intense black appearance. Its general appearance plus the serrate to pectinate antennae of the male and a bright white fringe on the hindwings should readily identify it. The male genitalia vary considerably within a single location and among localities, a property of many other species of *Cucullia.*

The larvae of *pulla* pose a problem. Crumb (1956) divided the larvae I refer to *pulla* into four species although there are two basic morphotypes in this material. The situation is confused because no adult voucher specimen was reared for one of Crumb's species, and the voucher for another is missing. The genital preparation for his "species 2" is missing. Both the voucher and slide are present only for his "species 4." The four "species" described by Crumb fall into two basic types. In one type the larval pattern is like that of *dammersi* (see following), a complicated pattern of black patches between the dorsal line and the spiracles. In the other type these black patches are reduced to small lines and dots and the pattern is dominated by dorsal, subdorsal, and subspiracular lines. The head varies in its markings from having conspicuous dark-brown or black freckles to almost totally lacking markings. The lined form usually does not have a central black spot on

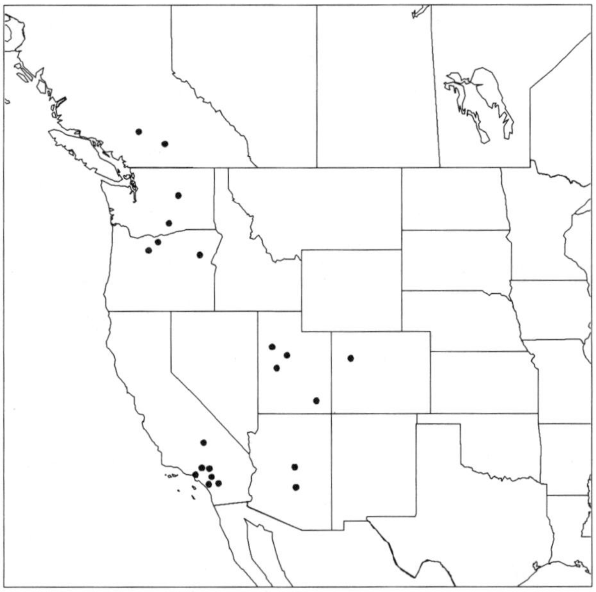

FIGURE 39: DISTRIBUTION OF SPECIMENS EXAMINED OF *CUCULLIA PULLA*

each proleg, but a spot is present (as in *dammersi*) in the more complicated form.

Three possible alternatives are; 1) more than one species is involved, 2) the larvae are polymorphic and highly variable, or 3) one or more of these "species" may be the larva of *Cucullia strigata* or *serraticornis.* I am adopting the second hypothesis but suggest that this problem needs to be further examined. Wing length: mean = 20.04 mm, standard deviation = 1.00 mm, n = 10.

Assuming Crumb's four "*Lathosea* species" are indeed all *pulla* (a voucher specimen for one Arizona larva is the only one available), the larva feeds on *Chrysothamnus* (rabbitbush) in Washington and Idaho and *Gutierrezia wrighti* Gray and "*Chrysoma laricifolia*" in Arizona, all shrubby composites (Asteraceae).

Cucullia pulla is distributed primarily in the low, dry montane areas of the Great Basin and Mohave Deserts. It has not been collected in the rest of California or Nevada. It occurs as far east as western Colorado and as far north as eastern and central Washington. The southern limit of its range is in central Arizona. *Cucullia pulla* is apparently sympatric with *dammersi* in the Mohave Desert area. There appears to be little significant geographical variation in the species, although the Colorado and Utah specimens are a bit browner and darker than other populations of the species. There is a single specimen from Los Angeles County, which is slightly smaller and grayer than other specimens from the area; but I believe it is only a dwarfed individual. The adults fly in early spring (January to March), and the larvae occur in the spring, pupating well before summer. The significance of this early flight period is unknown, but possibly the larvae feed on the newly developing foliage and avoid a significant proportion of the secondary plant chemicals found in many shrubby composites.

Cucullia dammersi (McDunnough), NEW COMBINATION

PL. 3, FIGS. 10, 11; PL. F, FIGS. 11, 12 (♂ gen.); PL. J, FIG. 1 (♀ gen.); TEXT FIG. 40 (map) (RWH 10181).

Lathosea dammersi McDunnough, 1935, *Can. Ent.,* **67**: 135.
Type locality: [USA], California, Riverside. [CNC]

Cucullia dammersi is morphologically very similar to *pulla* but superficially very distinct. The valve of *dammersi* (plate F, figure 11) tends to be slightly narrower and more elongate than in *pulla* (plate F, figure 9) although the difference is obscured by the variability within each species. *Cucullia dammersi* is far less robust and hairy than *pulla.* The forewing is dark gray with rusty brown shades in the subterminal area, not black as in *pulla.* The forewing costa is lighter than the rest of the wing. The hindwing is sexually dimorphic; dirty white with conspicuous veins in the males and dark brown in the females. Wing length: mean = 19.87 mm, standard deviation = 0.62 mm, n = 10.

The early stages have been described and figured by Comstock and Dammers (1935). The larva is described as blue gray with a complicated pattern of black patches. There is a dorsal yellow line, interrupted into approximately two yellow patches per abdominal segment and also a yellow band below the spiracles. A complicated mixture of black patches surrounded by the blue-gray background exists between the dorsal and subspiracular bands. Each proleg is green with a central black spot. The head is blue gray with numerous irregular black patches. The superficial impression is a pattern very much like that found in the *convexipennis* and *speyeri* groups. The recorded foodplants are *Acamptopappus sphaerocephalus* (Harv. & Gray) Gray, *Ericameria linearifolia* (DC.) Urbatsch & Wussow, *E. palmeri* (Gray) H.M. Hall, *E. pinifolia* (Gray) Hall, and *Gutierrezia sarothrae* (Pursh) Britton & Rusby, all shrubby composites (Asteraceae) of the Mohave Desert.

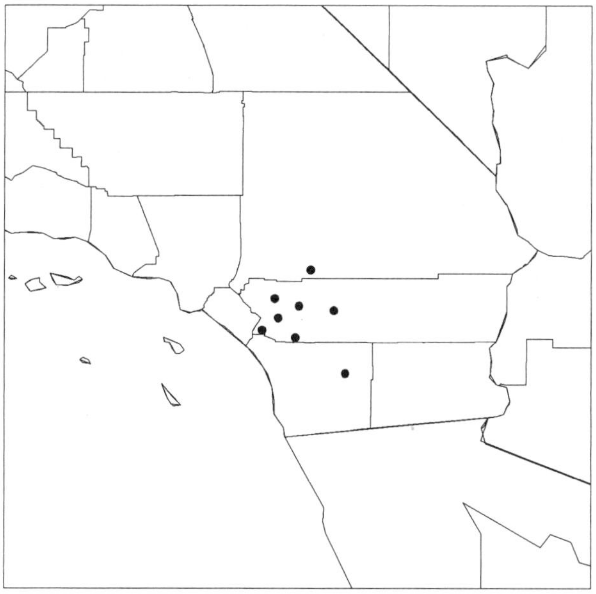

FIGURE 40: DISTRIBUTION OF SPECIMENS EXAMINED OF *CUCULLIA DAMMERSI*

Cucullia dammersi is known only from the Mohave Desert region of southern California and is sympatric with *pulla* there. The adults fly early in the year, December and January. Pupation takes place by late March.

GENUS

Dolocucullia Poole, NEW GENUS

Dolocucullia Poole.
Type species: *Xylina dentilinea* Smith, 1899.

Dolocucullia is the first of a group of three genera without cornuti in the vesica. The cornuti are replaced by sclerotized, often dentate, nodules. The other two genera are *Opsigalea* and *Emariannia.* These three genera are most diverse in Mexico and Central America although species of *Opsigalea* reach as far south as northern Argentina and one species of *Dolocucullia* is known from the mountains of Peru. This group of genera is probably an offshoot of the line that gave rise to *Cucullia* because they share with *Cucullia* the same basic structure of the valve, the same general shape of the vesica despite the absence of cornuti, and the same external tympanic structure. There are two major differences between the female genitalia of *Dolocucullia* and *Cucullia.* 1) The ductus seminalis arises from the caudal, not cephalic, end of the corpus bursae. 2) A sclerite is present between the two ovipositor lobes in *Dolocucullia.* This sclerite is found elsewhere in the trifid noctuids, but not in *Cucullia.*

Two species of *Dolocucullia* occur in the United States and a number of species, mostly undescribed, are in Central and South America. Its center of development is in the mountains of Mexico and Guatemala. One species, *Dolocucullia buddhae* Walker, NEW COMBINATION (*Lithomia buddhae* Walker,) reaches as far south as Peru. The other named, included species is *Dolocucullia emungens* Draudt, NEW COMBINATION (*Cucullia emungens* (Draudt).

Head: antenna simple, scaled dorsally at base, otherwise minutely hairy; tuft of scales on vertex of head projecting through antennal bases, second tuft from top of front; tuft of white scales from antennal scape present; strong set of lashes from below base of antenna; front smooth, unmodified, tightly scaled; lashes from posterior margin of eye; palpus normal for subfamily, but third segment a bit smaller than in *Cucullia,* first and second segments fringed with hair below; proboscis normal. Thorax: patagia apparently not capable of being raised into a hood; dorsal vestiture of long, hairlike scales; no significant thoracic tufts; venation as in subfamily. Prothoracic leg: tibia unspined and without terminal claw; tarsal segments with usual three rows of spines except fifth which has only two marginal rows; first segment approximately 2.5 × as long as second. Mesothoracic leg: first tarsal segment 3 × as long as second, otherwise as for prothoracic leg. Metathoracic leg: first tarsal segment slightly swollen, slightly arched, twice as long as second segment, otherwise as for prothoracic leg. External tympanic region: text figure 41 *b*; hood moderate in size followed by a large, vaguely triangular bulla; first tergum without a lateral lobe. Male abdomen: weak tufts on second and third terga; basal hair pencils and Stobbe's glands present or absent depending on species; eighth tergum not swollen; eighth sternum U-shaped with weak central setal hair brush and well-developed lateral apodemes. Male genitalia: text figures 41 *c, d*; uncus hairy, apex pointed; tegumen simple, without penicillus lobes, but hairy at base; free pleurite present, separate from vinculum, ovoid, articulated along inner margin of tegumen; valves simple, generally as in *Cucullia;* corona well developed, consisting of a single row of spines; clasper variously developed, basal process of sacculus usually well developed; vesica without cornuti, but with sclerotized, denticulate lobes. Female abdomen: seventh sternum without lateral specialized scale patches of *Cucullia*; membrane between seventh and eighth sterna glandular and tough as in *Cucullia.* Female genitalia: text figure 41 *a*; ovipositor lobes not specialized, but spines seated in well-developed areolae; a strong triangular sclerite between lobes; ovipositor short; eighth abdominal sternum with distinct lateral pockets; ductus bursae short, variously sclerotized depending upon species; appendix bursae greatly reduced or absent, not distinctly separated from corpus bursae; ductus seminalis arising from caudal end of corpus bursae, not cephalic end as in *Cucullia.*

None of the larvae or foodplants of the species in *Dolocucullia* are known.

Dolocucullia dentilinea (Smith), NEW COMBINATION

PL. 3, FIGS. 12–14; TEXT FIG. 41 *c, d* (♂ gen.); PL. F, FIGS. 13, 14 (♂ gen.); PL. J, FIG. 2 (♀ gen.); TEXT FIG. 41 *a* (♀ gen.); TEXT FIG. 42 (map) (RWH 10188).

Xylina dentilinea Smith, 1899, *Jour. New York Ent. Soc.,* **7**: 229.
Type locality: [USA], Colorado, Denver. [AMNH]; lectotype designated by Todd (1982).

Cucullia dentilinea form *mexicana* Draudt, 1924, *in* Seitz, *Gross-Schmetterlinge der Erde,* **7**:174.
Type locality: Mexico.
NOTE—Draudt's Noctuidae types from Mexico were apparently destroyed in World War II.

Dolocucullia dentilinea is a difficult species to characterize because of its obscure appearance. It is gray with elongate, but not pointed, forewings and strongly lashed eyes. The hindwing is suffused with gray brown. In contrast the hindwings of *Dolocucullia minor* are usually pure white (males), or with only a dark outer margin (females). The clasper is forked in *dentilinea* but single in *minor.* Wing length: mean = 16.84 mm, standard deviation = 0.48 mm, n = 10.

The larva and its foodplants are unknown.

Dolocucullia dentilinea occurs in California, particularly in the San Francisco Bay area, and has been taken as far south as the Los Angeles area. It has been collected in Oregon, middle Colorado, southern Arizona, and southern New Mexico. The species is fairly common in the prairie regions of Nebraska. It occurs in Mexico as far south as Mexico City. *Dolocucullia dentilinea* appears to have a very patchy distribution. Significant geographical variation is present. Specimens from Oregon and the San Francisco Bay area of California are more brightly marked than those from other localities. The valve of California specimens is narrower and the corona not as well developed as in material from Colorado and Arizona. There are small differences in the shape of the clasper. I was inclined to think two species might be involved. However, a study of Central and South American species appears to indicate that the species of *Dolocucullia* are prone to geographical variation in the male genitalia. Most specimens have been collected in April, May, and June.

Dolocucullia minor (Barnes and McDunnough), NEW COMBINATION

PL. 3, FIGS. 15, 16; PL. F, FIG. 15, PL. G, FIG. 1 (♂ gen.); PL. J, FIG. 3 (♀ gen.); TEXT FIG. 43 (map) (RWH 10193).

Cucullia minor Barnes and McDunnough, 1913, *Contrib. Nat. Hist. Lep. N. Am.,* **2**(3): 110, pl. 5, fig. 5.
Type locality: [USA], New Mexico, Deming. [USNM]
NOTE—This species was described from two males, one labeled type male, the other cotype male. The specimen marked "type male" was the specimen figured, and it is hereby designated and has been labeled as LECTOTYPE.

Dolocucullia minor bears a superficial similarity to *Cucullia incresa* with its white male hindwings and pointed, smooth, dark-gray forewing. This species is the most aberrant member of *Dolocucullia* and perhaps warrants its own generic name. The vesica is basically like that of *dentilinea,* but the clasper is single and is situated more toward the middle of the valve. The clasper is closer to the ventral edge of the valve in *dentilinea.* Wing length: mean = 15.40 mm, standard deviation = 0.64 mm, n = 10.

The larva and its foodplants are unknown.

Dolocucullia minor has been collected in western Texas, southern New Mexico, and southern Arizona but is best represented by specimens from southwestern Texas. Females are darker than males and the outer margin of the hindwing usually has a black tinge in females. Adults have been collected in May–June and August–September, indicating possibly two broods.

GENUS *Opsigalea* Hampson

Opsigalea Hampson, 1906, *Catalogue of the Lepidoptera Phalaenae in the British Museum,* **6**: 84.
Type species: *Nystalea ocellata* Walker, 1865. Original designation.

Opsigalea is an enigmatic genus. The species look like members of *Cucullia,* but there are significant

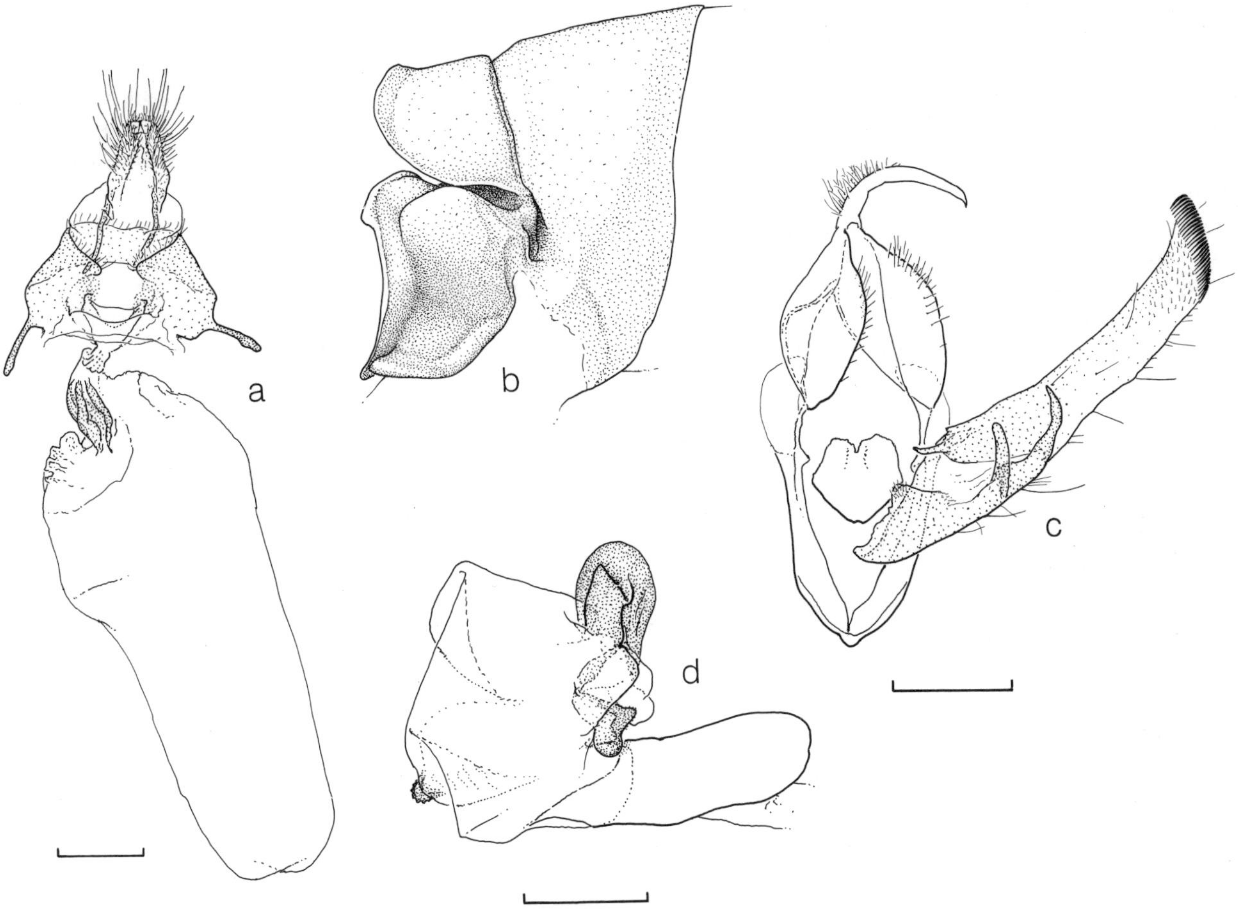

FIGURE 41: STRUCTURAL FEATURES OF *DOLOCUCULLIA DENTILINEA*
a. Female genitalia. *b.* Lateral view of abdominal segments one and two. *c.* Male genitalia. *d.* Aedoeagus. Scale = 1 mm.

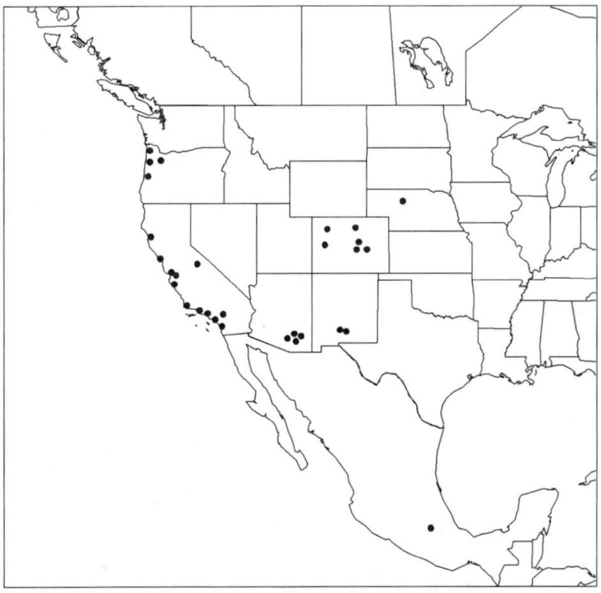

FIGURE 42: DISTRIBUTION OF SPECIMENS EXAMINED OF *DOLOCUCULLIA DENTILINEA*

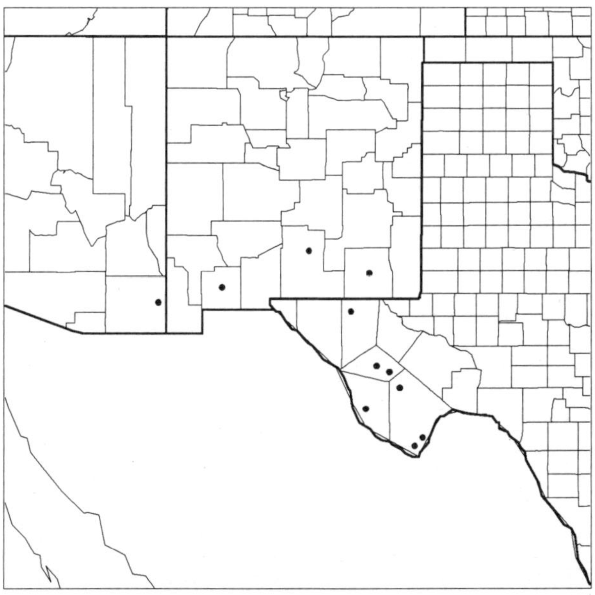

FIGURES 43: DISTRIBUTION OF SPECIMENS EXAMINED OF *DOLOCUCULLIA MINOR*

morphological differences. The costa of the forewing is bent downward at the apex giving the forewing a distinctly rounded appearance. The vesica is characterized by an absence of distinct cornuti, as found in *Cucullia,* and their replacement by one or more sclerotized patches as in *Dolocucullia.* The ductus seminalis arises from the top of the bursa, not the bottom as in *Cucullia*; the appendix bursae is absent, not the corpus bursae. On the other hand the external internal tympanal structure (text figure 44 *b*) is essentially the same as in *Cucullia.*

Opsigalea appears, despite the vast superficial dissimilarity, to be closely related to *Dolocucullia.* I am inclined to believe that these two genera are Neotropical derivatives of the ancestral line that gave rise to *Cucullia* despite the differences between them and *Cucullia.* I base this belief on the tympanic structure and the superficial similarities of *Opsigalea* and *Cucullia.* Admittedly, superficial similarity can be misleading, but in the absence of contradictory evidence, it is better than nothing.

The distribution of the species of *Opsigalea* is unusual. There are two species in Mexico and Central America in addition to the Texas species *Opsigalea blanchardi*; *Opsigalea ocellata* (Walker) and *Opsigalea phocylides* (Druce), NEW COMBINATION, (*Cucullia phocylides* Druce). These three species are separated by a large distributional hiatus from an undetermined number of species in southern South America, but including *Opsigalea hildana* (Schaus), *Opsigalea perlucida* (Jones), NEW COMBINATION, (*Cucullia perlucida* Jones), and *Opsigalea prima* (Köhler), NEW COMBINATION, (*Cucullia prima* Köhler).

Head: vestiture mostly of scales; palpus with second segment not fringed ventrally with hair; proboscis normal; antenna simple; eyes very weakly lashed; front smooth, clypeus not produced, viewed laterally ventral margin hardly projecting beyond margin of eye, not triangular; no distinct tufts on head except for two weak tufts from the bases of antennae. Thorax: patagia not capable of being produced into a hood, black lines forming circles, not straight lines; in type species, but not all species, a large woolly tuft followed by a small round tuft at the rear of dorsum of thorax. Forewing: elongate, costa curved downward at apex; apex rounded; outer margin curved inward, beginning of incurving occurring about midway between apex and inner margin; venation as in *Cucullia* except for distortion caused by wing shape. Prothoracic leg: without tibial claw; femur and tibia with long, strongly developed fringes of hair; three rows of spines on first four tarsal segments, two rows on fifth segment. Mesothoracic and metathoracic legs: femur and tibia fringed with long hair; tarsal spines as on foreleg. Internal tympanic structure as in *Cucullia.* External tympanic structure: text figure 44 *b*; much as in *Cucullia*; hood moderate, separated by a rift from a bullalike swelling; tympanal groove continued as a depression along lower margin of second abdominal tergum. Abdomen: without dorsal tufts; apophyses of first abdominal sternum short, thick, not elongate; eighth sternum of male not enlarged and sclerotized. Male genitalia: text figure 44 *c, d*; vesica of aedoeagus with no cornuti, but with two sclerotized patches, a basal patch with distinct teeth, and a large elongate apical patch; valve as in *Cucullia* with both basal processes and claspers, corona weakly developed; uncus simple. Female genitalia: text figure 44 *a*; ovipositor lobes unmodified in known species; ductus bursae well sclerotized; corpus bursae consisting of a larger elongate lobe and a smaller lobe (possibly the appendix bursae) near junction with ductus bursae; ductus seminalis arising from smaller lobe, not from bottom of larger lobe as in *Cucullia.*

Opsigalea blanchardi Todd

PL. 3, FIGS. 17, 18; TEXT FIG. 44 *c, d* (♂ gen.); TEXT FIG. 44 *a* (♀ gen.); TEXT FIG. 45(map) (RWH 10216).

Opsigalea blanchardi Todd, 1966, *Proc. Ent. Soc. Washington,* **68**: 149, figs. 1, 5–7.
Type locality: [USA], Texas, Brewster Co., Alpine. [USNM]

Opsigalea blanchardi is the only species of the genus known from the United States and cannot be mistaken for any other noctuid living north of the Mexican border. In particular the elongate forewings and the eyelike spot in the tornus are particularly diagnostic. *Opsigalea blanchardi* is clearly related to the Mexican *O. ocellata* (Walker) and looks like a miniature version. The modal length of the forewing is 17 mm in *blanchardi* and 23 mm in *ocellata.* Wing length: mean = 16.98 mm, standard deviation = 0.79 mm, n = 10.

The life history and foodplants are unknown. However *Opsigalea ocellata* has been reared from *Buddleja sessiliflora* H.B.K., a shrub or small tree in the family Loganiaceae.

This species has been collected only in western Texas.

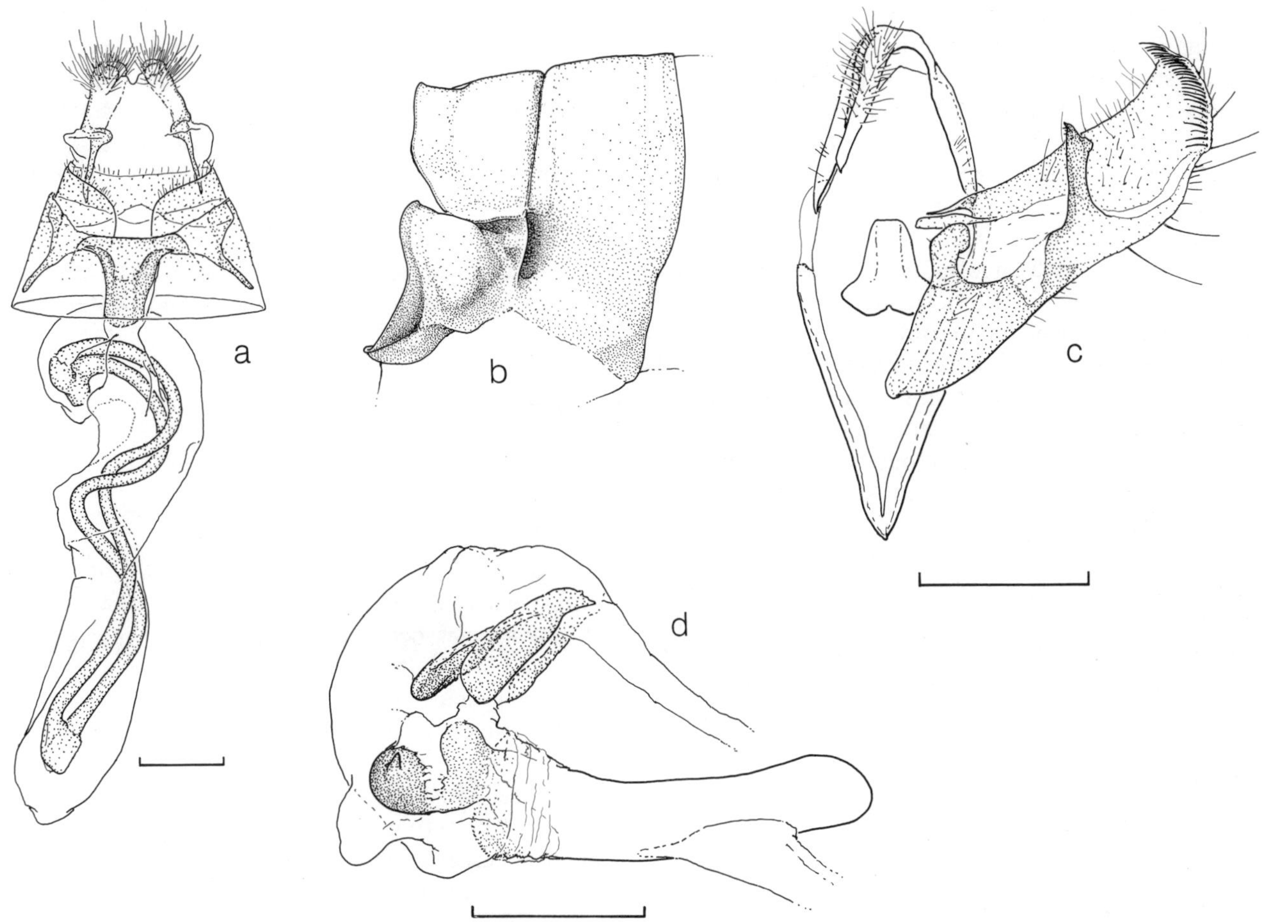

FIGURE 44: STRUCTURAL FEATURES OF *OPSIGALEA BLANCHARDI*

a. Female genitalia. *b.* Lateral view of abdominal segments one and two. *c.* Male genitalia. *d.* Aedoeagus. Scale = 1 mm.

GENUS

Emariannia Benjamin

> *Emariannia* Benjamin, 1933, *Pan-Pacific Ent.*, **9**: 2.
>
> Type species: *Emariannia cucullidea* Benjamin. Original designation.

Emariannia is the third and last of the group of three genera with sclerotized nodules in the vesica. In many ways it may be the most unusual. The following set of characters should identify it: 1) the eyes are lashed; 2) a strong tubular frontal process is present; 3) the prothoracic tibiae do not have an apical claw; and 4) the forewings are elongate. The most unusual features are found in the vesica. Sclerotized nodules are present, but the vesica is longer than in the other two genera and *Cucullia.* Minute denticules, which may be the remnants of the original cornuti, are present in the vesica. Assuming, as I have, that an elongate spined vesica is primitive

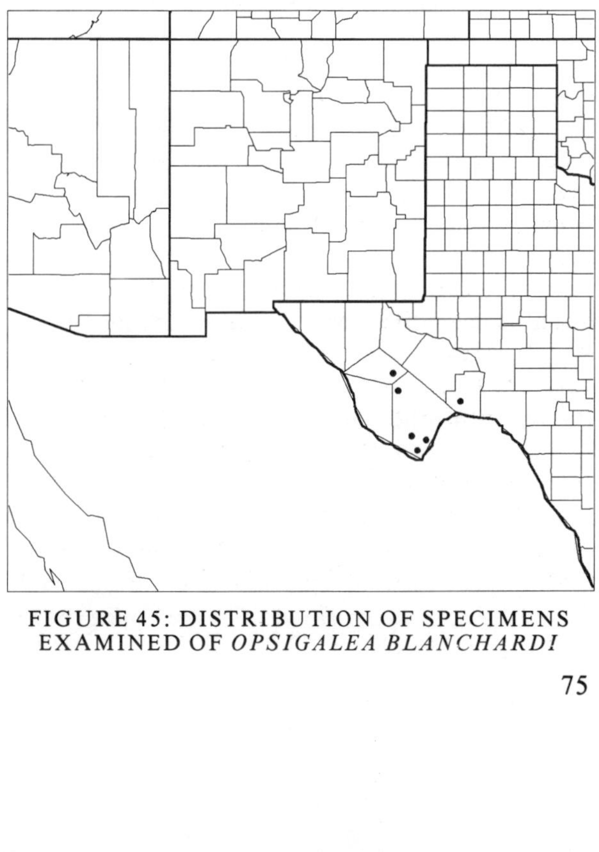

FIGURE 45: DISTRIBUTION OF SPECIMENS EXAMINED OF *OPSIGALEA BLANCHARDI*

for the Cuculliinae, this genus would appear to have retained more primitive characters than any other in the subfamily. The external structure of the tympanic region is not significantly different from that of *Cucullia.* The eighth sternum of the female is entire beneath the ostium and not membranous as it is in almost all other genera of the subfamily. The early stages are unknown.

Head: front with large, elongate, cylindrical frontal process, process without a central prominence; ridges extending from frontal process to edges of clypeus, clypeal margin protruding forward; antenna simple, scaled dorsally, minutely hairy otherwise; a tuft of white scales from scape of antenna and tuft of lashes from near antennal base; posterior margin of eye without lashes; usual tufts from vertex projecting through antennal bases and another tuft from top of frons, but both tufts bushy and disturbed by the frontal process; frons loosely scaled; palpus normal for subfamily; proboscis normal. Thorax: patagia apparently not capable of being raised in a hood; a moderately large tuft on rear margin of thorax (metathoracic scutellum); venation as in subfamily. Prothoracic leg: tibia without terminal claw; tarsal segments short, about as long as wide, and with three strong rows of spines on first four segments and two lateral rows on fifth. Mesothoracic and metathoracic legs: without the foreshortened tarsal segments, but otherwise as in prothoracic leg; first tarsal segment about twice as long as second. External tympanic structure: not significantly different from *Cucullia;* abdominal bulla kidney shaped, elongate dorsoventrally. Internal tympanic structure: not examined. Abdomen: male eighth sternum degenerate with only a faint hint of lateral apodemes and no hair comb; basal hair pencils and Stobbe's glands absent. Male genitalia: text figures 46 *b, c;* vesica rather elongate for group with only one sclerotized node and studded with denticules. Female genitalia: text figure 46 *a;* a double denticulate bullalike structure in bursa; ductus bursae sclerotized, but not ribbed; ovipositor lobes heavily sclerotized; last sternum entire beneath ostial mouth and margin with two lobes.

Emariannia cucullidea Benjamin

PL. 3. FIGS. 19, 20; TEXT FIG. 46 *b, c* (♂ gen.); TEXT FIG. 46 *a* (♀ gen.); TEXT FIG. 47(map) (RWH 10217).

Emariannia cucullidea Benjamin, 1933, *Pan-Pacific Ent.,* **9**: 3.
Type locality: [USA], Texas, Brewster Co., Alpine. [USNM]

Emariannia cucullidea has elongate, gray-brown forewings with small, white reniform and orbicular marks. The presence of lashed eyes, the long cylindrical frontal process, and the absence of a tibial claw should identify it immediately. Wing length: mean = 15.65 mm, standard deviation = 0.82 mm, n = 10.

The larva and its foodplants are unknown.

E. cucullidea is known only from western Texas. Adults have been collected in July and September.

SUBFAMILY
Stiriinae Grote

Stiriinae Grote, 1882, *New Check List of North American Moths,* p. 35.
Type genus: *Stiria* Grote, 1874.

The subfamily Stiriinae is a group of colorful genera and species confined to North and Central America. A few species occur in the northern, central, and eastern United States and Canada, but the vast majority of species live in a zone from the southwestern United States through Mexico.

The species in the Stiriinae have a distinctive appearance. Most are colorful, and some are among the most attractive noctuids in North America. They exhibit a great variety of external colors and shapes. Among them are the bright-yellow and brown pattern of species of *Stiria,* the metallic-green patterns of the species of *Chalcopasta,* the bright-orange and striate appearance of *Cirrhophanus,* and sharp-pointed wings and sometimes contrasting colors found in some of the species of the genus *Plagiomimicus.* The species differ in forewing shape from rounded to those with acute apices. Size varies from large species such as *Basilodes pepita* to small ones such as *Lineostriastiria hutsoni.* The superficial diversity among species contrasts with a conservative internal morphology.

Most species in the subfamily are denizens of desert and scrub habitats, with many of the remaining species occurring in the prairie regions of the United States. Relatively few species in the subfamily have been reared, but those species whose biologies are known are all feeders on the flowers and seeds of the plant family Asteraceae. The bright colors of some species are possibly cryptic coloration for adults sitting on the flowers of their foodplants. The bright-yellow wings of *Stiria rugifrons,* for example, matches the yellow petals of its foodplant *Helianthus microcephalis.*

No formal subfamily description is given. The generic description for *Plagiomimicus* may be used

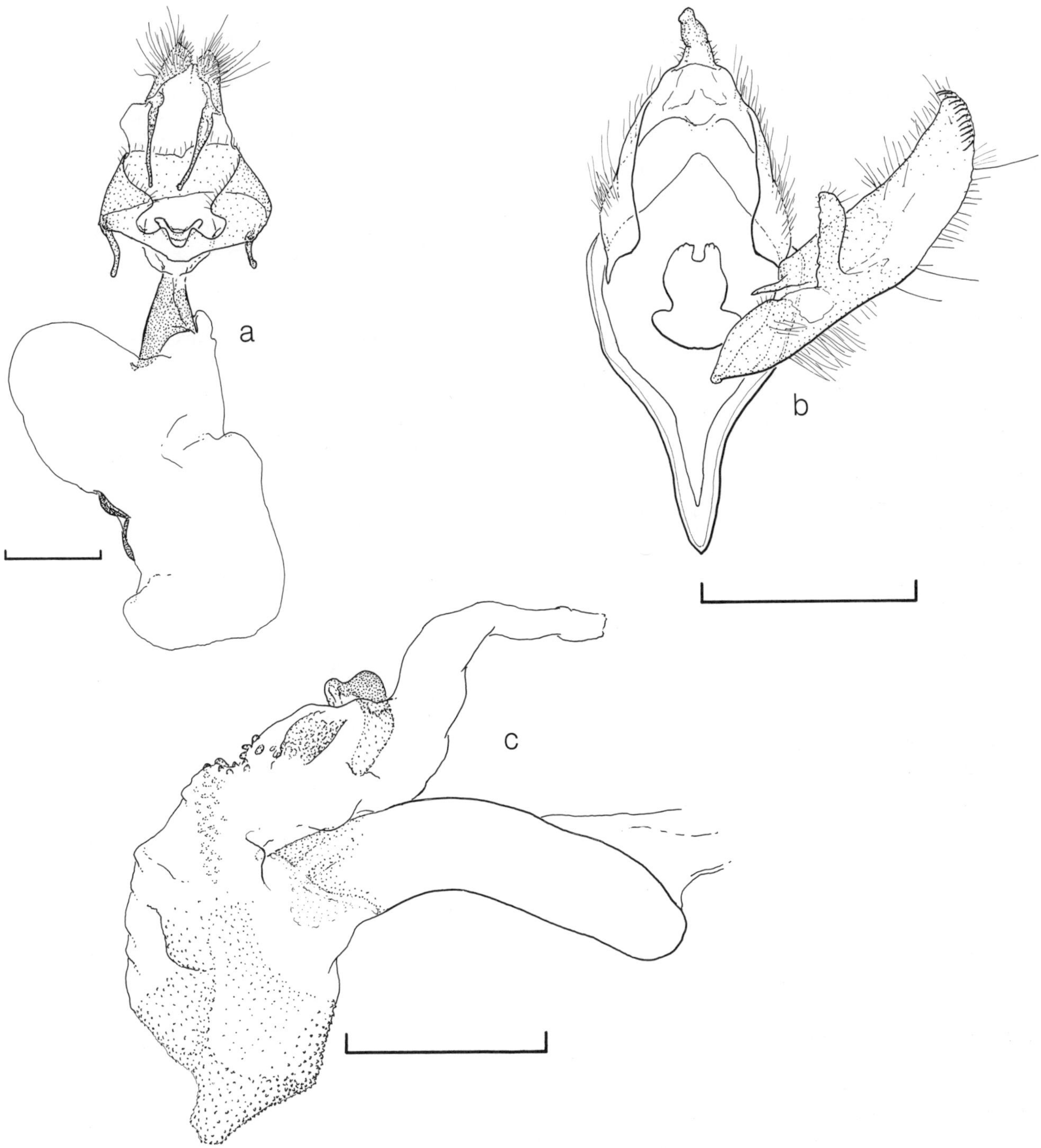

FIGURE 46: STRUCTURAL FEATURES OF *EMARIANNIA CUCULLIDEA*

a. Female genitalia. *b.* Male genitalia. *c.* Aedoeagus. Scale = 1 mm.

as a formal description if desired. Two characters appear to be synapomorphic for the subfamily. 1) The general shape and spination of the vesica: The vesica of the Stiriinae is generally ovate and slightly elongate. There are two major patches of cornuti in the vesica; a basal row of short, stubby cornuti and a more apical, diffuse patch of elongate cornuti. Variation occurs among the genera and species. For example, the basal row of short cornuti may be fused into a sclerotized nodule, or the basal row may be absent (e.g., *Basilodes*). However, all of the genera of the Stiriinae, as here defined, share this same basic

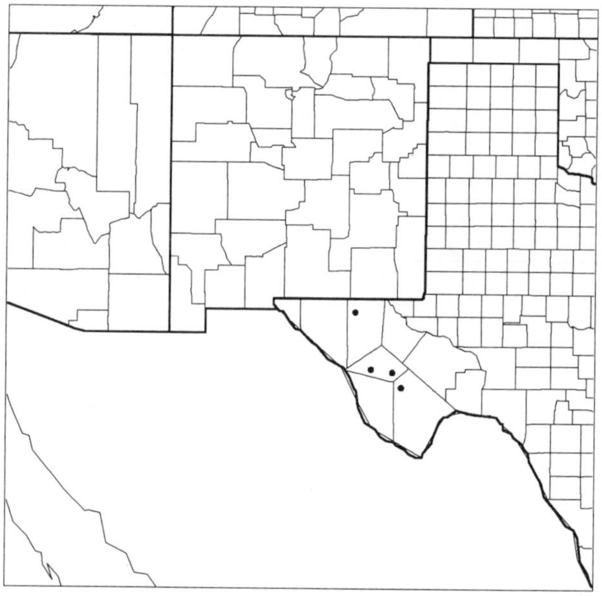

FIGURE 47: DISTRIBUTION OF SPECIMENS EXAMINED OF *EMARIANNIA CUCULLIDEA*

spining pattern. 2) Reduction of the larval spinneret to a small, scalelike object. This severe reduction in size of the spinneret is not unique to the Stiriinae and is also found in the Condicinae and in the European genus *Panameria* Hübner. The presence of seta SV3 on A1 in the Condicinae and its absence in the Stiriinae will separate these two distantly related subfamilies. I have not examined *Panameria* and cannot speculate on its phylogenetic position other than to note the absence of the typical stiriine spining pattern in the vesica.

Several characters are typical of the Stiriinae other than the two synapomorphies listed above. None of them is unique to the Stiriinae, and all occur scattered throughout other trifid subfamilies in the Noctuidae. Most of these characters seem to be adaptations to a desert life style. 1) There is a large, circular frontal process. Although variable in form, the process usually consists of a large outer ring and a central noselike protuberance. Such a prominence is found in several subfamilies of the Noctuidae (e.g., Heliothinae, Cuculliinae, Psaphidinae) and is also known from species in the Geometridae. A frontal protuberance is found elsewhere in the Lepidoptera but not in the exact form described above. 2) The prothoracic tibia has a terminal claw. This claw in the Stiriinae is composed of three or four enlarged, fused setae. Tibial claws are found commonly in many branches of the Noctuidae and are known in the Geometridae as well. The tibial claw is most commonly found in species from xeric environments. 3) The male genitalia of the Stiriinae are simplified and of the primitive "trifid" noctuid type. The valve has an erect clasper oriented perpendicularly to the long axis of the valve, the corona is weak and disorganized, and there is no digitus or differentiation between the sacculus and cucullus. 4) The female genitalia almost always have elongate, pointed, and heavily sclerotized ovipositor lobes. Such lobes are commonly found in unrelated groups of the Noctuidae and apparently are an adaptation for placing eggs within flower or leaf buds. The corpus bursae is usually elongate and the upper fourth of the corpus bursae and appendix bursae are usually sclerotized with internal ridges. The larvae of all known species feed on the flowers and seeds of composites. Flower and seed feeding is also typical of the Cuculliinae and Heliothinae, and to a lesser extent in the Oncocnemidinae. The larva belongs to Crumb's "larvae confluentae" section while most other flower-feeding groups occur in the "larvae liberae."

The Stiriinae were recognized as a natural grouping by the earliest workers on North American noctuids. Grote (1874) in his "Checklist of the Noctuidae of North America" described two new genera (*Stiria* and *Stibadium*) and placed them together with the previously described genus *Plagiomimicus*. Grote, however, failed to include in the group the previously described genera *Basilodes* and *Cirrhophanus*. In his 1882 revised list of the Noctuidae he united these three genera again and included three additional generic names (*Polenta, Fala* and *Acopa*) but again failed to include *Basilodes* and *Cirrhophanus*. He formally recognized the group by proposing the name Stiriinae. He placed this new subfamily between the Calpinae and Heliothinae; some stiriines were placed in the Heliothinae. At approximately the same time Smith (1882) stated that he believed the genera *Stiria, Stibadium,* and *Plagiomimicus* were not generically distinct from *Basilodes*. Smith associated his composite genus *Basilodes* with *Fala, Antaplaga,* and *Xanthothrix* but mixed them together with two genera now included in the Heliothinae. The other generic names published by that time (*Cirrhophanus* and *Polenta*) were placed elsewhere and were well mixed with other trifid noctuids. Grote's treatment of the Stiriinae was generally followed into the early years of the 20th century. His grouping of Stiriinae genera was based on the presence of a prothoracic tibial claw and a frontal ring. Neither character is unique or synapomorphic for the Stiriinae, as noted above. Hampson (1910), in his monumental revision of the

Noctuidae of the world, treated the Stiriinae known by that time, but scattered them throughout the groups treated in volume 9, the third volume dealing with his subfamily "Acronyctinae" (Amphipyrinae of the "Mona Checklist"). Hampson employed an artificial system of classification and in general lumped all genera with a frontal tubercle and prothoracic tibial claw. Barnes and McDunnough (1917) restored some order to the group, but in their check list the group was not given a name and the genera now considered to belong to the Stiriinae are mixed with genera now placed in other tribes and subfamilies. McDunnough (1938) followed the arrangement in the Barnes and McDunnough (1917) list.

Hogue (1963) completed a thorough and exemplary analysis of the higher classification of the entire subfamily (treated as a tribe but omitting *Fala* or *Xanthothrix*), including the Mexican, U. S., and Canadian species. He examined in detail the morphology of most of the species in the subfamily but was not concerned with specific limits in the species groups. He constructed a phylogeny dividing the Stiriinae into 27 phylogenetic lines grouped into eight genera.

Two authors subsequent to Hogue (1963) have tried to broaden the boundaries of the subfamily Stiriinae. Hardwick (1970), while recognizing the Stiriinae as a monophyletic group as here defined by the spining structure of the vesica, added a number of other genera to the subfamily. Most of these genera are members of tribes in the Psaphidinae in the classification proposed at the beginning of this fascicle. He did not give specific reasons for including these genera except for the statement "There are, however, many genera currently resident in the Amphipyrinae which are closely related to Hogue's Stiriini, and which in combination with the latter form a small and moderately well-defined subfamily." No subfamily definition is given or supporting characters discussed. Presumably, the simple, plesiomorphic male genitalia, the presence of a frontal process, and the existence of a prothoracic claw motivated this conclusion. I have argued above, however, that none of these characters can be safely used to make phylogentic inferences. Hardwick's added genera, in the order they were listed, belong to the following groups in the classification proposed at the beginning of this fascicle: *Azenia* Grote (Psaphidinae, Azeniini); *Oslaria* Dyar (Psaphidinae, Nocloini); *Stiriodes* Hampson (Psaphidinae, Azeniini); *Lythrodes* Smith (Psaphidinae, Nocloini); *Tristyla* Smith (Psaphidinae, Azeniini); *Chamaeclea* Grote (*incertae sedis* following the Stiriinae); *Nocloa* Smith (Psaphidinae, Nocloini); *Triocnemis* Grote (Psaphidinae, Triocnemidini); *Axenus* Grote (*incertae sedis*); *Xanthothrix* Hy. Edwards (Stiriinae); *Crimona* Smith (Psaphidinae, Triocnemidini); *Podagra* Smith (Psaphidinae, Grotellini); *Pseudacontia* Smith (Oncocnemidinae); *Walterella* Dyar (*incertae sedis*); *Prothrinax* Hampson (*incertae sedis,* but closely related to *Walterella*). Hardwick explicitly excluded other genera from the Stiriinae even though they are at times closely related to the genera listed above. *Euamiana* Barnes and Benjamin and *Ruacodes* Hampson both belong to the Nocloini and are very closely related to *Nocloa* and *Oslaria. Oxycnemis* Grote belongs to the Triocnemidini. *"Chalcopasta" koebelei* (Riley) is *incertae sedis* and is placed in this fascicle following the Stiriinae. *Grotella* Harvey, *Neogrotella* Barnes and Benjamin, and *Hemigrotella* Barnes and Benjamin belong to the Psaphidinae, Grotellini.

Matthews (1991) accepted Hardwick's broadened definition of the Stiriinae and included even more genera; a group of genera of largely African origin, and a small group of Palearctic genera including *Aegle* Hübner, *Panemeria* Hübner, and *Synthymia* Hübner. Again no reason for the inclusion of these genera was given except the statement "Dissection of the following Old World species left little doubt that they are indeed stiriines." I believe Matthews also was influenced by the primitive and simplified structure of the male genitalia. The larvae of the additional genera included as "stiriines" are all flower feeders, but none of them feed on composites. The Stiriinae, as far as we know, all feed on genera in the Asteraceae. None of these genera have the distinctive stiriine spining pattern of the vesica, nor many of the other characters typical of the Stiriinae. This is not the proper place to try to reconstruct the phylogeny of these Old World groups; however, the African genera listed do not seem to form a monophyletic group based on my survey of material in the USNM.

The goal of both Hardwick and Matthews was to establish an out-group for their studies of the phylogeny of the Heliothinae, and to this extent their broad definition of the Stiriinae is an out-group, although a heterogeneous one. I have tried to establish a first, admittedly crude and incomplete, phylogeny for this part of the "trifid" Noctuidae in the introduction to this fascicle. Unfortunately character analysis in this part of the Noctuidae evolutionary tree is very difficult because of the dominance of primitive character states, a tendency toward simplication of both the male and female genitalia, and the repetitive loss and reappearance of the most obvious external characters. The last word on this subject is far from being written.

PHYLOGENY OF THE STIRIINAE. Hogue (1963) used the traditional comparative method in practice at the time to construct a phylogeny of the Stiriinae. I have also attempted to construct a phylogeny for the subfamily using newer methods. My treatment, by necessity, is considerably less thorough than his and is limited to the United States and Canadian species of the subfamily. I included most of the lines defined in Hogue's study in my analysis including the groups found only in Mexico, later reducing the data set to those species or species-groups found north of Mexico. The tree was estimated using the "bb" (branch and bound) algorithm found in the computer program Hennig 86 (Farris, 1988). The data set was too large to run using the "ie" option (implicit enumeration). The species *Fala ptychophora* was used as the outgroup, even though it is included as part of the Stiriinae. Species in the *incertae sedis* group following the Stiriinae were used to polarize characters even though they were not explicitly included in the analysis. The consistency index was 51 percent and over 100 trees were generated, an indication of considerable instability in the estimated phylogeny. One of these trees is shown in text figure 48 and was chosen by the criterion that it best fit my own personal biases. Some of the characters considered are placed in text figure 48 and discussed individually below. The phylogeny does not, for the most part, differ materially from that constructed by Hogue (1963), but there are four major differences: 1) The genera *Fala* and *Xanthothrix* are recognized as part of the Stiriinae. *Fala ptychophora* was used as the sister species of the rest of the subfamily for the purposes of outgroup comparisons. There are no explicit outgroup comparisons in Hogue's analysis. I did not choose an explicit outgroup from outside the subfamily because it was not clear what genera and species would be a suitable sister group. Choosing an outgroup from within the subfamily is questionable, but seemed preferable to selecting a potentially unsuitable group from outside the subfamily. The genus *Fala* has many attributes of the species and genera found in the *incertae sedis* group discussed following the Stiriinae.

2) Many of the differences shown between the two hypotheses are attributable to differences in generic limits. For example the genus *Lineostriastiria* has been separated from *Plagiomimicus*. This analysis seems to indicate that *Lineostriastiria* is less related to *Plagiomimicus* than postulated by Hogue, although adjacent to it in the phylogeny. The genus *Cirrhophanus*, as defined by Hogue, is a much wider grouping of species than I have used. In particular I have segregated a group of species into the genus *Eulithosia*.

3) The major difference in the phylogeny is the separation of the species of *Stiria* from *Basilodes* within which they were included by Hogue.

4) Most of the differences in the upper reaches of the tree result from tree instability. The analysis seems to indicate that the relative positions of *Chalcopasta, Cirrhophanus, Basilodes, Stiria, Eulithosia,* etc. are relatively unstable given the data used in this analysis.

In general the cladogram shown in text figure 48 is not much different from the phylogeny postulated by Hogue given the general instability of the analysis. Therefore I have limited my discussion to characters and character states indicating major breaks in the estimated tree and refer the reader to Hogue's paper for a more detailed discussion. The convention is used that the primitive state of a character is indicated as (0) and subsequent steps in the evolution of a character as (1), (2), etc. Homoplasious character states such as reversals and parallelisms are not discussed even though they abound in the analysis.

1. Vesica globular with a basal row of cornuti and a wider, dispersed mass of cornuti in the body of the vesica. This basic structure of the vesica is characteristic of the Stiriinae (although sometimes modified) and is the derived state (1) versus the primitive (0) state (as exhibited in *Neumoegenia* for example) of a simple, globular vesica without this pattern of cornuti.

2. Ovipositor lobes unmodified (0) versus ovipositor lobes strongly sclerotized and usually pointed at the apices (1). This character can be further subdivided into lobes partially sclerotized (1a) and fully sclerotized (1b). All of the Stiriinae with the exception of *Fala ptychophora* have modified ovipositor lobes.

3. Eighth sternum of female abdomen without unsclerotized areas on either side of the ostium (0) versus (1) round clear areas present. In some species of *Plagiomimicus* there is an indication of a weakening of the sclerotization but never distinct clear areas. I have used this character more explicitly than Hogue did. Consequently a group of species included in *Plagiomimicus* by Hogue are separated into the genus *Lineostriastiria* in this analysis. A few Mexican species show an intermediate state between *Plagiomimicus* and *Lineostriastiria* as defined in this work. In these Mexican species (e.g., *Plagiomimicus argyropolius* Dyar and *Plagiomimicus phalaenoides* Dyar) there are small, irregular clear areas in the eighth sternum.

4. Clear areas, when present, bounded on the caudal margin of the eighth abdominal sternum of the female

abdomen (0) versus clear areas not bounded on the caudal margin (1). Characters 3 and 4 are the two two-state characters formed by subdividing a three-state character.

5. Minute hairs in the ventral membrane of the ovipositor neck of the female abdomen not arranged in distinct lines (0) versus minute hairs arranged in distinct lines. These lines of minute setae are sometimes found elsewhere in the subfamily (e.g., in some species of *Plagiomimicus*), particularly near the margin of the eighth sternum, but are never continuous over the length of the ovipositor neck.

6. Species not day flying and eye size normal (0) versus species day flying and with reduced eye size. A reduction in eye size is a common occurrence in day-flying noctuid species. This character defines the genus *Xanthothrix*. The position of *Xanthothrix* within the phylogeny is very unstable. Most features of both the male and female genitalia are difficult to code because of strong tendencies toward reduction in most of the structures.

7. Clear areas in eighth sternum of female genitalia (when present) with strong, pleated lines (1) versus pleated lines absent (0). I have given considerably more weight to this character than did Hogue. This character unites his genera *Chrysoecia, Gorgora,* and *Chichimeca* into the genus *Chrysoecia*.

8. Ostium sunk into a rounded, laterally edged, invaginated region of the eighth sternum (1) versus no such invaginated region (0).

9. Uncus relatively long and narrow, not swollen either at the base or middle (1) versus uncus short and generally swollen for part of its length (0). This is a difficult character to code and cannot always be determined. The sequence postulated may not be a true character transition, but a mixture of several different states. Running the character as unordered did not affect the series of trees produced.

Synapomorphies of some of the terminal taxa follow:

1. *Plagiomimicus pityochromus* group. The wings are rounded at the apex unlike the pointed apices found in most of the other species of *Plagiomimicus*. The frontal process is circular, strongly defined with a deep concavity, and without a central process. The ovipositor lobes are short and stubby as are the apophyses anteriores and posteriores.

2. *Basilodes*. The basal row of cornuti of the vesica is absent. The forewing is broad and usually has a golden, metallic sheen. The apex of the sacculus is produced into a long apical spine. The ostium of the female genitalia is more deeply invaginated than in any other stiriine genus.

3. *Hoplolythrodes*. The forewing has a distinctive pattern and shape. The male is unknown and the position of this genus is very uncertain.

4. *Cirrhophanus*. The forewing pattern and color is distinctive. The frontal process of the head is triangular, not ovate. The prothoracic tibial claw is weak. The ovipositor lobes are blunt tipped, not pointed as in the adjacent genera of the subfamily.

5. *Eulithosia*. The ovipositor lobes of the female genitalia are short, stubby, and splayed outward. The basal group of cornuti consists either of a group of large knobs with short terminal cornuti or the cornuti are grouped together on a sclerotized base.

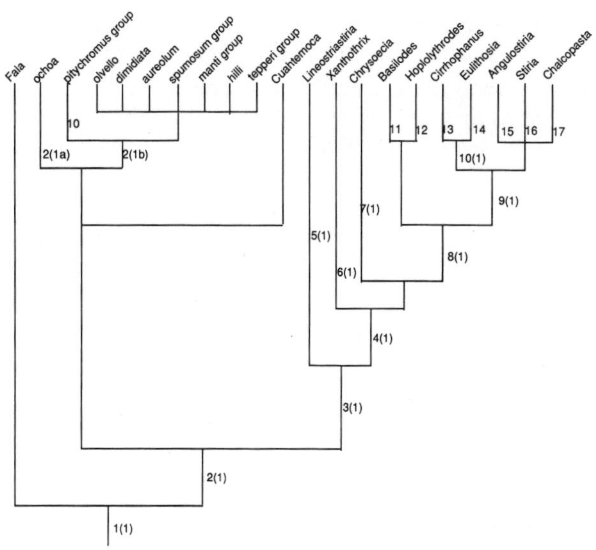

FIGURE 48: POSTULATED PHYLOGENY OF THE SUBFAMILY STIRIINAE

6. *Angulostiria*. The vesica has a basal diverticulum covered with a mat of small spinules.

7. *Stiria*. Forewing pattern is yellow with brown patches.

8. *Chalcopasta*. Areas of metallic-green scaling are on the forewing.

KEY TO THE SPECIES OF THE STIRIINAE

1. Forewing with patches of metallic-green or yellow-green scales on a tan or dark-brown background 2
— Forewing not as above 7

2. Forewing with metallic-green scaling confined to basal and subterminal areas, absent from median area .... *Cuahtemoca unicum* p. 105
— Metallic-green scaling present in forewing median area 3

3. Forewing dark brown; green patch a single diagonal slash from basal ⅓ of costa to outer part of inner margin; no metallic-green scaling in subterminal area .... *Chalcopasta acema* p. 148
— Forewing light brown; green patch as above but continued into subterminal area to apex of wing 4

4. Extensive areas of brown scaling in basal area of forewing, particularly along inner margin . . . 5
— Forewing basal area almost entirely filled with metallic-green scales except for small brown spot at wing articulation . . . . *Chalcopasta territans* p. 145

5. Forewing dusted with white scales giving it a frosted appearance . . . . *Chalcopasta fulgens* p. 147
— Forewing not dusted with white scales, not appearing frosted 5

6. Hindwing entirely brown . . . . *Chalcopasta sinuata* p.145
— Hindwing with some areas of white, not entirely brown . . . . *Chalcopasta howardi* p. 147

7. Forewing orange with a striate appearance (plate 4, figure 19) 8
— Forewing not as above 11

8. Forewing with postmedial line strongly bowed outwardly at discal cell toward outer margin; pattern appearing complex (plate 4, figure 19) 9
— Forewing postmedial line not strongly bowed outwardly at discal cell; pattern not appearing complex (plate 4, figure 25) . . . . *Eulithosia papago* p. 132

9. Distinct triangular indentation of postmedial line of forewing at CuA_2; southwestern United States . . . . *Cirrhophanus dyari* p. 126
— No distinct triangular indentation of postmedial line of forewing at CuA_2; eastern or Great Plains regions of United States 10

10. Larger species (16–18 mm from forewing base to apex); postmedial line on ventral surface of forewing present . . . . *Cirrhophanus triangulifer* p. 127
— Smaller species (13–15 mm from forewing base to apex); postmedial line on ventral surface of forewing absent . . . . *Cirrhophanus pretiosa* p. 129

11. Forewing with all maculation absent and with a strong violet tinge to basal area . . . . *Eulithosia plesioglauca* p. 132
— Forewing not as above 12

12. Forewing entirely off-white, maculation completely absent . . . . *Chrysoecia thoracica* p. 119
— Forewing not entirely off-white, some maculation present 13

13. Forewing with median area white or off-white, reniform and orbicular marks absent; postmedial and antemedial lines black, contrasting with median area 14
— Median area not as above; postmedial and antemedial lines not as above 16

14. Basal and subterminal areas of forewing yellow orange . . . . *Lineostriastiria hachita* p. 106
— Basal and subterminal areas of forewing white or off-white 15

15. Hindwing suffused with gray or brown; forewing subterminal areas with series of black dots . . . . *Lineostriastiria sexseriata* p. 108
— Hindwing not suffused with gray or brown; forewing subterminal area usually (but not always) without a series of black dots . . . . *Lineostriastiria biundulalis* p. 108

16. Forewing pure orange or yellow orange; all maculation absent . . . . *Eulithosia composita* p. 130
— Forewing not as above 17

17. Forewing yellow or yellow orange with a brown patch along outer margin and usually a brown patch in median area along inner margin 18
— Forewing not as above 24

18. Forewing with patch in median area along inner margin 19
— Forewing without patch in median area along inner margin . . . . *Stiria blanchardi* p. 141

19. Forewing without a brown patch at base of wing 20
— Forewing with a brown patch at base of wing 21

20. Forewing color and maculation vague, washed with a white film . . . . *Stiria consuela* p. 143
— Forewing bright yellow, maculation distinct, not washed with a white film . . . . *Stiria dyari* p. 141

21. Basal brown patch large and confluent with median patch; hindwing dark brown ...... *Stiria satana* p. 143

– Basal brown patch not as large, never confluent with median patch; hindwing generally, but not always, white or lightly suffused with brown 22

22. Smaller species; forewing length from base to apex 13–15 mm; median brown patch of forewing usually taller than wide; hindwing usually suffused with light brown .... *Stiria sulphurea* p. 142

– Larger species; forewing length from base to apex 16–18 mm; median brown patch of forewing either wider than tall or square; hindwing usually, but not always, white with at most brown along outer margin. 23

23. Median patch of forewing usually wider than tall, suffused with light-gray scales; eastern and central North America .... *Stiria rugifrons* p. 138

– Median patch of forewing usually square in shape, not suffused with light-gray scales; southwestern United States and Great Basin region .... *Stiria intermixta* p. 140

24. Eye reduced in size; very small (forewing length less than 10 mm) day flying moths from southern California 25

– Eye not reduced in size; forewing length greater than 10 mm; not strictly day flying 26

25. Forewing bright rusty orange with definite maculation; dark-brown discal dot in hindwing .... *Xanthothrix neumoegeni* p. 112

– Forewing yellow green, olive green, or brown green; absolutely no maculation in either forewing or hindwing .... *Xanthothrix ranunculi* p. 113

26. Forewing ovate (plate 4, figures 6, 7); forewing with subterminal area cream white, contrasting with dark median area; forewing median area with blue tints; outer ring of frontal process weak, not clearly defined 27

– Forewing not as above; outer ring of frontal process usually clearly defined 29

27. Forewing with a white patch or spot between reniform and orbicular marks 28

– Forewing without a white patch or spot between reniform and orbicular marks .... *Chrysoecia scira* p. 115

28. Forewing with white patch between reniform and orbicular marks square, filling discal cell .... *Chrysoecia gladiola* p. 115

– White patch between reniform and orbicular marks smaller, comma shaped or triangular, not reaching costal margin of forewing .... *Chrysoecia requies* p. 115

29. Forewing brown, tan, or dark brown with a contrasting dark-brown triangular patch on the costa following postmedial line 30

– Forewing not as above; if with this triangular patch, forewing not a shade of brown 32

30. Forewing with the orbicular mark or parts of it black, contrasting with background color; black may be reduced to small dots in orbicular mark 31

– Forewing with orbicular mark not black, not contrasting with rest of forewing .... *Plagiomimicus curiosum* p. 95

31. Forewing light brown; a strong, contrasting dark-brown subterminal shade in forewing running from inner margin to about middle of subterminal area; southwestern U.S. and Mexico .... *Plagiomimicus triplagiatus* p. 101

– Forewing dark brown; subterminal shade in forewing present but weak, not contrasting with rest of forewing; eastern and central United States .... *Plagiomimicus pityochromus* p. 101

32. Forewing light brown with a broad, diffuse, bright-yellow shade following postmedial line .... *Plagiomimicus aureolum* p. 90

– Forewing not as above 33

33. Forewing light gray brown with a shining ovate patch in the position of the claviform mark and a shining white lunate line representing the reniform mark .... *Plagiomimicus olvello* p. 89

– Forewing not as above 34

34. Forewing with a large, shining white basal area and a grayish-green median area .... *Plagiomimicus dimidiata* p. 90

– Forewing not as above 35

35. Forewing yellow orange; maculation consisting entirely of fine black lines; postmedial line a double black line, continuous near costa and inner margin, breaking into dots in middle of wing; frontal process of head laterally elongate, central tubercle absent .... *Chrysoecia atrolinea* p. 117

– Forewing not as above 36

36. Forewing dull white with olive-green antemedial, postmedial, and subterminal lines .. .. *Lineostriastiria olivalis* p. 109

– Forewing not as above 37

37. Forewing with dull olive-green ground color; costa of forewing near apex with a darker olive-green triangular patch; postmedial and antemedial lines white; foretibial claw very small 38

– Forewing not as above; foretibial claw various 39

38. Cubital vein in median area of forewing accented with white or basal area significantly lighter than median area; outer half of postmedial area before subterminal line darker than inner half; reniform mark composed of both white and dark-green components; clasper of valve may be elongate but not exceeding outer margin .... *Plagiomimicus tepperi* p. 97

– Cubital vein in median area of forewing not accented with white and basal area not significantly lighter than median area; outer half of postmedial area before subterminal line not darker than inner half; reniform mark a single faint white line without dark-green components; clasper of valve in male genitalia very elongate, apex exceeding outer margin .... *Plagiomimicus mimica* p. 98

39. Forewing with a yellow, golden, or brassy sheen 40

– Forewing without golden or brassy sheen, or if present limited to subterminal area 42

40. Forewing with a frosted appearance .... *Basilodes chrysopis* p. 123

– Forewing without a frosted appearance 41

41. Postmedial line of forewing straight and with only one crenulation in it on vein 1A near inner margin; basal area with patch of brown from articulation of forewing along lower third of area to antemedial line .... *Basilodes pepita* p. 120

– Postmedial line of forewing vaguely crenulate for its entire length; basal area without patch of brown from articulation of forewing along lower third of area to antemedial line .... *Basilodes straminea* p. 122

42. Forewing elongate, dull white; veins in subterminal area accented with brown; area between median line and postmedial line in lower half of median area contrastingly darker than rest of wing; orbicular mark small, white, surrounded by a thin black line .... *Hoplolythrodes arivaca* p. 125

– Forewing not as above 43

43. Forewing color light blue gray with a suffused appearance .... *Plagiomimicus caesium* p. 94

– Forewing not as above 44

44. Forewing ground color yellow with rust-brown striations running along veins; postmedial and antemedial lines and all other maculation absent; hindwing orange .... *Eulithosia discistriga* p. 130

– Forewing and hindwing not as above 45

45. Forewing light brown; orbicular and reniform marks present, large, surrounded by a dark-brown ring; median line present, distinct, ventral surface of forewing and hindwing with a distinct postmedial line ..... *Plagiomimicus ochoa* p. 87

– Forewing and hindwing not as above 46

46. Forewing brown with apices produced into points; orbicular and reniform marks present, large, surrounded by a fine white line .... *Plagiomimicus astigmatosum* p. 96

– Forewing without this combination of characters 47

47. Forewing dull olive green; forewings elongate; small species (forewing length 7–9 mm); foretibial claw very strong and tibia about as long as wide .... *Lineostriastiria hutsoni* p. 111

– Forewing not as above; forewing length greater than 10 mm; foretibial claw not extraordinarily strong and tibia longer than wide 48

48. Forewing appearing rectangular; forewing color light clay brown to gray brown; forewing maculation consisting solely of dull, broad white bands .... *Plagiomimicus hilli* p. 96

– Forewing triangular, often with apex produced into a point; color of forewing dark brown, brown, or gray brown; postmedial and antemedial lines, if present, thin, not diffuse and broad 49

49. Forewing apex not produced into a point (plate 3, figure 45); ground color of forewing gray brown; orbicular mark often a small black dot .... *Plagiomimicus expallidus* p. 103

– Forewing apex produced into a point (plate 3, figure 29); ground color of forewing dark brown or brown; orbicular mark absent 50

50. Forewing coloration brown; postmedial line yellow white, distinct against background color .... *Plagiomimicus spumosum* p. 92

– Forewing coloration dark brown; postmedial line, if present, obscure, not distinct against background color 51

51. All maculation of forewing indistinct; smaller species (forewing length 11–12 mm); hindwing light brown; midwestern U. S. species .... *Plagiomimicus heitzmani* p. 91

– Not as above 52

52. Subterminal region of forewing without metallic yellow scales; median area without violet tint; hindwing dark brown .. *Plagiomimicus manti* p. 95

– Subterminal region of forewing with metallic yellow scales; median area with violet tint; hindwing light brown ... *Angulostiria chryseochilus* p. 134

GENUS
Fala Grote

> *Fala* Grote, 1875, *Proc. Acad. Nat. Sci. Philadelphia,* **1875**: 425.
> Type species: *Fala ptycophora* Grote, 1875. Monotypy.

Fala is a monotypic genus containing the southern Californian species *ptycophora* Grote. This genus is clearly off the main branch of stiriine evolution and appears to exhibit a number of primitive features. Therefore, I have used *Fala* as the outgroup for the rest of the subfamily. *Fala ptycophora* does not look like a stiriine, as unscientific a statement as that is. However, the male vesica has the basic oblong shape characteristic of the Stiriinae with the main field and the basal row of cornuti. *Fala ptycophora* is a robust, almost woolly, moth without bright colors or metallic reflections. The male and female genitalia exhibit a number of unique features. The ovipositor lobes are unmodified and membranous with scattered hairs and small spines (the primitive condition in the "trifid" noctuids) with none of the modifications found in the rest of the subfamily. Otherwise the female genitalia are typical of the Stiriinae, although the appendix bursae is somewhat larger than is typical for the subfamily. The valve is unusually robust for a species in the Stiriinae. The tympanic region has a large hood and a well-developed abdominal bulla located toward the distal end of the intersegmental membrane. Both the hood and the abdominal bulla are reduced in size in the main branch of stiriine evolution, and an increasing degree of reduction is apparently a transition from primitive to advanced.

Head: antenna of both male and female simple, slightly ciliate ventrally; eye naked; eye unlashed either from base of antenna or rear margin of eye; a rough row of bushy scales between antennal bases; front with strong, circular frontal process; ventral margin of process includes clypeal margin; dorsal margin interrupted; concavity with strong central prominence, dorsoventrally flattened, bladelike, located closer to dorsal than ventral margin; palpus with third segment small, ovate, projecting forward; second segment fringed outwardly with coarse hair; proboscis strong. Thorax: dorsal vestiture of rough, elongate hairlike scales; tufting obscure with obscure metathoracic tuft; vestiture of forewing of elongate ovate scales only weakly toothed; no sexual modifications either dorsally or ventrally. Prothoracic leg: femur and tibia without large tufts or fringes of hairs; tibia about 3 × as long as wide; tibial claw present, strong, blade shaped with both margins confluent with apex of tibia; first tarsal segment about 1.5 × as long as tibia, as long as remaining four segments; first four tarsal segments with three rows of spines; fifth segment with only scattered spinules; tarsal claw with basal tooth. Mesothoracic and metathoracic legs: femur weakly fringed ventrally with hair; first four tarsal segments with three rows of spines, fifth segment with scattered small hairlike setae. External tympanic region: hood ovate, relatively large, conspicuously free of intersegmental membrane; abdominal bulla relatively large, positioned toward

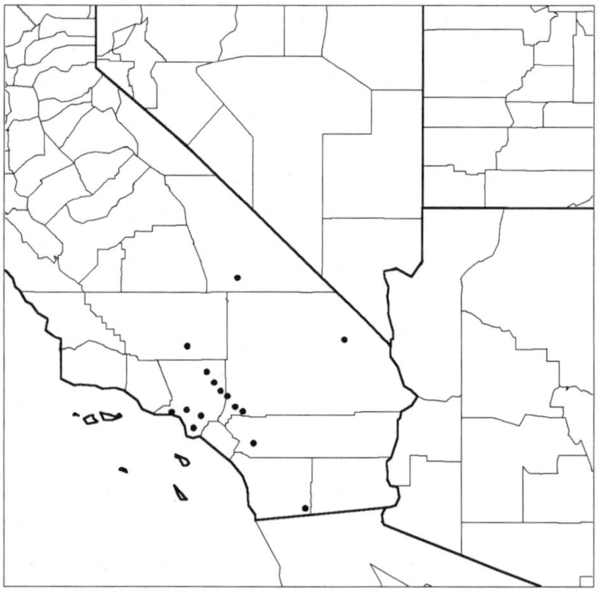

FIGURE 49: DISTRIBUTION OF SPECIMENS EXAMINED OF *FALA PTYCHOPHORA*

distal margin of segment. Internal tympanic region: not examined. Abdomen: no noticeable tufts; male without basal hair pencils; sterna with heavy deposition of melanin; male eighth sternum U-shaped but with sclerotization indistinct; eighth tergum with distal margin enlarged, sclerotized, forming a small cup-shaped container. Male genitalia: plate J, figures 4, 5; uncus about 4 × as long as wide, apex with a small tooth and a slight ventral ridge; uncus moderately swollen, hairy dorsally; valve strong with simple clasper; corona present; juxta with dorsal triangular indentation; vesica typical of subfamily. Female genitalia: plate P, figure 1; ovipositor lobes unmodified; ovipositor neck without distinct lines of microspinules; eighth sternum without unsclerotized regions; ostium broadly triangular; ostium region not indented; appendix bursae as long as wide.

Fala ptycophora Grote

PL. 3, FIG. 21; PL. J, FIGS. 4, 5 (♂ gen.); PL. P, FIG. 1 (♀ gen.); TEXT FIG. 49 (map) (RWH 9788).

Fala ptycophora Grote, 1875, *Proc. Acad. Nat. Sci. Philadelphia,* **1875**: 426.
Type locality: [USA], California. [BMNH]

Fala ptycophora is atypical of the Stiriinae in superficial appearance and biology. Unlike the majority of the Stiriinae the adults fly in the spring, not in the autumn. The other notable exception to this rule are the two species of *Xanthothrix.* The adults have a robust, slightly shaggy appearance, possibly because of the spring flight period. The forewing coloration is clay brown. The reniform and orbicular marks are filled with dull-white scales. The postmedial line is heavily crenulate. The frontal prominence has a vertical, blade-shaped central prominence, and the ring is open at the top. Wing length: mean = 15.13 mm, standard deviation = 0.37 mm, n = 10.

The larva and its foodplants are unknown.

This species appears to be limited to the Mohave Desert region of southern California. It occurs as far north as southern Inyo County and as far south as the Mexican border in San Diego County. Its eastern most boundary appears to be eastern San Bernardino County. There is some variation in the amount of dull-white scaling in the forewing. A single male specimen appears to be a dark chocolate-brown melanic. Adults have all been collected in March and April.

GENUS

Plagiomimicus Grote

Plagiomimicus Grote, 1873, *Bull. Buffalo Soc. Nat. Sci.,* **1**: 182.
Type species: *Plagiomimicus pityochromus* Grote, 1873. Monotypy.

Stibadium Grote 1874, *Bull. Buffalo Soc. Nat. Sci.,* **2**: 32, 74.
Type species: *Stibadium spumosum* Grote, 1874. Original designation.

Antaplaga Grote, 1877, *Can. Ent.,* **9**: 70.
Type species: *Antaplaga dimidiata* Grote, 1877. Monotypy.

Polenta Morrison, 1880, *Proc. Boston Soc. Nat. Hist.,* **18**: 124.
Type species: *Schinia tepperi* Morrison, 1875. Original designation.

Neophaeus Dyar 1918, *Proc. U. S. Natl. Mus.,* **54**: 350.
Type species: *Neophaeus chalcospilans* Dyar, 1918. Original designation.

Plagiomimicus is the largest genus of the subfamily Stiriinae. The genus consists of stiriines with modified ovipositor lobes, but without large clear areas in the eighth sternum or an invagination of the ostium in the female genitalia. *Plagiomimicus* could be split into a number of smaller genera more in keeping with the average genus size in the rest of the subfamily. However, I have been reluctant to change the concept of the genus as established by Hogue (1963) for reasons of stability. Almost all of the species, with the exception of the *pityochromus* group, have pointed apices in the forewings. Although species in other genera may have pointed apices, the condition is never as distinct as for most species of *Plagiomimicus,* particularly in the *spumosum* and *tepperi* groups.

Head: antenna of male and female simple, scaled dorsally, finely ciliate ventrally; no lashes from base of antenna or rear margin of eye; eye naked; ocelli present; proboscis normal; palpus with second segment roughly scaled outwardly, third segment short and stubby; frontal process present, large, variable among groups, but generally large and round with strongly raised rim; process smaller and elliptical in *ochoa*; process flattened ventrally and with ventral fingerlike projection in *tepperi* group; vestiture of dorsal region of head of large, flattened scales, not arranged in rows. Thorax: forewing with accessory cell; R_3 and R_4 stalked about ⅓ the distance from end of accessory cell to outer margin; R_{3+4} and R_5 connate at apex of accessory cell; M_2 of hindwing well developed, arising closer to M_1 than M_3; cross-vein mdc slightly triangular; M_3 and CuA_1 connate at their interception of cross-vein ldc; venation otherwise as in the Noctuidae as a whole; wing shape variable, without specialized scales; patagia not raised into a hood, vestiture of large, flattened scales; vestiture of dorsum of thorax of large, flattened scales, scales caudally with long, thin stalks; thorax without tufts except for vague tuft on metathorax. Prothoracic leg: femur 3–4 × longer than wide, closely covered with square scales; no tufts or fringes of hair present; tibia about twice as long as wide; spines absent; strong tibial claw present, degree of development dependent on species; first two tarsal segments scaled dorsally and ventrally; first tarsal segment about 6 × longer than wide, as long as segments two and three together; tarsal spines on first four segments in three irregular rows; tooth on tarsal claw absent. Mesothoracic and metathoracic legs: tibia and femur approximately equal in length; tibia without spines; femur covered with flat scales with at most a weak ventral ridge of long scales; tarsal segments as in prothoracic leg. External tympanic region: hood moderate in size with spiracle located cephalad and toward the ventral margin of hood; no bullae in intersegmental membrane of abdominal segments one and two; alula weak. Internal tympanic structure: as in *Cucullia*; pocket four with a large, circular opening dorsally. Male abdomen: dorsum of abdomen without tufts; basal hair pencils and Stobbe's glands absent, but sternum of form for potential hair pencils; eighth sternum with U-shaped sclerotization; eighth tergum with V- to U-shaped sclerotization; no line of long hairs in eighth sternum. Male genitalia: uncus variable in length and width, sometimes with medial swellings; uncus covered with many long hairs; tegumen single, unmodified; free pleurite present, a simple elongate strap; saccus of vinculum rounded to pointed; valve with corona (except in *tepperi* group); clasper of various shapes and positions depending on species, but elongate and not arising from a large plate in most species; clasper oriented perpendicularly to long axis of valve; basal process of sacculus present; no differentiation between sacculus and cucullus; digitus absent; vesica elongate, globular with basal row of cornuti and a large distal patch of diffuse cornuti. Female genitalia: ovipositor lobes elongate, strongly sclerotized, and with apices usually pointed (however, see *ochoa*); weakly sclerotized ridges sometimes existing between ovipositor lobes; apophyses anteriores and posteriores normal to slightly elongate depending on species; microspinules in ovipositor neck not arranged in rows, or arranged in rows only near margin of eighth abdominal sternum; ostium not invaginated; clear areas on either side of ostium in eighth abdominal segment absent, or represented at most by small, scattered clear patches; ductus bursae short, usually no more than 3 × longer than wide; ductus bursae sclerotized or not depending on species-group; corpus bursae elongate, about 4–6 × longer than wide; upper fourth of corpus bursae and appendix bursae sclerotized, with internal ridges.

Plagiomimicus ochoa (Barnes)

PL. 3, FIG. 22; TEXT FIG. 50 *b, c, e* (♂ gen.); TEXT FIG. 50 *a* (♀ gen.); TEXT FIG. 50 *d* (head); TEXT FIG. 51 (map) (RWH 9740).

Stibadium ochoa Barnes, 1904, *Can. Ent.*, **36**: 241.
Type locality: [USA], Arizona, Wilgus. [USNM]
NOTE—*Stibadium ochoa* was described from one male and one female. The type male has been labeled and is hereby designated the LECTOTYPE.

Plagiomimicus dollii Smith, 1908, *Ann. New York Acad. Sci.*, **18**: 118.
Type locality: [USA], Arizona, Cochise Co., Palmerly [=Palmerlee]. [USNM]; lectotype designated by Todd (1982).

Plagiomimicus ochoa has many characters considered primitive for the Stiriinae. In particular the

FIGURE 50: STRUCTURAL FEATURES OF *PLAGIOMIMICUS OCHOA*
a. Female genitalia. *b.* Lateral view of male genitalia. *c.* Right valve of male genitalia. *d.* Head. *e.* Aedoeagus.

ovipositor lobes are only slightly sclerotized and only weakly elongate. This state is intermediate between the primitive, unmodified lobes of *Fala ptychophora* and the strongly sclerotized ovipositor lobes found in most of the Stiriinae. The adult is light to medium brown with a slightly yellow tint. The forewing is slightly broader than in most species of the genus (but not greatly so), and the apex of the wing is pointed. The thin, medium-brown postmedial line runs diagonally from the inner margin to a point opposite the reniform mark and is then angled back to the outer margin. The reniform mark is a large, but inconspicuous, circle. The outer third of the forewing is distinctly lighter than the basal and median areas. The frontal prominence is a dorsoventrally elongate ellipse. The clasper of the valve is medially placed and not modified. The uncus is long (about 5 × as long as wide). The vesica contains the wide apical field and basal row of cornuti typical of the Stiriinae. The spicules in the ovipositor neck are arranged in rough rows toward the distal margin of the eighth sternum, but randomly placed nearer the ovipositor lobes. Wing length: mean = 16.18 mm, standard deviation = 0.74 mm, n = 10.

The sister species of *ochoa* is *Plagiomimicus resoluta* Dyar from central and south-central Mexico. The maculation and genital morphology are similar in the two species, but *resoluta* is a darker species with a distinct olive rather than yellow tinge as in *ochoa.*

The larva and its foodplants are unknown.

Plagiomimicus ochoa has been collected in the Chiricahua, Huachuca, and Santa Rita Mountains of southeastern Arizona and the White Mountains of east-central Arizona. It also occurs in the Davis Mountains of western Texas and has been taken in the central part of Chihuahua in northern Mexico. The species presumably occurs in southern New Mexico, but I have not seen any specimens from there. There is no discernible variation among or within populations. Adults have been collected in August and September.

Plagiomimicus olvello (Barnes)

PL. 3, FIG. 23; PL. J, FIGS. 6, 7 (♂ gen.); PL. P, FIG. 2 (♀ gen.); TEXT FIG. 52 (map) (RWH 9742).

Stibadium olvello Barnes, 1907, *Can. Ent.,* **39**: 94.

Type locality: [USA], southern New Mexico. [USNM]

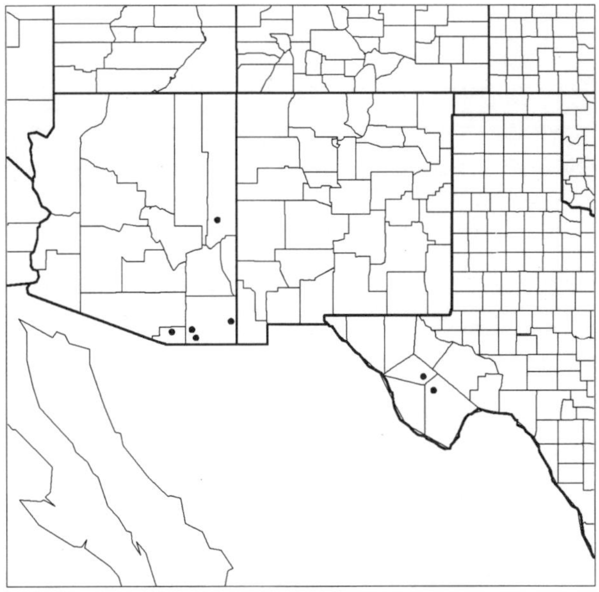

FIGURE 51: DISTRIBUTION OF SPECIMENS EXAMINED OF *PLAGIOMIMICUS OCHOA*

Plagiomimicus olvello is one of the most distinctive noctuids in North America. The forewing is medium tan. The claviform mark is represented by a huge round silver-white spot. The reniform mark is a silver-white lunule with its points directed toward the costa. All other maculation is absent except for a vague, dull-white postmedial line. The frontal process is round, and its lower margin is slightly concave. The central prominence is represented by a small vague knob. The male clasper is a double structure: the distal part is large and slightly curved distally; the basal process is small and represented by a small knob. The uncus is 4–5 × as long as wide and not swollen medially. The vesica is unremarkable for the subfamily. The ovipositor lobes are highly sclerotized with long, pointed apices projecting from wide bases. The ovipositor neck is short and heavily covered with hairlike spicules. The spicules are not arranged in clear-cut rows. The appendix bursae is large relative to the *spumosum* group and not convoluted into sclerotized wrinkles as in that species complex. *Plagiomimicus olvello* does not appear to have any obvious close relatives. Wing length: mean = 14.40 mm, standard deviation = 0.53 mm, n = 10.

The larva and its foodplants are unknown.

Plagiomimicus olvello occurs in Cochise County in southern Arizona, southern New Mexico, and southwestern Texas. The species undoubtedly also occurs in the corresponding regions of northern Mexico, but no specimens have been seen. There is

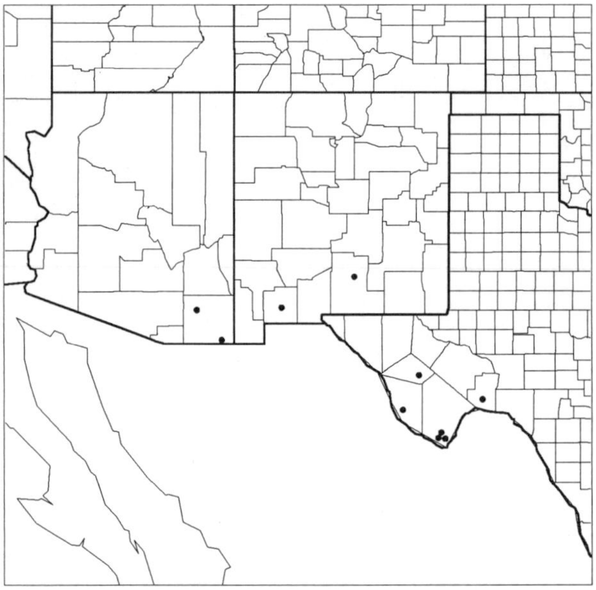

FIGURE 52: DISTRIBUTION OF SPECIMENS EXAMINED OF *PLAGIOMIMICUS OLVELLO*

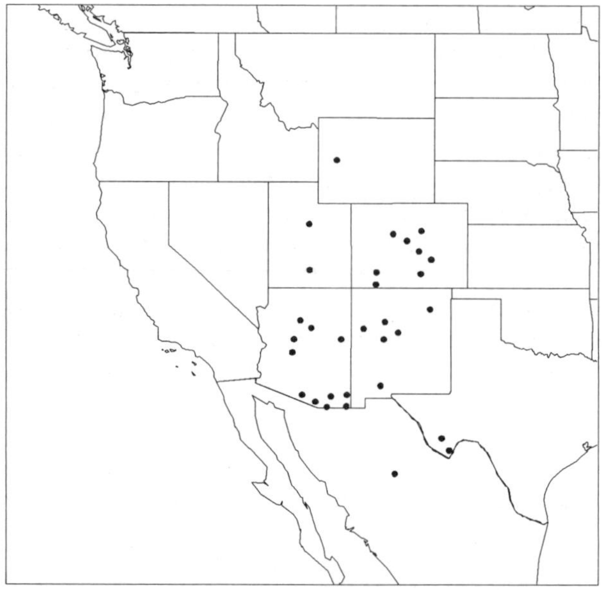

FIGURE 53: DISTRIBUTION OF SPECIMENS EXAMINED OF *PLAGIOMIMICUS DIMIDIATA*

no obvious variation in the species. Adults have been collected in September and October.

Plagiomimicus dimidiata (Grote)
PL. 3, FIG. 24; PL. J, FIGS. 8, 9 (♂ gen.); PL. P, FIG. 3 (♀ gen.); TEXT FIG. 53 (map) (RWH 9745).

Antaplaga dimidiata Grote, 1877, *Can. Ent.*, **9**: 71.
Type locality: [USA], Colorado. [BMNH]

Plagiomimicus dimidiata is another highly distinctive species and is unlikely to be confused with any other North American noctuid. The forewing base is silvery white and the outer ⅔ from the median line to the outer margin is greenish bronze. The postmedial line is represented by a diffuse, white shade. There is a vague, small, triangular olive-green apical triangle. The male genitalia are like those of *olvello*; however, the distal part of the clasper is strongly hooked. The ovipositor lobes of *dimidiata* are uniform in size from base to apex and lack the greatly expanded bases of *olvello*. Wing length: mean = 14.02 mm, standard deviation = 0.94 mm, n = 10.

The larva and its foodplants are unknown.

Plagiomimicus dimidiata has a wide distribution in western North America. It has been collected in southwestern Texas, New Mexico, Colorado, most of Arizona (except for the Mohave Desert region of the western part of the state), and at scattered localities in Utah. The northern boundary of its range is the Grand Teton region of western Wyoming. The species has also been collected in Chihuahua in northern Mexico. The strength of the orbicular and reniform marks in the forewing varies a little, but not greatly so. Adults have been collected in September and October.

Plagiomimicus aureolum (Hy. Edwards)
PL. 3, FIG. 25; PL. J, FIGS. 10, 11 (♂ gen.); PL. P, FIG. 4 (♀ gen.); TEXT FIG. 54 (map) (RWH 9741).

Stibadium aureolum Hy. Edwards, 1882, *Papilio*, **2**: 126.
Type locality: [USA], Arizona, Prescott. [USNM]

NOTE—Edwards states that this species was described from four specimens from the Neumoegen collection. There are currently one male and one female from the Neumoegen collection in the USNM. The male is hereby designated and has been labeled as the LECTOTYPE.

Plagiomimicus aureolum is characterized by a bright-yellow band between the forewing postmedial and subterminal lines. The forewing ground color is tan. The only other obvious markings are fine, white antemedial and postmedial lines. The orbicular and reniform marks are absent. The frontal process is ovate. The male genitalia are similar to those of *olvello*. However the distal process of the two-pronged clasper is smaller and has a sharp hook at its apex. The uncus is swollen medially. The female ovipositor lobes are short and triangular. The ovipositor neck is short and densely covered with hairlike spicules, but the spicules are not arranged in obvious rows. The appendix bursae is longer than in any other species in the genus, is ovate at its apex, and has internal rows of spinules. Wing length: mean = 16.65 mm, standard deviation = 0.92 mm, n = 10.

Plagiomimicus aureolum is superficially most similar to *astigmatosum. Plagiomimicus astigmatosum* is about half the size of *aureolum* and lacks the bright-yellow band in the forewing.

The larva and its foodplants are unknown.

This species occurs in the southwestern United States and probably northern Mexico. *Plagiomimicus aureolum* is widely distributed in Arizona except for the northern and western parts of the state. It occurs as far north as Williams in Coconino County and the White Mountains in the central-eastern part of the state. In the southeast it is commonest in the mountains of Cochise, Pima, and Santa Cruz Counties. The species has been collected in the western and southern parts of New Mexico. In western Texas it occurs in Brewster and Jeff Davis Counties. The species undoubtedly occurs in the corresponding regions of northern Mexico, but I have not seen specimens from there. Some individual variation in size exists, but otherwise *aureolum* is not significantly variable. Adults have been collected in August and September.

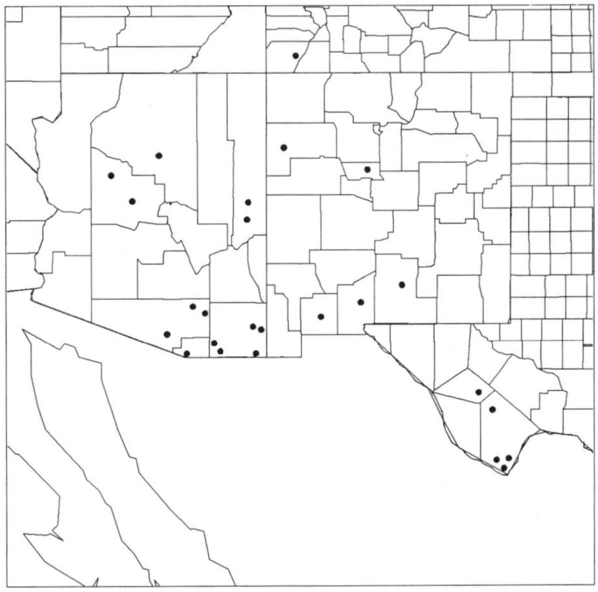

FIGURE 54: DISTRIBUTION OF SPECIMENS EXAMINED OF *PLAGIOMIMICUS AUREOLUM*

spumosum SPECIES COMPLEX

The *spumosum* complex consists of three species, *spumosum, heitzmani,* and *caesium.* They have a similar morphology although they are superficially abundantly distinct. The three species have a peculiar biogeography. *Plagiomimicus spumosum* is wide spread in the United States and southern Canada. In contrast *caesium* appears to be limited to Padre Island along the eastern coast of Texas. *Plagiomimicus heitzmani* is a denizen of the long grass prairie region of the USA midwest. All three species have distinctly pointed forewing apices. The male clasper is large, curved outward, and located near the base of the valve. The frontal process is nearly circular, large, and has the ventral margin slightly pointed.

Plagiomimicus heitzmani Poole, NEW SPECIES

PL. 3, FIG. 26; PL. J, FIG. 12; PL. K, FIG. 1 (♂ gen.); PL. P, FIG. 5 (♀ gen.); TEXT FIG. 55 (map).

Plagiomimicus heitzmani Poole.
Type locality: [USA], Missouri, Clay Co., E. of Birmingham. [USNM]

Plagiomimicus heitzmani is restricted to a narrow region of long-grass prairie in Illinois, Missouri, and Kansas. It is much smaller than *spumosum,* about half the size, and the forewing is a darker brown. The forewing is devoid of all maculation except for a poorly marked postmedial line. The hindwing is dull white in the male and lightly suffused with brown in the female. The male genitalia (plate J, figure 12; plate K, figure 1) are virtually identical to those of *spumosum* except for size. The clasper tends to be slightly thinner. The clasper is not recurved as in *manti* and its relatives. The apices of the female ovipositor lobes (plate P, figure 5) are slightly longer and more sharply pointed than those of *spumosum.* The ostium of *heitzmani* is laterally rectangular but is distally to proximally rectangular in *spumosum.* The above diagnosis and the figures of the adults, male, and female genitalia are given instead of a formal description. Wing length: mean = 12.40 mm, standard deviation = 0.72 mm, n = 5.

The larva and its foodplants are unknown, although I speculate it may be some sunflowerlike composite. This hypothesis is based on the known plant hosts of *spumosum* and the predominance of sunflowerlike composites in the flora of the region.

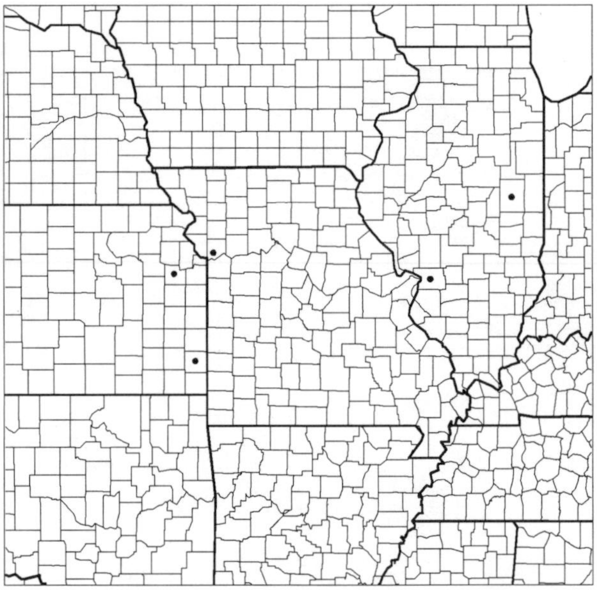

FIGURE 55: DISTRIBUTION OF SPECIMENS EXAMINED OF *PLAGIOMIMICUS HEITZMANI*

TYPES. Holotype: ♂. [USA], Missouri, Clay Co., near Missouri River, East of Birmingham, September 2, 1962, J. R. Heitzman, genitalia on slide USNM 43,692. USNM. Paratypes: 2 ♂, 4 ♀. [USA], Illinois, Urbana, August 20, 1886 (1 ♀). Same locality, August 24, 1886 (1 ♀). Illinois, Madison Co., S. Illinois Campus, September 3, 1972, R. S. Funk (1 ♂). Kansas, Crawford Co., 4 mi E Pittsburg, Sept. 1–15, 1972, E. L. Todd (1 ♂, 1 ♀). Kansas, Lawrence, August 3, 1933, W. D. Field (1 ♀). CNC and USNM. The species is named in honor of the collector of the holotype, J. R. Heitzman.

Plagiomimicus heitzmani is extraordinarily rare in collections and is known only from central and southern Illinois, western Missouri, and eastern Kansas. It may be restricted to areas of long-grass prairie. Adults have been collected in August and September.

Plagiomimicus spumosum (Grote)

PL. 3, FIGS. 27–30; TEXT FIG. 56 *b, c, e* (♂ gen.); TEXT FIG. 56 *a* (♀ gen.); TEXT FIG. 56 *d* (head); TEXT FIG. 57 (map) (RWH 9748 9749).

Stibadium spumosum Grote, 1874, *Bull. Buffalo Soc. Nat. Sci.*, **2**: 74.
Type locality: [USA], Kansas. [BMNH]

NOTE—*Plagiomimicus spumosum* is based on one male and one female. The male type has been labeled and is hereby designated as the LECTOTYPE.

Stibadium mavina Barnes and McDunnough, 1910, *Can. Ent.*, **42**: 250. NEW SYNONYMY.
Type locality: [USA], Utah, Provo. [USNM]

The forewing of *spumosum* is gray brown and has a slightly frosted appearance in the typical eastern and midwestern populations. The only significant forewing marking is a light postmedial line. The subterminal area is usually slightly lighter than the median and basal areas. Specimens from Utah and westward have a propensity to become dull, pallid yellow with yellowish and brownish specimens occurring within the same populations. In a few specimens the antemedial line is present as well. The frontal process is an ovate ring with a deep central depression and only a slight indication of a central prominence near the ventral part of the depression. The ovipositor lobes are elongate and apically pointed. Both pairs of apophyses are moderately elongate. The ostium is either quadrate or slightly distoproximally elongate. Wing length: mean = 16.62 mm, standard deviation = 1.19 mm, n = 10.

The sister species of *spumosum* are *caesium* (restricted to the coastal region of southeast Texas), and *heitzmani* (known only from a small area in the midwestern United States). Neither species is likely to be confused with *spumosum. Plagiomimicus spumosum* is almost twice as large as *heitzmani, heitzmani* is slightly darker, and the postmedial line is not as well developed. The color of *caesium* is light blue gray, and is unlike any other stiriine.

The larva has been described by Crumb (1956) and Forbes (1954). Both record the foodplant as the seedheads of sunflowers (*Helianthus annuus* L.) (Asteraceae) the larva burrowing in the heads and eating the seeds. There are also reared specimens in the LACM and USNM from this foodplant. The larva is indistinctive. In the preserved material I have seen the body is white without markings of any kind except for a slightly heavier sclerotization of the dorsal surfaces of the last four abdominal segments. The head is orange yellow. The only marking on the head is a slight reticulation caused by the underlying muscles and fat, and is not due to color patterns in the head capsule itself. *Plagiomimicus spumosum* has not yet been recorded as a pest on commercial sunflower crops.

This species has, perhaps, the widest geographical range of any stiriine. It occurs throughout most of the United States except the northeast. *Plagiomimicus spumosum* appears to be rare in the east and highly sporadic in its distribution. The northern most specimen I have seen is one from Hunterdon

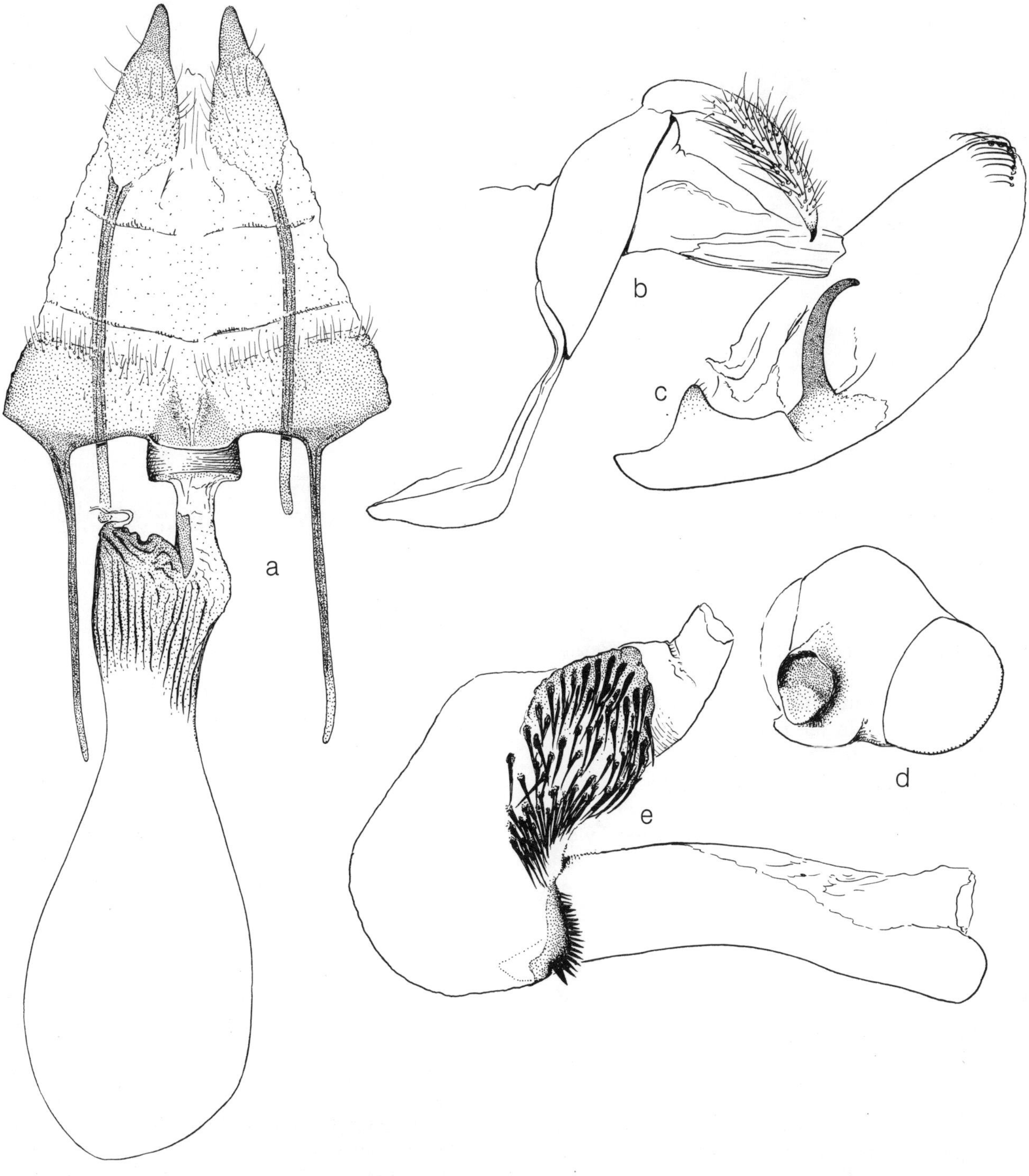

FIGURE 56: STRUCTURAL FEATURES OF *PLAGIOMIMICUS SPUMOSUM*
a. Female genitalia. *b.* Lateral view of male genitalia. *c.* Right valve of male genitalia. *d.* Head. *e.* Aedoeagus.

County, New Jersey. It has been collected in western South Carolina, northwestern Florida and west-central Mississippi. The species is much commoner in the midwestern United States ranging throughout the plains states from eastern North Dakota in the north to the Mexican border in the south. Although common in northern Texas, it has not been collected in the eastern and central parts of the state. *Plagiomimicus spumosum* has been commonly collected in the west from Colorado, Utah, New Mexico,

FIGURE 57: DISTRIBUTION OF SPECIMENS EXAMINED OF *PLAGIOMIMICUS SPUMOSUM*

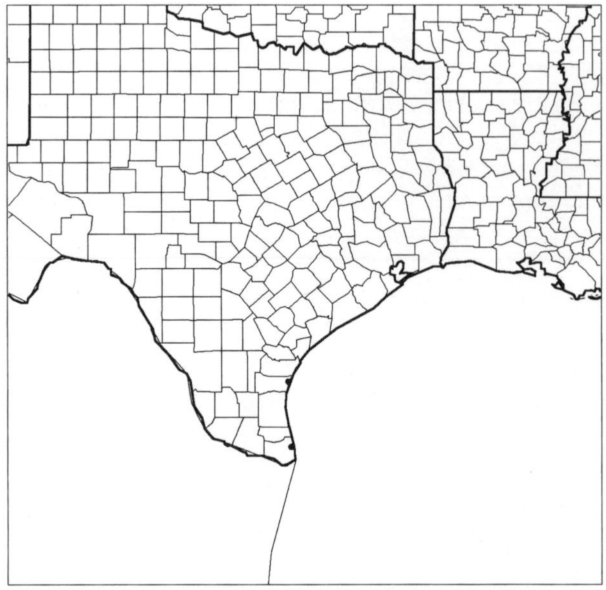

FIGURE 58: DISTRIBUTION OF SPECIMENS EXAMINED OF *PLAGIOMIMICUS CAESIUM*

western Texas, and Arizona. The species also occurs in southern California in Los Angeles, San Bernardino, and Riverside Counties, but has not been collected in either central or northern California or the state of Nevada. In the northwest the species has been collected in eastern and central Washington and eastern Alberta. I have not seen any specimens from Saskatchewan or Manitoba. *Plagiomimicus spumosum* is a variable species geographically and individually. In the eastern and midwestern parts of its range, the species is constant in color, as described above, but specimens vary considerably in size and distinctness of the forewing postmedial line. In the western part of its range, particularly in California and Washington, there is considerable variation in color as well. Many, if not most, specimens are a dull, washed-out yellowish color, although all gradations from almost typical eastern brown forewings to pale yellow exist within the same populations. The variation does not appear to be sex related. Both forms occur in a reared series from southern California in the LACM. This yellowish form has gone under the name *mavina.* The incidence of the pale forms rises as one goes west from western Colorado-Utah, to the West Coast. Adults fly from July to September.

Plagiomimicus caesium (Blanchard and Knudson), NEW COMBINATION.

PL. 3, FIGS. 31, 32; PL. K, FIGS. 2, 3 (♂ gen.); PL. P, FIG. 6 (♀ gen.); TEXT FIG. 58 (map).

Stibadium caesium Blanchard and Knudson, 1984, *Proc. Ent. Soc. Washington,* **86**: 346. Type locality: [USA], Texas, Cameron Co., South Padre Island. [USNM]

Plagiomimicus caesium is an elegant blue-gray species unique both in its coloration and distribution. Superficially it appears to be a blue-gray version of *spumosum*; in fact, the male and female genitalia of the two are virtually identical. The median area of the forewing of *caesium* is slightly darker than the median and subterminal areas, and the orbicular and reniform marks are only vaguely hinted at. The subterminal area is particularly lighter in color following the postmedial line. The outer margin of the hindwing is tinged with black. Wing length: mean = 14.93 mm, standard deviation = 0.90 mm, n = 8.

The immature stages are unknown.

Plagiomimicus caesium is very closely related to *spumosum* despite its distinctive appearance. It was described from a single male from South Padre Island in coastal Texas but has subsequently been collected in the North Padre Island region of San Patricio County by E. Knudson as well. Its curious coloration and restricted distribution as well as its relationship to the wide spread *spumosum* is the stuff of speculation. Adults have been collected in September and October.

manti GROUP

The three species *manti, astigmatosum,* and *curiosum* are superficially disparate, but morphologically similar, species characterized by a single recurved, distally pointed male clasper. The basal position of the clasper and its recurved shape is shared with the species of the *spumosum* group. The two groups are probably closely related. *Plagiomimicus corazona* Schaus from the Mexico City area also belongs to the *manti* group. The female ovipositor lobes are elongate, pointed, with the apices shaped like narrow spades.

Plagiomimicus manti (Barnes)

PL. 3, FIG. 33; PL. K, FIGS. 4, 5 (♂ gen.); PL. P, FIG. 7 (♀ gen.); TEXT FIG. 59 (map) (RWH 9744).

Stibadium manti Barnes, 1904, *Can. Ent.,* **36**: 243.
Type locality: [USA], Texas, Kerrville. [USNM]

NOTE—*Plagiomimicus manti* was described from a number of specimens. Three identically labeled syntypes are in the USNM, two males and one female. One of the male syntypes is in much better condition than the other. This specimen has been labeled and is hereby designated the LECTOTYPE.

Plagiomimicus manti is a dull species by the standards of the Stiriinae. The forewing is dark brown. The only obvious markings consist of dull, but discrete antemedial and postmedial lines. The frontal process is circular with a vague knob representing the central prominence. The male clasper is single and recurved toward the apex of the valve. The uncus is weakly swollen and 5–6 × longer than wide. The vesica is unremarkable. The ovipositor lobes are elongate and pointed with the apices resembling the blade of a rounded spade. The appendix bursae is reduced and has the sclerotized wrinkles found in the *spumosum* group of species. Wing length: mean = 14.07 mm, standard deviation = 0.56 mm, n = 10.

The larva and its foodplants are unknown.

Plagiomimicus manti is mostly known from south-central and southwestern Texas. Its distribution in Texas appears to be spotty with a concentration in Uvalde and surrounding counties (near San Antonio) and a second concentration in the mountains of the western region of the state. The species has also been collected in southeastern Arizona in the Huachuca and Chiricahua Mountains, and near Douglas, all in Cochise County. A single specimen in the USNM is from Tehuacan in the state of Puebla in south-central Mexico, a huge range disjunction that probably results from a lack of collecting in most of Mexico. There is no apparent variation in the species. Adults have been collected in September and October.

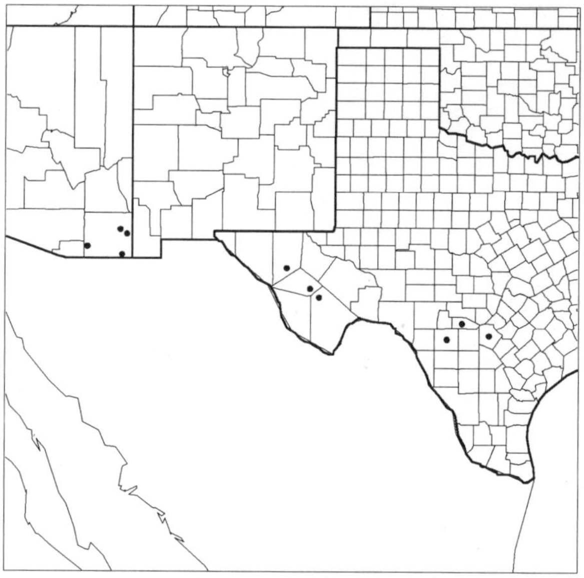

FIGURE 59: DISTRIBUTION OF SPECIMENS EXAMINED OF *PLAGIOMIMICUS MANTI*

Plagiomimicus curiosum (Neumoegen)

PL. 3, FIG. 34; PL. K, FIGS. 6, 7 (♂ gen.); PL. P, FIG. 8 (♀ gen.); TEXT FIG. 60 (map) (RWH 9747).

Stibadium curiosum Neumoegen, 1883, *Papilio,* **3**: 141.
Type locality: [USA], southern Arizona. [USNM]

The forewing of *curiosum* is medium brown. The most distinctive feature of *curiosum* is the large, semi-circular, dark-brown patch at the forewing apex following the postmedial line. This dark patch is outlined with white on both its inner and outer sides. The antemedial and postmedial lines are fine and white, and the reniform and orbicular marks are surrounded by fine, white lines. The male uncus is medially swollen and only about ⅔ as long as that of *manti.* The valve tends to be short and stubby and only about ⅔ as long as in *manti.* The apices of the ovipositor valves are turned outward; they are not turned outward in *manti.* The appendix bursae region is not nearly as heavily sclerotized as in *manti.* Wing length: mean = 14.20 mm, standard deviation = 0.77 mm, n = 10.

FIGURE 60: DISTRIBUTION OF SPECIMENS EXAMINED OF *PLAGIOMIMICUS CURIOSUM*

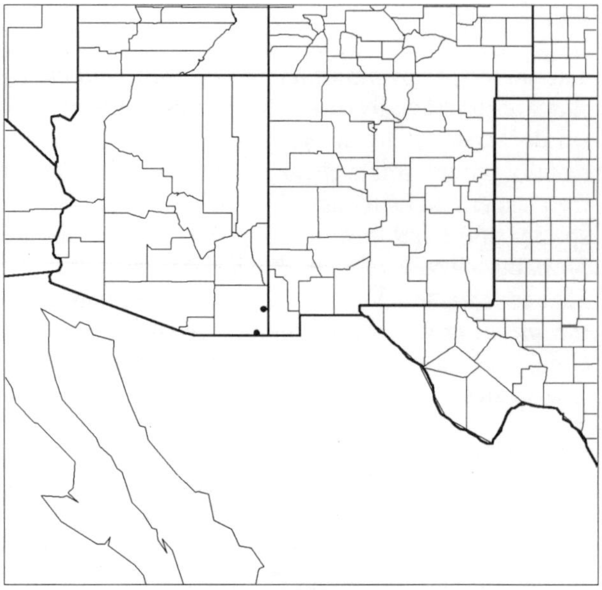

FIGURE 61: DISTRIBUTION OF SPECIMENS EXAMINED OF *PLAGIOMIMICUS ASTIGMATOSUM*

The larva and it foodplants are unknown.

Plagiomimicus curiosum has been collected in the Huachuca, Chiricahua, and Santa Rita Mountains of southeastern Arizona, New Mexico in Otero and Bernalillo Counties, and the Davis Mountains region of southwestern Texas. The species has been collected in the Mexican states of Veracruz, Puebla, Jalisco, and the Federal District. There is no significant variation in the species. Adults have been collected in August and September.

Plagiomimicus astigmatosum (Dyar)
PL. 3, FIG. 35; PL. K, FIGS. 8, 9 (♂ gen.); TEXT FIG. 61 (map) (RWH 9746).

Stibadium astigmatosum Dyar, 1921, *Insecutor Inscitiae Menstruus,* **9**: 42.
Type locality: [USA], southern Arizona. [USNM]

Plagiomimicus astigmatosum looks like a miniature version of *aureolum* but lacks a yellow band following the forewing postmedial line. The male genitalia are structurally a small version of *manti* with no other significant differences. The forewing is tan and elongate with pointed apices. The subterminal area is distinctly lighter than the median and basal areas. The reniform and orbicular marks are outlined by thin, white lines, and the postmedial line is thin and white. A small, white shade is present at the apex of the forewing. The frontal process is circular. The female genitalia have not been examined. Wing length from base to apex: mean = 12.79 mm, standard deviation = 0.28 mm, n = 3.

The larva and its foodplants are unknown.

This species is very rare in collections. It is known only from Paradise in the Chiricahua Mountains and Douglas, both in Cochise County in southeastern Arizona. Adults have been collected in September.

Plagiomimicus hilli (Barnes and Benjamin)
PL. 3, FIGS. 36, 37; PL. K, FIGS. 10, 11 (♂ gen.); PL. P, FIG. 9 (♀ gen.); TEXT FIG. 62 (map) (RWH 9750).

Stiria hilli Barnes and Benjamin, 1923, *Contrib. Nat. Hist. Lep. N. Am.,* **5**: 83.
Type locality: [USA], California, Riverside Co., Indio. [USNM]

Antaplaga caliente Hill, 1923, *Bull. So. California Acad. Sci.,* **22**: 17, pl. [unnumbered], fig. [unnumbered]
Type locality: [USA], California, Riverside Co., Indio. [USNM]

Plagiomimicus hilli is possibly the most disparate appearing species of the genus *Plagiomimicus.* Its morphology possesses some features unique to the genus. The forewing is dull, light brown gray and is heavily dusted, in most cases, with a frosting of white scales. The maculation is obscure, and the

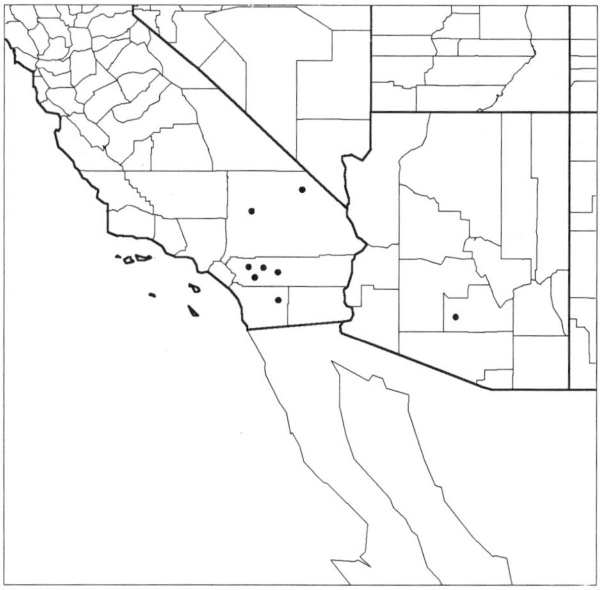

FIGURE 62: DISTRIBUTION OF SPECIMENS EXAMINED OF *PLAGIOMIMICUS HILLI*

only evident markings consist of white antemedial and postmedial lines. The subterminal line is present in some specimens. The forewing has an elongate, slightly rectangular appearance with an upswept appearance that should be recognizable in the figures (plate 3, figures 36, 37) if not easily described in words. One of the unique features of this species is the heavy sclerotization of the sixth, seventh, and eighth terga in the male abdomen. In particular the caudal apex of the seventh sternum is rounded and heavily sclerotized. The male genitalia are unremarkable for the genus. The clasper is basal and is slightly recurved indicating a relationship with species in the *spumosum* and *manti* groups. The uncus is elongate, not short and swollen, a condition slightly atypical for the genus. The vesica has the usual apical and basal groups of cornuti. The ductus bursae and appendix bursae are unsclerotized. A small sclerotized plate exists at the junction of the ductus bursae with the corpus and appendix bursae. Otherwise the female genitalia are not unlike those of the species of the previous two species groups. The spinules in the ovipositor neck at the margin of the eighth sternum are arranged in vague rows, but not as strongly or clearly as in *Lineostriastiria.* The frontal tubercle of the head is ovate-triangular with a large vertical central process. Wing length: mean = 13.04 mm, standard deviation = 0.76 mm, n = 10.

The larva and its foodplants are unknown.

Plagiomimicus hilli is a species of the Mohave Desert region. In California it has been collected in San Bernardino, Riverside, and San Diego Counties. Almost all of the specimens I have seen were collected in the general vicinity of Palm Springs in Riverside County. There are two known specimens from Arizona; one from Maricopa in Pinal County and a second from "Beaver Dam," a locality I have been unable to locate. Intrapopulation variability exists in the strength of the forewing markings and in the amount of yellow coloration in the forewing. Adults have been collected in September and October.

tepperi GROUP

The *tepperi* group consists of three species, *Plagiomimicus tepperi* (Morrison) and two undescribed species. One of the unnamed species is *Plagiomimicus mimica* and is described below. The second undescribed species is known from two males in the USNM from the Mexico City region. The two U. S. species, *tepperi* and *mimica,* are approximately equally common in the United States and have approximately equivalent ranges (although not in Mexico).

The species in this group are among the smallest in the Stiriinae; only the species in *Xanthothrix* and *Lineostriastiria hutsoni* are smaller. The valve is quadrate rather than rectangular, and the corona is totally absent. The uncus is short and swollen. The clasper of the valve is located near the outer margin and the basal process of the sacculus is well developed. The basal group of cornuti in the vesica is absent. The ductus bursae is short and sclerotized with pleats along its entire length. The upper ¼ of the corpus bursae is sclerotized, and there is a strong bulla between the entrance of the ductus bursae and the appendix bursae. The forewing pattern is a dull gray green with characteristic white lines.

Plagiomimicus tepperi (Morrison)

PL. 3, FIGS. 38–40; TEXT FIG. 63 *b, c, e* (♂ gen.); TEXT FIG. 63 *a* (♀ gen.); TEXT FIG. 63 *d* (head); TEXT FIG. 64 (map) (RWH 9755).

Schinia tepperi Morrison, 1875, *Proc. Acad. Nat. Sci. Philadelphia,* **1875**: 68.
Type locality: [USA], Texas. [MSU]

Plagiomimicus richii Grote, 1886, *Can. Ent.,* **18**: 99.
Type locality: [USA], Texas. [BMNH]

NOTE—The exact number of syntypes is unknown, but Grote lists the localities Texas and Florida. Florida is outside the range of this species and may be an error on Grote's part. For consistency, the female syntype from Texas in the BMNH is labeled and is hereby designated the LECTOTYPE.

Plagiomimicus tepperi and *mimica* have been confused in collections under the name *tepperi.* The two are superficially very similar, but the differences in male genitalia are considerable. The forewing color in *tepperi* is gray green without yellow-green tints. The forewing color of *mimica* is usually duller and has a distinct yellowish cast to it. The cubital vein in the median area of the forewing is accented with white in most populations of *tepperi*; it is not accented in *mimica* or only weakly so. The outer half of the postmedial area before the subterminal line is darker than the inner half in *tepperi,* but there is no significant difference in the two halves of the postmedial region in *mimica.* The reniform mark is composed of both white and dark-green components in *tepperi* but is an obscure thin white line in *mimica* with dark-green spots. *Plagiomimicus mimica* is a consistently smaller species than *tepperi.* The size difference is best seen in the color photographs of the adults. Size variation in *tepperi* is considerable, and the size difference will not always separate small individuals of *tepperi* from *mimica.* The male clasper of the right valve in *tepperi* is at right angles to the long axis of the valve (approximately) and does not exceed the outer margin of the valve. The clasper, however, is variable in length within the species. In contrast the clasper on the right valve of *mimica* is elongate, and runs in the approximate direction of the long axis of the valve. The apex of the clasper usually exceeds the outer margin of the valve. Differences in the female genitalia are more subtle. The ostium tends to be globular in *tepperi* but rectangular in *mimica.* The ductus bursae is approximately as long as wide in *tepperi* but about 1.5 × longer than wide in *mimica.* A distinct sclerotized plate is present at the junction of the ductus and corpus bursae in *mimica.* This plate is absent or very small in *tepperi.* Wing length: (southeastern Arizona) mean = 13.40 mm, standard deviation = 0.85 mm, n = 10; (western Texas) mean = 12.56 mm, standard deviation = 0.72 mm, n = 10.

The larva has not been described. There is a single individual in the USNM from near Flagstaff, Arizona reared from *Chrysothamnus nauseosus* (Blake) N. C. Anders. (Asteraceae).

Plagiomimicus tepperi has a moderately wide distribution in the southwestern United States and Mexico. It occurs as far east as Kerrville in south-central Texas and westward to southern California. It has been collected in the drier parts of Colorado and New Mexico and has been collected as far north as north-central Colorado. A disjunct and very distinctive population of *tepperi* occurs in south-central Washington. This population is discussed below. The species is variable in size and is illustrated in the color plate. Females tend to be larger than males, but size is not totally sex linked.

Variation in superficial appearance and the male genitalia is common in *tepperi.* A very distinctive, disjunct population of *tepperi* is known from Yakima County in south-central Washington (plate 3, figure 40). The status of this population is problematical. The forewing has a slightly more rounded appearance and the postmedial and antemedial lines are not as obliquely slanted as in Arizona and Texas populations of *tepperi.* No white accenting is present on the cubital vein in the median area of the forewing. The subterminal and basal areas of the forewing are more distinctly lighter than the median area in this population than in Arizona and Texas specimens. The male clasper is shorter and the ductus bursae longer than in Arizona and Texas populations. All of these distinguishing characters could be interpreted as specific differences. However a variety of observations have led me to treat it as a distinctive and disjunct population of *tepperi.* First, a series of specimens from the Charleston Mountains in southeastern Nevada are superficially nearly identical to the Washington population, but the male and female genitalia fall well within the variability found within Arizona and Texas populations. Second, the male genitalia of *tepperi* are extremely variable, particularly the size and length of the clasper. This variation is complicated but apparently occurs both within and among populations in Arizona and Texas. Individuals from Arizona exist with claspers as short as those found in the Washington population. Similar variation is found in the relative length of the female ductus bursae. This Washington population may represent a distinct species, but the evidence is inconclusive.

Adults have been collected primarily from late July to mid-September. A few specimens are dated April and May.

Plagiomimicus mimica Poole, NEW SPECIES
PL. 3, FIG. 41; PL. K, FIGS. 12, 13 (♂ gen.); PL. P, FIG. 10 (♀ gen.); TEXT FIG. 65 (map).

Plagiomimicus mimica Poole.
Type locality: USA, Arizona, Cochise Co., Paradise [USNM]

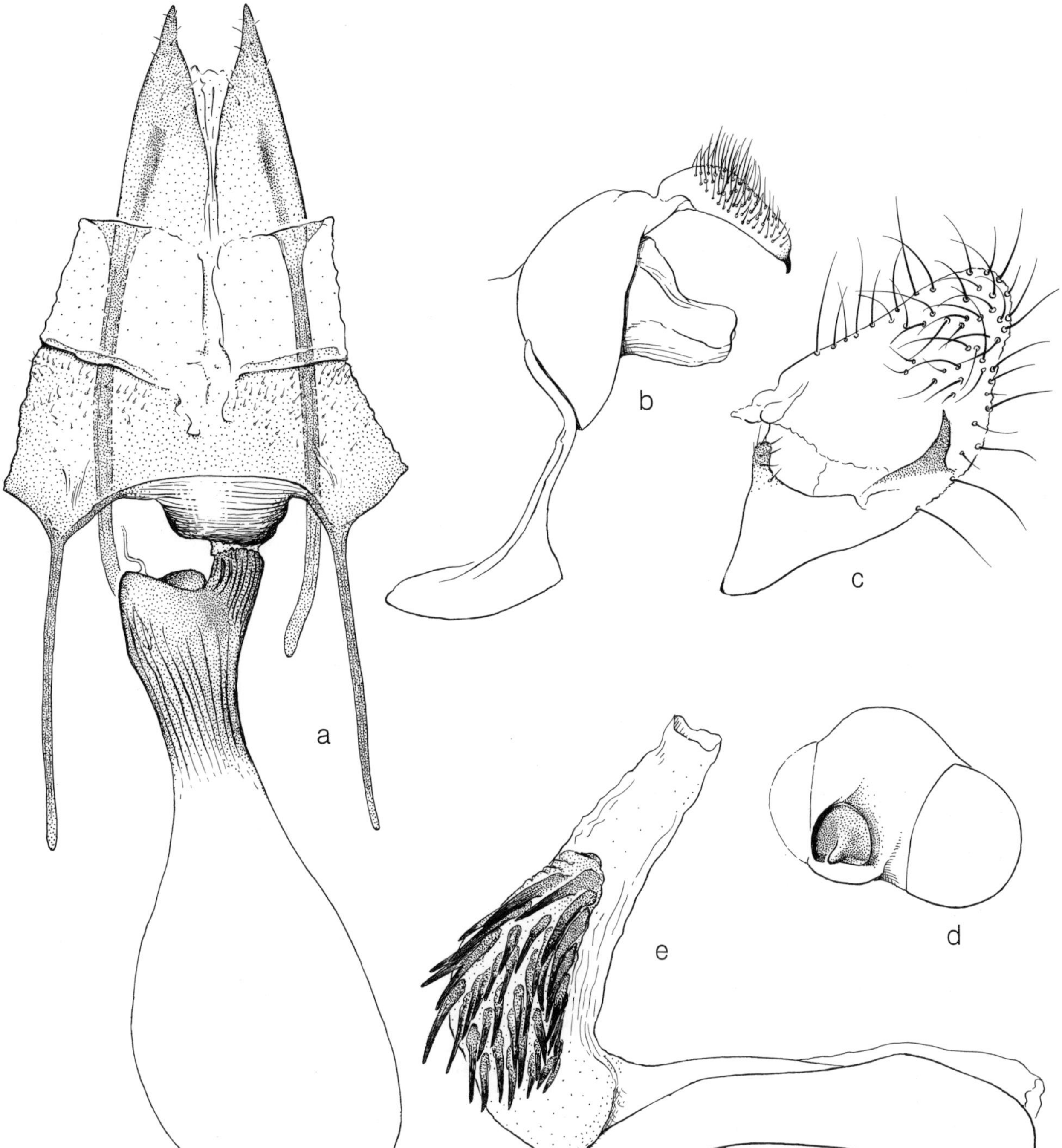

FIGURE 63: STRUCTURAL FEATURES OF *PLAGIOMIMICUS TEPPERI*
a. Female genitalia. *b.* Lateral view of male genitalia. *c.* Right valve of male genitalia. *d.* Head. *e.* Aedoeagus.

The characters differentiating *mimica* from *tepperi* are discussed under *tepperi.* In general *mimica* is smaller than *tepperi* and has a distinct olive or yellow tint. Although *mimica* has been confused with *tepperi,* its closest relative is an undescribed species from the Mexico City region. This new Mexican species is twice as large as *mimica,* but the basic pattern and coloration are identical. The male gen-

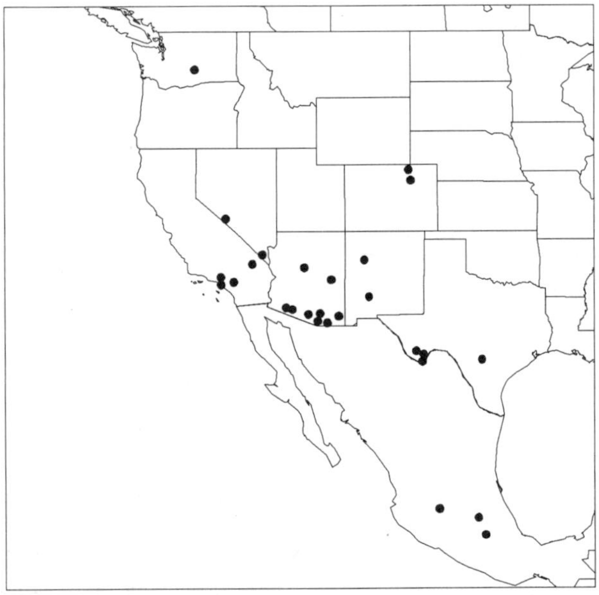

FIGURE 64: DISTRIBUTION OF SPECIMENS EXAMINED OF *PLAGIOMIMICUS TEPPERI*

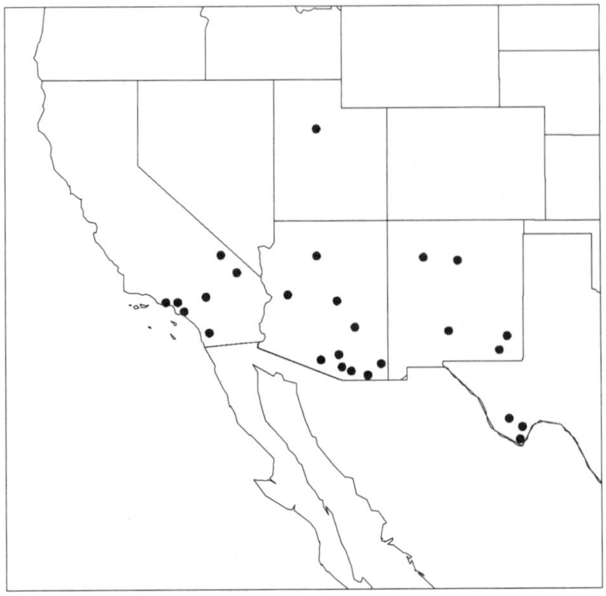

FIGURE 65: DISTRIBUTION OF SPECIMENS EXAMINED OF *PLAGIOMIMICUS MIMICA*

italia are also much closer to those of *mimica* than to those of *tepperi*. Wing length: mean = 11.82 mm, standard deviation = 0.77 mm, n = 10.

Head: frontal tubercle as in *tepperi* (text figure 63 *d*). Forewing maculation and coloration as in plate 3, figure 41; color gray green with a distinct olive tint; no white streak on the median vein through median area or if present, extremely weak; basal area olive to gray green, lighter than median area; antemedial line white, distinct, straight; reniform mark a simple, obscure white streak, without green patches; postmedial line white, distinct, excurved toward bottom of apical patch, and then sharply angulate to costa; apical patch dark olive green, distinct, although perhaps not as distinct as in *tepperi*; subterminal area with proximal half not lighter than distal half; subterminal line white, distinct, straight; upper ⅓ of terminal area slightly darker than lower ⅔. Hindwing: uniformly dull olive green. Male genitalia: plate K, figures 12, 13; clasper of right valve elongate, pointed, running along long axis of valve and with apex exceeding outer margin of valve; clasper of left valve as in right valve but shorter; basal process of sacculus long, well developed; corona absent; vesica without basal row of cornuti. Female genitalia: plate P, figure 10; ostium more nearly rectangular than in *tepperi*; ductus bursae about 1.5 × longer than wide, sclerotized over entire length and plicate; upper fourth of corpus bursae sclerotized; a sclerotized plate located near junction of corpus and ductus bursae; ovipositor lobes somewhat shorter than in *tepperi* but length within range of variation of *tepperi*.

The larva and its foodplants are unknown.

TYPES. Holotype: ♂. [USA], Arizona, Cochise Co., [Chiricahua Mountains], Paradise, genitalia on slide USNM 41,805. USNM. Paratypes: 55 ♂, 34 ♀. Same locality as for holotype, some with the months July, August, or September. AMNH, CNC, LACM, USNM. The type series is restricted to the type locality.

Plagiomimicus mimica has a wide distribution in the southwestern United States. It occurs as far east as western Texas, running westward throughout New Mexico, Arizona, and southern California. I have also seen a specimen from central Utah. The species has not been collected in Mexico although it undoubtedly occurs in the northern states. There is little or no variation in *mimica* and none of the size variation of *tepperi*. Adults have been collected from July to September.

pityochromus GROUP

The *pityochromus* group contains the three closely related U. S. species: *pityochromus, expallidus,* and *triplagiatus*. A fourth species belongs to the species group, *Plagiomimicus concinnum* Dyar from Mexico. These three species have rounded forewings and lack the apical elongation found in *spumosum* and its allies. The frontal process is circular, strongly defined with a deep concavity, but without a central protuberance. The lower margin of the circle protrudes much more than the upper margin. The ovipositor lobes are short and apically pointed, and both pairs of apophyses are short and stubby. No distinct linear striae are present in the ovipositor membrane. The ostium is short and linear. The distal end of the bursa is extensively sclerotized and tends to be folded into weak plates. The clasper is smaller and situated much nearer the apex of the valve than in *spumosum* and its allies. The base of the uncus is swollen into two lateral lobes, most strongly in *pityochromus,* less so in *expallidus* and *triplagiatus.*

These three species are readily separable on superficial appearance. *Plagiomimicus triplagiatus* is bright brown with a jet-black orbicular mark, a strong apical triangular patch, and a dark-brown, diffuse patch in the lower half of the subterminal area. The antemedial and postmedial lines are white and distinct. *Plagiomimicus triplagiatus* is the brightest appearing of the three species. *Plagiomimicus pityochromus* is dark brown. The orbicular mark may be filled with black. The antemedial and postmedial lines are not pure white and generally do not contrast with the rest of the wing. Although a subterminal line is present in the lower half of the subterminal area, it does not contrast with the rest of the wing's coloration. The forewing of *expallidus* is drab yellow brown with no contrasting maculation. The orbicular mark is never filled with jet black. The postmedial and antemedial lines are not white. The triangular apical patch is indistinct, although vaguely present, and does not contrast with the rest of the wing.

Plagiomimicus pityochromus Grote

PL. 3, FIGS. 42, 43; TEXT FIG. 66 *b, c, e* (♂ gen.); TEXT FIG. 66 *a* (♀ gen.); TEXT FIG. 66 *d* (head); TEXT FIG. 67 (map) (RWH 9754).

Plagiomimicus pityochromus Grote, 1873, *Bull. Buffalo Soc. Nat. Sci.,* **1**: 182.
Type locality: [USA], Alabama. [BMNH]

NOTE—Grote described *pityochromus* from a number of syntypes from different localities in the United States. The syntype in the BMNH from Alabama and bearing Grote's type label has been labeled and is hereby designated the LECTOTYPE.

Schinia media Morrison, 1875, *Proc. Boston Soc. Nat. Hist.,* **18**: 123.
Type locality: [USA], Kansas, Leavenworth. [FMNH]

Plagiomimicus pityochromus is a dark-brown species. The forewing orbicular mark is often filled or partially filled with jet black. The amount of black filling in the orbicular mark is variable ranging from a large bilobed blotch to a complete absence of black scales. The two sister species of *pityochromus* are *triplagiatus* and *expallidus.* The three are easily separable by superficial appearance as indicated in the species-group statement. The differences between the three species in the male and female genitalia are minute. The male clasper of *expallidus* is slightly thinner and the ductus bursae slightly shorter than in the other two species. The cornuti in the basal row of the vesica are 3 × as long as wide in *pityochromus* and *triplagiatus,* but only 2 × as long as wide in *expallidus.* Wing length: mean = 14.05 mm, standard deviation = 0.68 mm, n = 10.

The larva has been described by Crumb (1956) who records *Ambrosia trifida* L. (giant ragweed) (Asteraceae) as the foodplant, the larva feeding on the seeds and seed capsules. The larva is indistinctive. I am unable to determine the larva's original color from the preserved material I have seen. All the setae are on well-defined pinacula, but the pinacula do not differ in color from the rest of the body. There are large, triangular areas on the dorsum of each abdominal segment that are vaguely marked with brown. The head is slightly speckled with brown. The cervical shield of the prothorax has a vague, white, W-shaped pattern with brown located between the three points of the "W." The larva is almost identical to that of *expallidus* except the cervical shield is slightly darker brown.

Plagiomimicus pityochromus occurs widely in eastern North America from New York City and Long Island, New York south to central Florida. It is commonest in the midwest ranging as far north as Wisconsin, southwestern Ontario and Iowa south to south-central Texas. It has a spotty distribution in the southeastern United States. The western limit of *pityochromus* is in eastern Colorado. There is no discernible geographical variation in the species. Within-population variation in size exists. The major within-population variation, as noted above, is the amount of black scaling in the orbicular mark and the degree of definition of the reniform mark. Adults have been collected from August to early October.

Plagiomimicus triplagiatus Smith

PL. 3, FIG. 44; PL. K, FIGS. 14, 15 (♂ gen.); PL. P, FIG. 11 (♀ gen.); TEXT FIG. 68 (map) (RWH 9753).

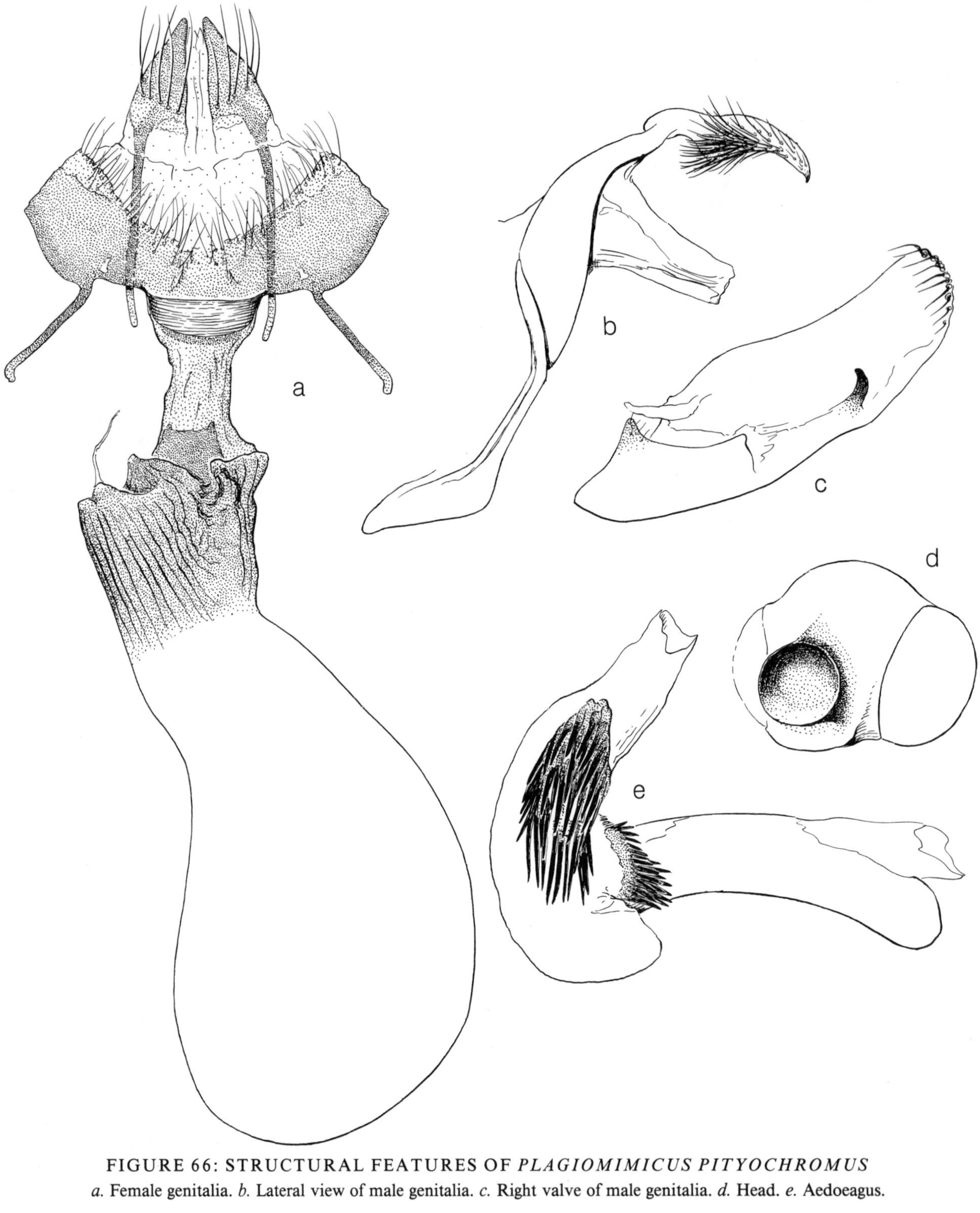

FIGURE 66: STRUCTURAL FEATURES OF *PLAGIOMIMICUS PITYOCHROMUS*
a. Female genitalia. *b.* Lateral view of male genitalia. *c.* Right valve of male genitalia. *d.* Head. *e.* Aedoeagus.

Plagiomimicus triplagiatus Smith, 1890, *Ent. Americana,* **6**: 139.
Type locality: [USA], New Mexico, Las Vegas. [USNM], lectotype designated by Todd (1982).

Plagiomimicus triplagiatus is essentially a slightly smaller and more brightly colored version of *pityochromus.* Except for size, there do not appear to be any significant differences in either the male or

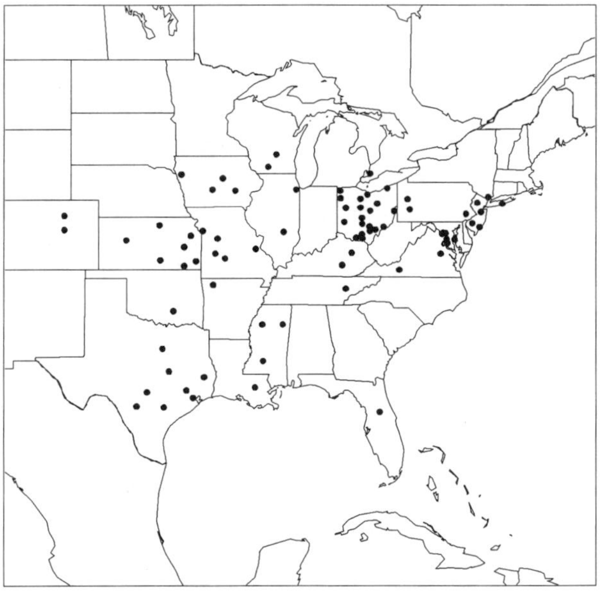

FIGURE 67: DISTRIBUTION OF SPECIMENS EXAMINED OF *PLAGIOMIMICUS PITYOCHROMUS*

FIGURE 68: DISTRIBUTION OF SPECIMENS EXAMINED OF *PLAGIOMIMICUS TRIPLAGIATUS*

female genitalia. *Plagiomimicus triplagiatus* is characterized by a contrasting dark brown lower half of the forewing subterminal line. The antemedial and postmedial lines are white and much better defined than in *pityochromus.* The two species are closely related and no known overlap of their ranges exists, although both species have been collected in Colorado. A very closely related sister species of *triplagiatus* exists in Mexico, *Plagiomimicus concinnum* Dyar. This Mexican species has a large, round orbicular mark in the forewing, and the reniform mark is clearly marked. The orbicular mark is L-shaped in *triplagiatus,* and the reniform mark is consistently absent. Wing length: mean = 12.38 mm, standard deviation = 0.92 mm, n = 10.

The larva and its foodplants are unknown.

Plagiomimicus triplagiatus is a rare species in collections and has been most widely collected in New Mexico and along the front of the Rocky Mountains in Colorado. The species has also been taken in Santa Cruz County in southeastern Arizona. Little variation is found in this species. The amount of jet black in the orbicular mark is variable although perhaps not as much as in *pityochromus.* Dated adults have been collected in September.

Plagiomimicus expallidus Grote

PL. 3, FIG. 45; PL. L, FIGS. 1, 2 (♂ gen.); PL. P, FIG. 12 (♀ gen.); TEXT FIG. 69 (map) (RWH 9752).

Plagiomimicus expallidus Grote, 1882, *Papilio,* **2**: 185.

Type locality: [USA], Montana. [USNM]

NOTE—Grote states that the types of this species were from Montana and in the Neumoegen collection. There are two specimens in the USNM fitting this description and with Grote type labels. In addition there appears to be a pseudotype with a Grote type label but of a different style and not from the Neumoegen collection. The specimen bearing the USNM red type label and from the Neumoegen collection has been labeled and is hereby designated the LECTOTYPE.

Plagiomimicus expallidus is the third, and most distantly related, of the species in the *pityochromus* group. The coloration of the forewing is dull yellow brown with a slight olive tint. The forewing also has a slightly frosted appearance. All the maculation is indistinct and the orbicular mark is, at most, represented by a small, dull-black dot, never by the jet-black blotches sometimes found in the other two species. There are slight differences in the male and female genitalia between *expallidus* and the other two species. The male clasper is slightly thinner and the swelling of the base of the uncus not quite as pronounced as in the other two species. The ductus bursae is slightly shorter than in either *pityochromus* or *triplagiatus.* Wing length: mean = 13.64 mm, standard deviation = 0.76 mm, n = 10.

The larva has been described by Crumb (1956)

FIGURE 69: DISTRIBUTION OF SPECIMENS EXAMINED OF *PLAGIOMIMICUS EXPALLIDUS*

who records *Iva xanthifolia* Nutt. (marsh elder) (Asteraceae) as a larval foodplant. The genus *Iva* is presumed to be closely related to *Ambrosia.* The larvae is virtually identical with that of *pityochromus.* The cervical shield is perhaps a slightly lighter brown based on preserved material.

Plagiomimicus expallidus has a mostly midwestern prairie distribution and is commonest east of the Rocky Mountains. In the midwest it occurs from North Dakota in the north to northern Texas in the south. The most easterly known specimen was collected near Chicago, Illinois. The western limit of the species is in central Utah, southwestern Idaho, and southern Alberta. There is no apparent geographical and very little individual variation in the species. Adults have been collected from July to September.

GENUS
Cuahtemoca Hogue

Cuahtemoca Hogue, 1963, *Contrib. Sci.*, **64**: 70. Type species: *Chalcopasta chalcocraspedon* Dyar, 1913. Original designation.

The genus *Cuahtemoca* was established for a Mexican species that does not occur in the United States. The North American species *Cuahtemoca unicum* is known from a single female specimen from Cochise County in southeastern Arizona. The male genitalia are unknown, and the placement of *unicum* in *Cuahtemoca* is tentative. *Cuahtemoca unicum* does share some diagnostic features with the type species of the genus. The forewing pattern of *unicum* is essentially the same as in *chalcocraspedon,* the type species of *Cuahtemoca,* except for the shade of brown in the forewing median area. Both species have metallic-green scaling filling the basal area and following the postmedial line, although the metallic green is not as distinctive against the dark-brown background in *chalcocraspedon* as it is against the light-brown color of the median area in *unicum.* The eighth sternum of the female abdomen lacks clear areas on either side of the ostium, a character treated as plesiomorphic in this work and by Hogue (1963). The ovipositor lobes are strongly sclerotized and dorsally recurved, a character also found in the type species of the genus. The corpus bursae has an L-shaped bend at approximately its middle. This bend is present in *unicum* although not as well developed as in the type species. The sternal-ostial opening is depressed with a spiny, swollen lamella antevaginalis slightly covering the opening of the ostium. Both *chalcocraspedon* and *unicum* have the setae in the intersegmental membrane between the ovipositor lobes and the eighth sternum arranged in vague, linear, striate rows. This combination of characters is somewhat confusing, combining both putative primitive and derived characters. The depressed opening of thc ostium and the linear arrangement of setae in the membrane caudal to the eighth sternum are both considered to be derived characters. The combination of the depressed ostium opening and the metallic-green patches in the forewing argue for a relationship with *Chalcopasta,* but the absence of clear patches in the eighth sternum does not. The linear arrangement of minute setae in the membrane caudal to the eighth sternum appears to argue against a connection to *Chalcopasta.* These striae do appear in *Cuahtemoca,* but only just posterior to the eighth sternum while the rest of the membrane has a random dispersion of spinules. Certain species in *Plagiomimicus* also have this pattern, although only weakly so. *Cuahtemoca* may be a relict group ancestral to *Plagiomimicus, Chalcopasta,* and *Chrysoecia.*

There are two important differences between *unicum* and the type species of *Cuahtemoca.* First, the frontal process of the head is vertically ovate in *unicum* with three horizontal ridges, a shape not found elsewhere in the Stiriinae. In contrast the frontal process of *chalcocraspedon* is ovate with a ventral raised point more typical of the subfamily. Second, the wing shape of *unicum* is elongate with a pointed apex. The wing is more ovate in *chalcocraspedon,* and the apex of the wing is not pointed.

I suspect *unicum* represents a relict group deserving of its own generic name. A new generic name seems premature without a male. The species is placed in *Cuahtemoca* with the understanding that its placement is only temporary. The diagnostic characters of this species are listed above and no formal description of *unicum* or the type species of *Cuahtemoca* is given.

Cuahtemoca unicum (Barnes and Benjamin), NEW COMBINATION
PL. 3, FIG. 46; PL. Q, FIG. 1 (♀ gen.); TEXT FIG. 70 (map) (RWH 9743).

Stibadium unicum Barnes and Benjamin, 1926, *Insecutor Inscitiae Menstruus,* **14**: 4.
Type locality: [USA], Arizona, Cochise Co., Paradise. [USNM]

The light-tan color of the forewing, the metallic-green basal area, and the metallic-green scaling following the postmedial line will easily identify this species. Wing length: mean = 14.91 mm, n = 1.

The larva and its foodplants are unknown.

Cuahtemoca unicum is known only from the female holotype from the Chiricahua Mountains in southeastern Arizona. The single specimen is not dated.

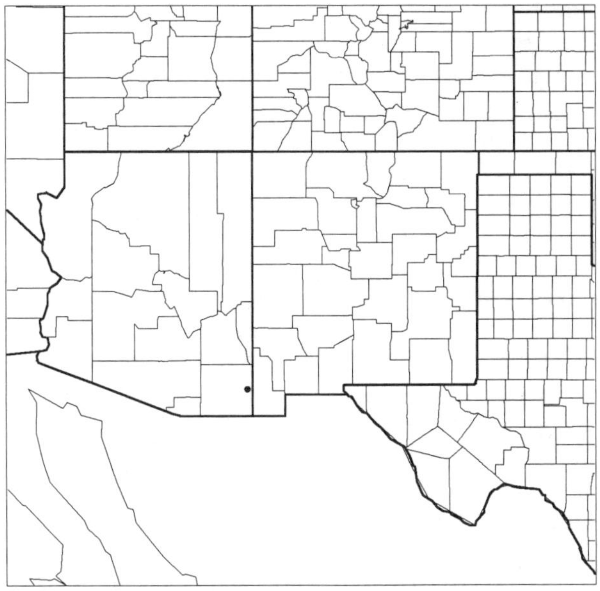

FIGURE 70: DISTRIBUTION OF SPECIMENS EXAMINED OF *CUAHTEMOCA UNICUM*

GENUS
Lineostriastiria Poole, NEW GENUS

Lineostriastiria Poole.
Type species: *Antaplaga hachita* Barnes, 1904.

Lineostriastiria is proposed for five North American species; *biundulalis, hachita, sexseriata, olivalis,* and *hutsoni.* There are no exclusively Mexican species in the genus. *L. biundulalis, hachita,* and *sexseriata* form a close, highly distinctive group (the *hachita* group as here treated). *L. olivalis* and *hutsoni* appear to be closely related because of similarities in their male and female genitalia despite great differences in superficial appearance.

Lineostriastiria is defined by two synapomorphic characters of the female genitalia. First, ovate clear areas are present in the eighth sternum caudal to the apophyses posteriores. These clear areas are bounded in both the caudal and cephalic directions by sclerotized regions. These clear areas in other genera (if present) are open caudally into the membrane of the ovipositor neck. Much smaller, partially unsclerotized areas are found in the two Mexican species *argyropolius* and *phalaenoides.* In this work these clear areas are treated as transitional between their absence in *Plagiomimicus* and the large clear areas found in *Lineostriastiria* and the other derived genera following it.

The second derived character used to establish this genus is the arrangement of the fine hairs or spinules in the ventral region of the neck of the ovipositor. These hairs are arranged in distinct lines from the caudal margin of the eighth sternum to the bases of the ovipositor lobes in *Lineostriastiria.* This linear arrangement of hairs is also found in a partial form in some species of *Plagiomimicus,* for example *Plagiomimicus curiosum,* and in *Cuahtemoca unicum,* but never as distinctly and only along the caudal margin on the eighth sternum, never running the entire length of the ovipositor neck. Some species of *Chalcopasta* also have a partial, weak, linear arrangement as described above. The male clasper is apical in this genus. The basal group of cornuti in the vesica is often situated on a sclerotized base (e.g., *olivalis* and *hutsoni*), although this sclerotized base is weak to almost absent in the *hachita* group. The uncus is relatively long and narrow, not short and swollen as in *Plagiomimicus.* The ductus bursae is unsclerotized and 3–4 × longer than wide. The apical claw of the foretibia is weak. The forewing of all the species is ovate without the pointed apex found in most of the species of *Plagiomimicus.* A rounded forewing apex is plesiomorphic for the subfamily but is one of the distinctive features of the species in *Lineostriastiria*.

Lineostriastiria is an intermediate step between *Plagiomimicus* and the remaining genera of the Stiriinae. The unsclerotized, but bounded, areas in the eighth sternum of the female abdomen are considered to be transitional to the large, unsclerotized regions in such genera as *Chrysoecia, Chalcopasta,* and *Stiria.* The technical description follows.

Head: male and female antennae simple; eye unmodified; frontal process (text figure 71 *d*) either ovate with a ventral tubercle (*olivalis* and *hutsoni*), or circular without a central projection (*hachita* group) but ventral margin protruding more than dorsal margin; front covered with long, dentate scales, scales not arranged in tufts; eye naked, unlashed; proboscis unmodified; palpus unmodified. Forewing: apex rounded, wing either ovate or rectangular-ovate; scales of forewing ovate, slightly dentate, closely appressed to wing. Prothoracic leg: tibial claw weak compared to other genera in tribe; leg without tufting; spines on ventral surface of first tarsal segment arranged in three rows, but central row weak and irregular; tarsal spines of fifth tarsal segment in two strong rows; first tarsal segment approximately ¾ × length of remaining four. Mesothoracic and metathoracic legs: scales closely appressed to leg segments, without tufting; tarsal spining as in prothoracic leg; first tarsal segment about as long as remaining four segments. External tympanic region: hood small; abdominal bulla rectangular, weak. Male abdomen: basal hair pencils and Stobbe's glands absent; dorsal tufts absent, scales closely appressed to abdomen. Male genitalia: text figure 71 *b, c, e*; uncus long and narrow, with apical tooth; clasper of valve apical, generally small; no backward projecting secondary clasper; basal process of sacculus weak; spines at apex of valve strong, but diffuse, not arranged in corona; cornuti arranged in usual two groups; basal group on a sclerotized base, but base weak in *hachita* group. Female genitalia: text figure 71 *a*); ovipositor lobes either strongly sclerotized, short, and triangular (*hachita* group) or longer and not as strongly sclerotized (*olivalis*); distinct, clear, ovate areas present in eighth sternum caudal to apophyses posteriores, not open caudally; a few weak spines in open areas; fine hairs (spinules) in ovipositor neck arranged in distinct rows from caudal margin of eighth sternum to bases of ovipositor lobes; ostium weakly developed, not contained in an evaginated region; ductus bursae unsclerotized, about 3–4 × longer than wide; appendix bursae simple, slightly sclerotized and ribbed (*olivalis*) or almost unsclerotized (*hachita* group); corpus bursae unsclerotized, elongate, about 3–4 × longer than wide.

The larvae and foodplants are unknown for all the species of this genus.

hachita GROUP

The three species *hachita, biundulalis,* and *sexseriata* form a distinctive, close-knit group. They are relatively small with rounded forewing apices. The forewing ground color is cream white and the antemedial and postmedial lines are black contrasting with the rest of the wing. The female genitalia are notable because of the shortness of both pairs of apophyses and the narrowness of the eighth sternum and tergum. The ductus bursae and the appendix bursae are both unsclerotized. The male clasper is short, stout, uncomplicated, and located near the apex of the valve. The vesica is unremarkable with the typical apical and basal groups of cornuti. The basal group is located on a very weakly sclerotized base.

Lineostriastiria hachita is the most distinctive species of the three and has orange basal and subterminal areas that contrast sharply with a white median area. The entire forewing is white in *biundulalis* and *sexseriata* without orange areas. The best character to separate *biundulalis* and *sexseriata* is the presence of a series of spots in the forewing subterminal area in *sexseriata* and its usual absence in *biundulalis* (but see below). The hindwing of *sexseriata* is suffused with brown, to varying degrees, but is unicolorous white in *biundulalis.*

Lineostriastiria hachita (Barnes), NEW COMBINATION

PL. 3, FIG. 47; TEXT FIG. 71 *b, c, e* (♂ gen.); TEXT FIG. 71 *a* (♀ gen.); TEXT FIG. 71 *d* (head); TEXT FIG. 72 (map) (RWH 9758).

Antaplaga hachita Barnes, 1904, *Can. Ent.*, **36**: 241.
Type locality: [USA], Arizona, Santa Catalina Mts. [USNM]

The forewing of *Lineostriastiria hachita* has yellow-orange basal and subterminal areas that contrast strongly with the white median area. A series of black dots represents the subterminal line. The hindwing is yellow orange with a little black suffusion near the apex of the wing in some specimens. The ventral side of the forewing is suffused with black except along the costa and outer margin. The

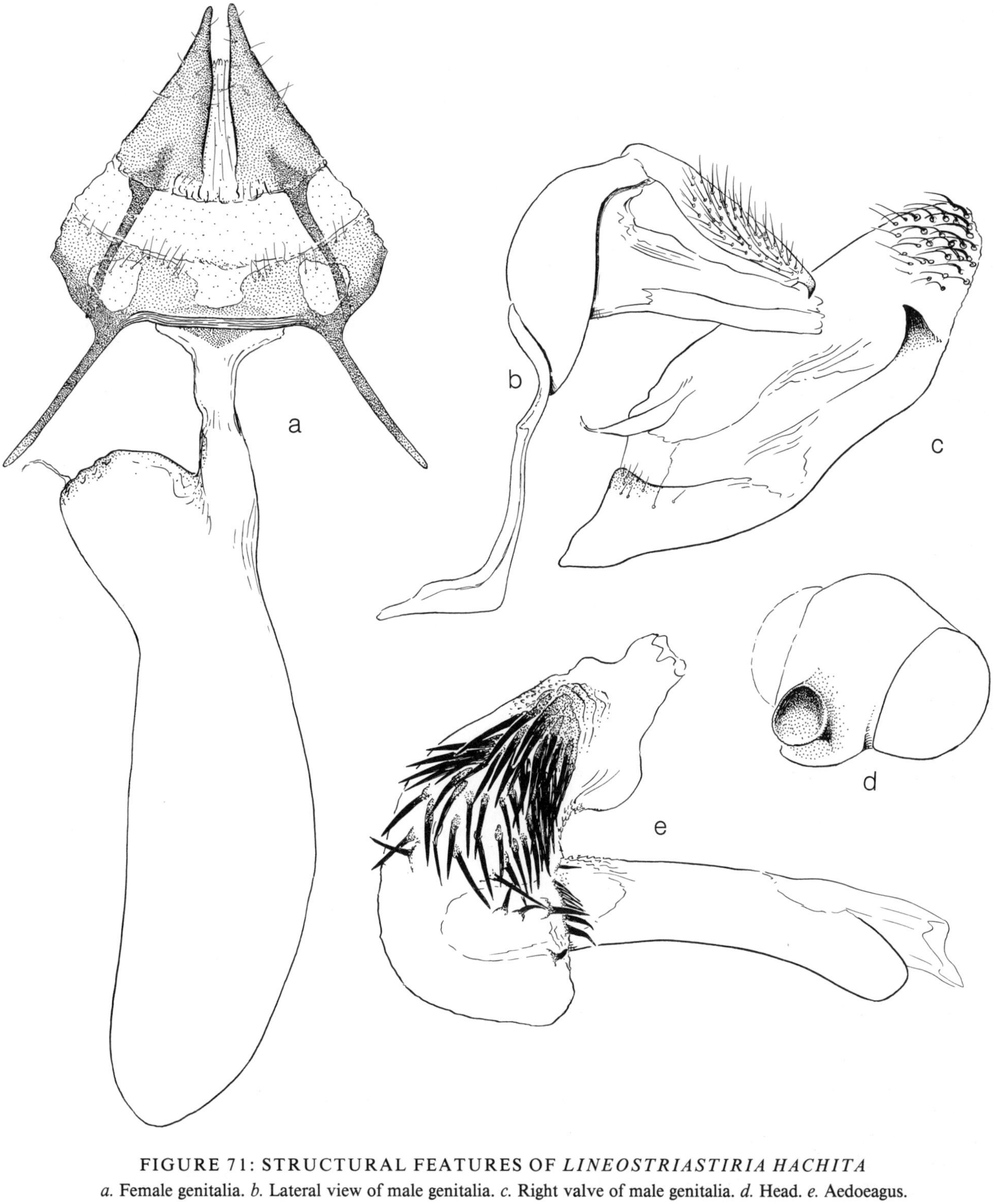

FIGURE 71: STRUCTURAL FEATURES OF *LINEOSTRIASTIRIA HACHITA*
a. Female genitalia. *b.* Lateral view of male genitalia. *c.* Right valve of male genitalia. *d.* Head. *e.* Aedoeagus.

clasper is located well away from the inner margin of the valve, at least by its own length. The lower margin of the juxta does not have a triangular invagination. The corpus bursae is more elongate and not as globular as in *sexseriata.* The ratio of the greatest length of the corpus bursae to its greatest width is 3.60 in *hachita* and 2.25 in *sexseriata,* (the higher number indicating a more elongate shape). The female genitalia differ from those of *biundulalis* by the differing lengths of the apophyses anteriores.

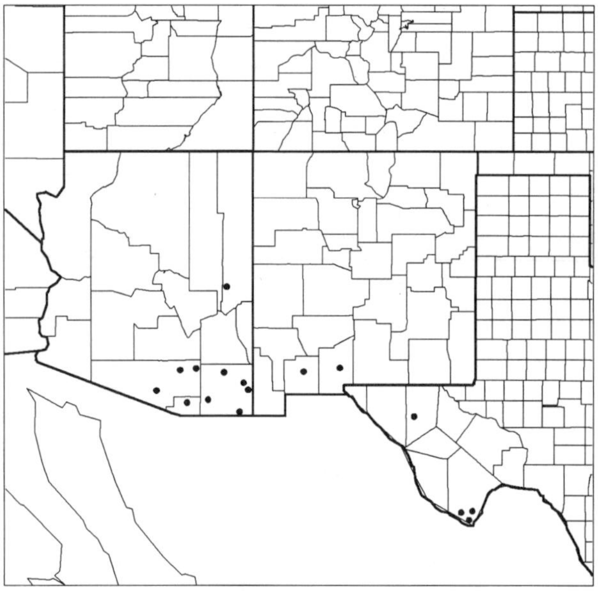

FIGURE 72: DISTRIBUTION OF SPECIMENS EXAMINED OF *LINEOSTRIASTIRIA HACHITA*

They reach the bottom (cephalic margin) of the eighth sternum in *hachita,* but do not reach it in *biundulalis.* Wing length: mean = 11.91 mm, standard deviation = 0.63 mm, n = 10.

L. hachita occurs in western Texas, southern New Mexico, and the southern half of Arizona. In Arizona it ranges as far north as the White Mountains. Some individual variation exists in the strength of the black dots in the forewing subterminal area. The orange of the forewing varies slightly in intensity. Adults have been collected in late summer and early autumn.

Lineostriastiria sexseriata (Grote), NEW COMBINATION

PL. 3, FIG. 48; PL. L, FIGS. 3, 4 (♂ gen.); PL. Q, FIG. 2 (♀ gen.); TEXT FIG. 73 (map) (RWH 9759).

Grotella sexseriata Grote, 1881, *Papilio,* **1**: 155.
Type locality: [USA], Arizona. [USNM]

NOTE—There is some confusion about the type of *sexseriata.* Grote states that the type is in the Neumoegen collection, a collection now in the USNM. Two specimens bear type labels in the USNM and one does in the BMNH. The type labels on the two specimens in the USNM are different, one in Grote's handwriting (Prescott, Arizona) and one in Neumoegen's (Arizona). The type locality cited in the original description is "Arizona." I suspect the specimen in the USNM with Grote's type label and the specimen in the BMNH with Grote's type label are pseudotypes. However, to be sure, the specimen labeled "Arizona" and with USNM type number 33995 has been labeled and is hereby designated the LECTOTYPE. All three putative syntypes are this species.

Lineostriastiria sexseriata has a cream-white forewing with contrasting black antemedial and postmedial lines. The forewing, in general, has a slightly polished appearance. *Lineostriastiria sexseriata* is most likely to be confused with *biundulalis.* However, a strong series of black dots representing the subterminal line is present in *sexseriata.* These black dots are almost always absent in *biundulalis.* In specimens of *biundulalis* with some black dots in the subterminal area, the postmedial line is finely dentate; the postmedial line of *sexseriata* is not finely dentate. The hindwing of *sexseriata* is always suffused, to some degree, with gray brown, but is pure white in *biundulalis.* The clasper of *sexseriata* is near the inner margin of the valve. The juxta of *biundulalis* has a triangular indentation along its ventral margin that is missing in *sexseriata.* The apophyses anteriores reach the the lower (cephalic) margin of the eighth sternum, but do not reach it in *biundulalis.* Wing length: mean = 12.16 mm, standard deviation = 0.39 mm, n = 10.

Lineostriastiria sexseriata is known only from Arizona. Most specimens are from southern Arizona, but the species occurs northward at least as far as the Mogollon Rim in Yavapai, Gila, and southern Navajo counties. The strength of the black dots in the forewing subterminal area varies a little, but not greatly. Adults have been collected in late summer and early autumn.

Lineostriastiria biundulalis (Zeller), NEW COMBINATION

PL. 3, FIGS. 49–51; PL. L, FIGS. 5, 6 (♂ gen.); PL. Q, FIG. 3 (♀ gen.); TEXT FIG. 74 (map) (RWH 9760).

Sedenia biundulalis Zeller, 1872, *Verh. Zool.-Bot. Ges. Wien,* **22**: 502.
Type locality: [USA], Texas. [BMNH]

The characters separating this species from *sexseriata* are given under that species. The forewing is cream white with black antemedial and postmedial lines. A row of black dots is almost always missing from the forewing subterminal area but some black dots occasionally are weakly present in some females. The hindwing is concolorous with the forewing. Wing length: mean = 11.49 mm, standard deviation = 0.46 mm, n = 10.

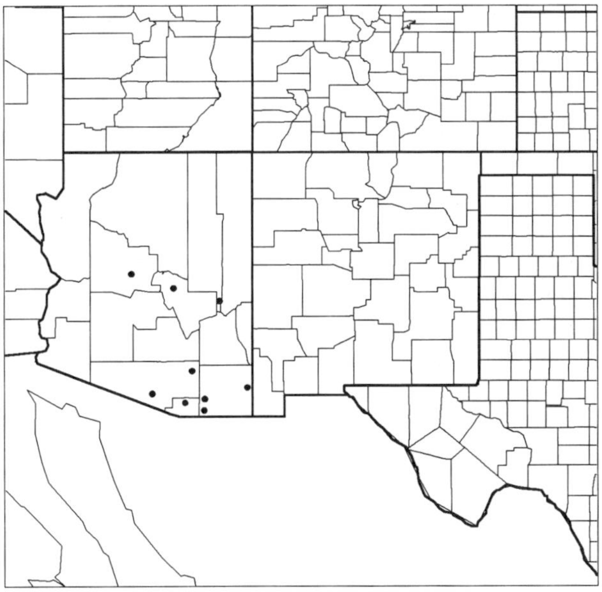

FIGURE 73: DISTRIBUTION OF SPECIMENS EXAMINED OF *LINEOSTRIASTIRIA SEXSERIATA*

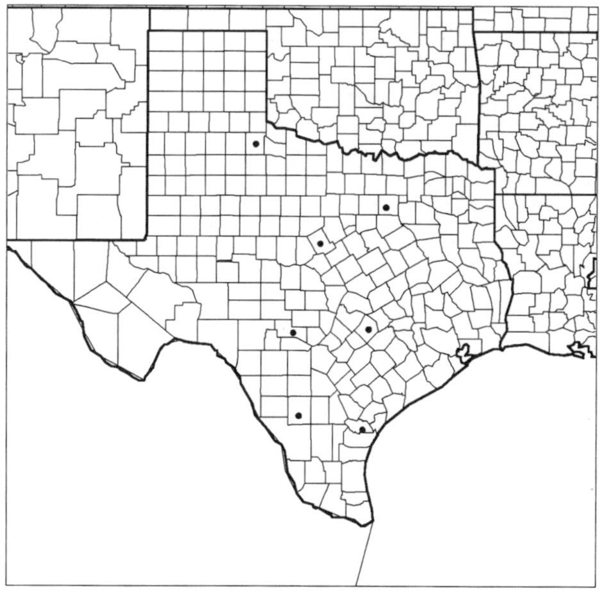

FIGURE 74: DISTRIBUTION OF SPECIMENS EXAMINED OF *LINEOSTRIASTIRIA BIUNDULALIS*

Lineostriastiria biundulalis is widely distributed in eastern and central Texas. It is found in Cottle county in the north and ranges southward through the central region of the state. The species is apparently not rare, but the specimens seen have a curiously spotty distribution. What these localities have in common is unknown. *L. biundulalis* does not seem to occur in western Texas, but a couple of specimens have been collected in southeastern Texas. A moderate amount of variability exists in *biundulalis.* Some females have very thick antemedial and postmedial lines, and the subterminal area is lightly suffused with gray. Likewise a few females have a series of weak, black dots in the subterminal area. Females tend to be slightly darker than males. Adults have been collected in May and then again in late summer and early autumn. Two generations possibly are present.

Lineostriastiria olivalis (Barnes and McDunnough), NEW COMBINATION
PL. 3, FIG. 52; TEXT FIG. 75 *b, c, e* (♂ gen.); TEXT FIG. 75 *a* (♀ gen.); TEXT FIG. 75 *d* (head); TEXT FIG. 76 (map) (RWH 9757).

Stiria olivalis Barnes and McDunnough, 1916, *Contrib. Nat. Hist. Lep. N. Am.*, **3**(1): 12, pl. 3, fig. 9.
Type locality: [USA], California, San Bernardino Co., Loma Linda. [USNM]

NOTE—*Lineostriastiria olivalis* was described from 2 males and 3 females. The specimen in the USNM with the label "type ♂" has been labeled and is hereby designated the LECTOTYPE.

Lineostriastiria olivalis is distinctive and unlike any other stiriine. It is not similar superficially to the other species in the genus. However, *olivalis* shares the striate intersegmental membrane between the ovipositor lobes and the eighth sternum synapomorphy used to define this genus. Both the male and female genitalia are basically the same as those found in the *hachita* group. The forewing ground color is dull white with diffuse olive-green antemedial, postmedial, and subterminal lines. The hindwing is suffused with brown. The frontal process is round with a thin, elongate tubercle near the ventral margin of the process. The clasper of the valve is short and stubby and located near the apex of the valve as in *hachita* and its relatives. The basal group of cornuti of the vesica is strong and located on a weakly sclerotized bulb. The ovipositor lobes are longer and not as triangular as in the *hachita* group. Both pairs of apophyses are normal in length, not short and stubby as in the *hachita* group, and the eighth sternum is not as narrow. The appendix bursae is sclerotized, not membranous. The ductus bursae is not sclerotized as is also true of the *hachita* group. Wing length: mean = 14.05 mm, standard deviation = 0.69 mm, n = 10.

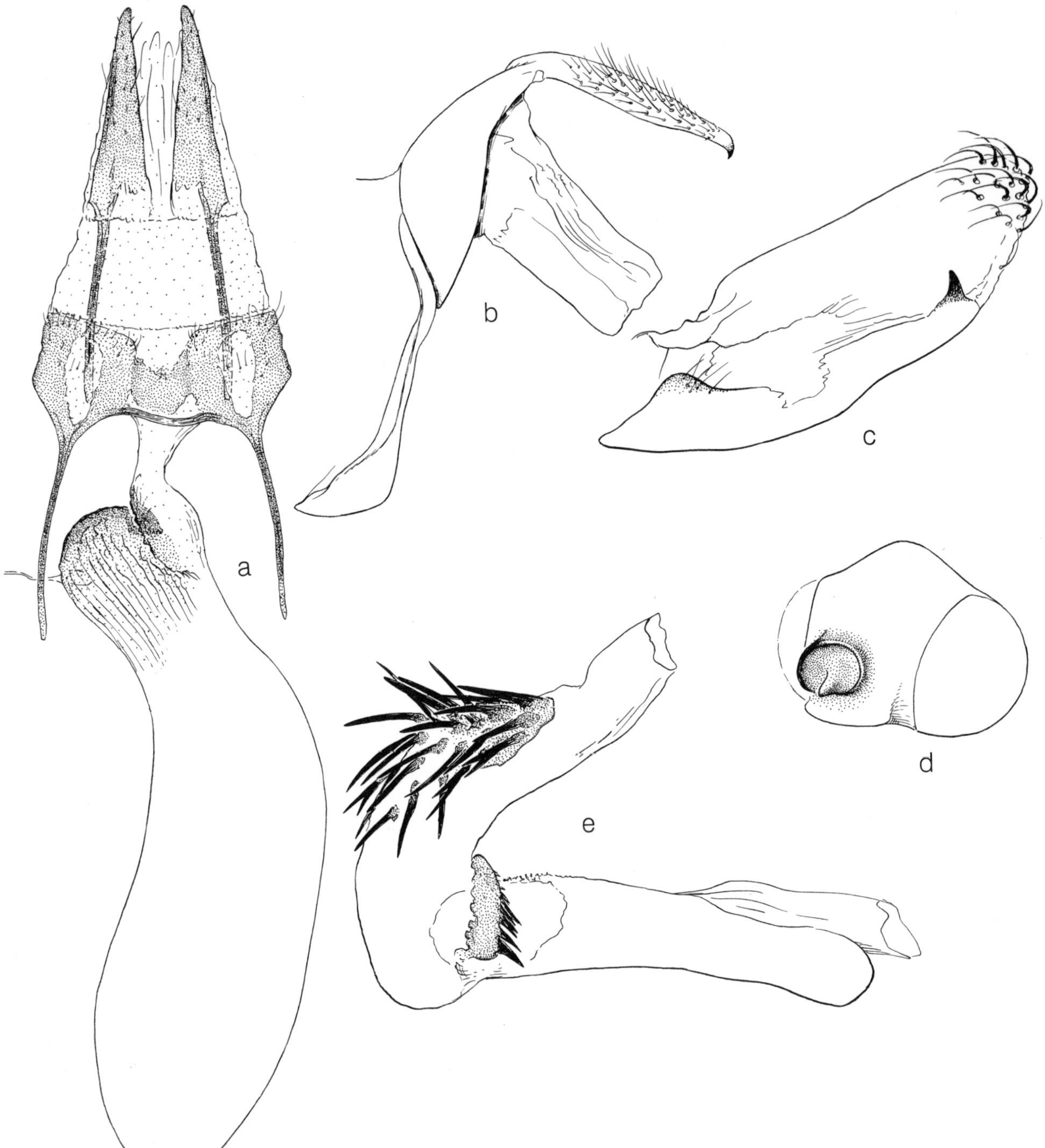

FIGURE 75: STRUCTURAL FEATURES OF *LINEOSTRIASTIRIA OLIVALIS*
a. Female genitalia. *b.* Lateral view of male genitalia. *c.* Right valve of male genitalia. *d.* Head. *e.* Aedoeagus.

L. olivalis occurs primarily in southern California and northern Baja California. There is a single specimen in the LACM from the Santa Catalina Mountains in Pima County, Arizona. Some variation exists in the shade of the olive markings of the forewing. Some specimens have a distinctive yellow tint, but others are slightly brownish. Adults have been collected between March and early May. This flight

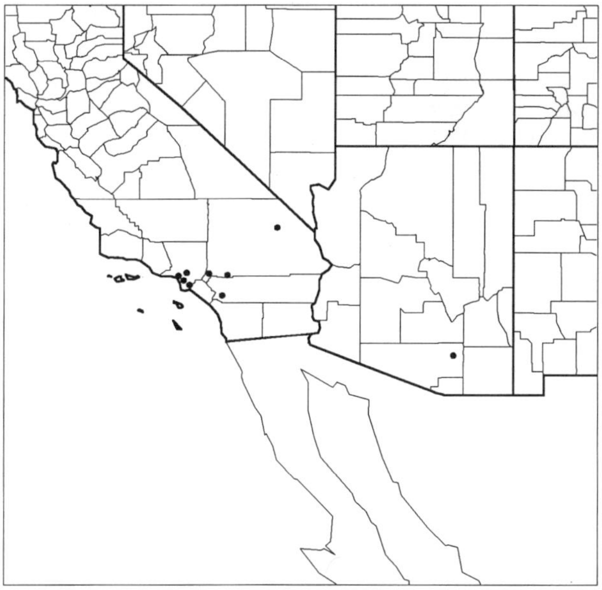

FIGURE 76: DISTRIBUTION OF SPECIMENS EXAMINED OF *LINEOSTRIASTIRIA OLIVALIS*

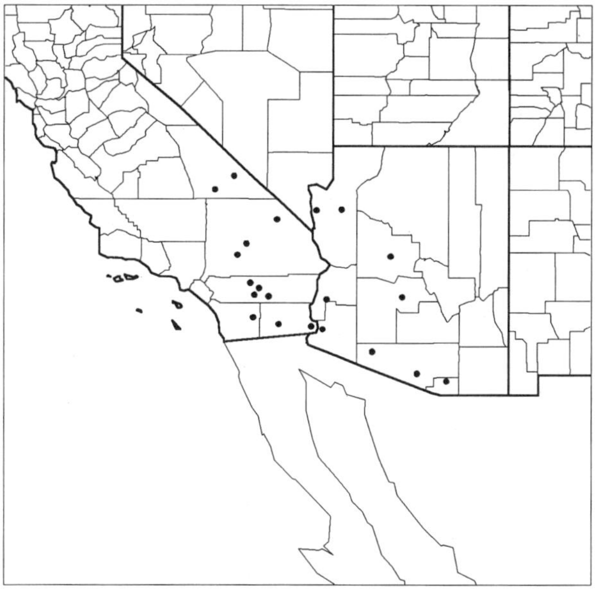

FIGURE 77: DISTRIBUTION OF SPECIMENS EXAMINED OF *LINEOSTRIASTIRIA HUTSONI*

period contrasts with the autumn flight period of the species in the *hachita* group and indeed the subfamily as a whole.

Lineostriastiria hutsoni (Smith), NEW COMBINATION

PL. 3, FIGS. 53, 54; PL. L, FIGS. 7, 8 (♂ gen.); PL. Q, FIG. 4 (♀ gen.); TEXT FIG. 77 (map) (RWH 9756).

Stibadium hutsoni Smith, 1907, *Trans. Am. Ent. Soc.*, **33**: 140.
Type locality: [USA], Arizona, Yuma Co. [AMNH]; lectotype designated by Todd (1982).

Stibadium fuliginosa Smith, 1907, *Trans. Am. Ent. Soc.*, **33**: 140.
Type locality: [USA], California, Walters Station. [AMNH]; lectotype designated by Todd (1982).

Lineostriastiria hutsoni is easily the dullest species of the Stiriinae. *Lineostriastiria hutsoni* and *olivalis* are structurally similar despite the deep superficial differences between them. *L. hutsoni* possesses all of the synapomorphies defining the genus, particularly two distinct, clear, ovate but delimited areas above the apophyses anteriores. The clasper is located near the apex of the valve. The basal group of cornuti in the vesica are strong and located on a sclerotized, bulbous region as in *olivalis*. Both patches of cornuti in the vesica are reduced in size, possibly because of the small size of the moth. Wing length: mean = 9.58 mm, standard deviation = 0.54 mm, n = 10.

Lineostriastiria hutsoni is the smallest species in the Stiriinae (although close in size to *Plagiomimicus mimica*). The forewing is elongate and the apex is slightly pointed unlike the round apex found in other species of the genus. The forewing coloration is dull green. The only distinctive forewing marking is an oblique, diffuse white postmedial line shaded along its inner side with dark green. The antemedial line, when present, is diffuse and darker green than the wing's ground color. The antemedial line may be followed by a slightly whitish band in the lower half of the median area. The subterminal area of the forewing, in a few specimens, is slightly lighter than the rest of the wing.

This species appears to be common in the driest regions of southern California and western Arizona. It occurs throughout the Mojave Desert regions of southern California and ranges as far north as southern Inyo County. Specimens are common from most of the western part of Arizona, and the species has been taken as far east as Santa Cruz County (Santa Rita Mountains). There are specimens in the USNM labeled only S. Nevada, probably from the Las Vegas region of Clark County. The species undoubtedly also occurs in Chihuahua and Baja California del Norte, but I have not seen specimens from Mexico. There is no discernible geographical variability. Individual variation was described above. The adults fly in March and April as in *olivalis*.

GENUS
Xanthothrix Hy. Edwards

Xanthothrix Hy. Edwards, 1878, *Pacific Coast Lepidoptera,* **29**: 7.
Type species: *Xanthothrix ranunculi* Hy. Edwards, 1878. Monotypy.

Euedwardsia Grote, 1882, *Papilio,* **2**: 122.
Type species: *Xanthothrix neumoegeni* Hy. Edwards, 1881. Original designation.

The genus *Xanthothrix* contains two day-flying species, *ranunculi* and *neumoegeni,* from the Mojave Desert region of southern California. *Xanthothrix* was not included in the Stiriinae by Hogue (1963). The genus is largely characterized by features that often occur in day-flying noctuids. The eyes and palpi are reduced in size and both the palpi and the dorsum of the thorax are covered with long, hairlike scales. The abdominal sclerites and the male genitalia are melanized, another feature common in day flying moths. The eighth sternum of the female abdomen has two lateral, distally open clear areas. This feature places the genus somewhere in the general vicinity indicated on the subfamily tree, but its exact position is unstable. Therefore, it has been placed arbitrarily in this part of the phylogenetic sequence.

The two species of *Xanthothrix* are superficially distinct. Both are among the smallest of the Stiriinae. *Xanthothrix neumoegeni* is bright orange with brown streakings in the forewing and a square shape to the wing. *Xanthothrix ranunculi* is smaller than *neumoegeni,* has an ovate wing, and a uniform light yellow color.

Head: male and female antennae simple, minutely ciliate ventrally; long hairs interpretable as lashes from base of antenna; long hairlike scales covering vertex of head; frontal process rounded dorsally, flat ventrally; ventral margin of frontal process with large, slightly pendulant, nose-shaped central tubercle; eye reduced in size with large anterior and posterior clear areas (may be partially covered with scales); ocellus present; palpus reduced, third segment minute; all palpal segments covered with long hairlike scales; proboscis normal. Thorax: dorsum of thorax covered with long, hairlike yellow scales underlain with dark-brown, ovate, closely appressed scales; wings with basal covering of ovate, dark-brown scales overlain with longer, dentate yellow scales. Prothoracic leg: tibial claw long, thin; femur closely scaled, with ventral furrow for reception of tibia; femur with two lateroventral rows of long hairs on either side of furrow, strongest distally; tibia closely scaled, about 3 × longer than wide; epiphysis large; first tarsal segment about 3 × longer than second; tarsal spining confined to apical ¼ of first segment; tarsal segments two-four with three ventral, weak, rows of spines; fifth tarsal segment with only weak hairs ventrally; dorsum of all tibial segments scaled, without spines or hairs. Mesothoracic leg: femur unmodified, with two lateroventral rows of long hairlike scales; tibia long, closely scaled, with usual one pair of apical spurs; tibia slightly triangular at apical end; tarsal segments as in prothoracic leg. Metathoracic leg: as in mesothoracic leg except for usual two pairs of tibial spurs. Internal tympanic region: not examined. External tympanal region: extremely reduced; hood totally absent; no bulla present; a slight lip of membrane cephalad of spiracle; alula present; other usual structures of tympanum present. Male abdomen: dorsal surface of abdomen without tufts, vestiture of rectangular, closely appressed scales; basal hair pencils and Stobbe's glands absent; sternum 1+2 about as long as wide; sclerites of abdomen melanized; eighth sternum and tergum typical for the Stiriinae. Male genitalia: plate L, figures 9, 10; structure simple; uncus short, swollen at middle (*ranunculi*) or basal third (*neumoegeni*); uncus with apical tooth, and large hairs on dorsal surface of apical half; free pleurite present, elongate, simple; saccus pointed, triangular; juxta trapezoidal, with two dorsal projections; anellus dentate; valve simple, with costal edge, particularly near articulation with tegumen, strongly melanized; corona or strong spines at apex of valve absent; clasper absent (*ranunculi*) or present as a small tooth (*neumoegeni*); basal process of sacculus weak; shaft of aedoeagus about 2.5 × longer than wide; vesica of normal stiriine type with linear basal group of cornuti and diffuse apical group of cornuti. Female genitalia: plate Q, figures 5, 6; ovipositor lobes either flattened and paddle shaped (*ranunculi*) or elongate with conical apices (*neumoegeni*); no perceptible pattern to fine hairs in membrane of ovipositor neck; eighth sternum with large clear ventral areas lateral to ostium, areas open distally, without striae, and with small setae; ostium simple, either not in a concavity (*ranunculi*) or in a small concavity (*neumoegeni*); ductus bursae short, about twice as long as wide, unsclerotized; appendix bursae present, sclerotized, ovate, slightly more heavily sclerotized than corpus bursae; corpus bursae elongate, about 4 × longer than wide, with an angulate bend about two-thirds the distance toward the cephalic end.

Xanthothrix neumoegeni Hy. Edwards
PL. 4, FIGS. 1, 2; PL. L, FIGS. 9, 10 (♂ gen.);

PL. Q, FIG. 5 (♀ gen.); TEXT FIG. 78 (map) (RWH 9771).

Xanthothrix neumoegeni Hy. Edwards, 1881, *Papilio,* **1**: 101.
Type locality: [USA], California, Santa Barbara. [USNM]

NOTE—*Xanthothrix neumoegeni* was described from 1 male and 1 female in the Neumoegen collection, now in the USNM. The male syntype currently bears the label "Was. T." and lacks a Neumoegen collection label. The female syntype lacks a locality label but does bear a Neumoegen collection label. The female syntype in the USNM has been labeled and is hereby designated the LECTOTYPE.

The combination of the orange forewing and hindwing, the frontal process, tibial claw, and reduced eyes, and day-flying habits should distinguish this species from all other noctuids in North America. The forewing is orange with brown markings on the veins. The antemedial, postmedial, and subterminal lines are all brown and slightly dentate. The postmedial line is shaded with brown on its inner side. A distinctive discal lunule is present in the hindwing. The apex of the forewing is shaded with dark brown and a dark line continues along the outer margin. The male valve has a small clasper, a structure absent in *ranunculi.* Wing length: mean = 9.80 mm, standard deviation = 0.54 mm, n = 10.

The larva has been reared by Charles Hogue (Los Angeles County Museum of Natural History) who provided me with photographs of the larva and some notes on general coloration. The foodplant is not listed in the notes but is probably some species of *Coreopsis* (Asteraceae) by analogy with *Xanthothrix ranunculi.* The dorsum of the larva is light yellow to cream yellow with a dorsal, segmented cream-colored line. A caudally pointed, gray-brown chevron is present on the dorsum of each abdominal segment. A light-yellow subspiracular line is present.

Xanthothrix neumoegeni has a restricted distribution in southern California. Almost all specimens seen are from the north side of the San Gabriel Mountains in Los Angeles County. A single specimen is known from the Argus Mountains in southern Inyo County. The type series was stated to be from Santa Barbara, but no specimens have been seen from there nor are the syntypes labeled as such. Nevertheless, I have plotted Santa Barbara on the distribution map (text figure 78). No specimens have been seen from San Bernardino, Riverside, San Diego, or Orange Counties. The current distribution suggests a species highly restricted to the Mojave Desert region. The species is not common. Adults fly from late April to early June and are day fliers.

FIGURE 78: DISTRIBUTION OF SPECIMENS EXAMINED OF *XANTHOTHRIX NEUMOEGENI*

Xanthothrix ranunculi Hy. Edwards

PL. 4, FIGS. 3–5; PL. L, FIGS. 11, 12 (♂ gen.); PL. Q, FIG. 6 (♀ gen.); TEXT FIG. 79 (map) (RWH 9772).

Xanthothrix ranunculi Hy. Edwards, 1878, *Pacific Coast Lepidoptera,* **29**: 7.
Type locality: [USA], California, Kern Co., Havilah. [AMNH]

NOTE—*Xanthothrix ranunculi* was based on 19 syntypes. There are three clearly recognizable syntypes in the AMNH. The male bearing Edwards' type label has been further labeled and is here designated as LECTOTYPE.

Xanthothrix ranunculi should be readily recognized by the combination of the generic characters, its small size, and yellow or yellow-green coloration. It is a day-flier. The forewing scaling consists of a lower layer of dark gray-green scales overlain with longer yellow scales. The yellow scales tend to be easily lost or to curl when wet. Therefore, fresh specimens are yellow or yellow orange, but old or worn specimens become darker olive green with a slight yellowish tinge. There also seems to be some inherent variation in how much of the olive green is visible even in fresh specimens. The forewing lacks any maculation. The hindwing is strongly tinged with dull brown, particularly along the apex and inner

FIGURE 79: DISTRIBUTION OF SPECIMENS EXAMINED OF *XANTHOTHRIX RANUNCULI*

margin, and somewhat less so along the outer margin. A discal mark is absent. The clasper is represented by, at most, two spinose knobs. The ovipositor lobes are wide and paddle shaped. The adaptive significance of this modification of the ovipositor lobes, if any, is unknown. Wing length: mean = 8.71 mm, standard deviation = 0.85 mm, n = 10.

The larva has been described and figured by Comstock and Henne (1940). The larva was reared on *Coreopsis douglasii* (DC.) H. M. Hall (Asteraceae). Charles Hogue (personal communication) has reared the species and provided me with color slides and written notes. The larva has a cream-colored dorsal line flanked on either side by a trapezoidal red-brown area tapering caudally. An extensive cream-colored region is present lateral to this trapezoidal area. A subdorsal cream-colored line is present followed by a supraspiracular dull-brown region. The ventral half of the body is dull white and the spiracular region is lighter than the rest of the ventral area. The larva lives within and eats the flower heads of its host plant. This species has a slightly earlier adult flight period than *neumoegeni.*

Xanthothrix ranunculi is known only from the Mojave Desert region of Los Angeles County and western San Bernardino County, both in southern California. Most specimens of *neumoegeni* are from the vicinity of the San Gabriel Mountains, but *ranunculi* seems to be more a species of the deep desert. Adults have been collected from March to April. In contrast *neumoegeni* flies from April to early June.

GENUS
Chrysoecia Hampson

Chrysoecia Hampson, 1908, *Catalogue of the Lepidoptera Phalaenae in the British Museum,* **7**: 9.
Type species: *Anthoecia scira* Druce, 1889. Subsequent designation by Hampson, 1910, ibidem, **9**: 182.

Gorgora Dyar, 1914, *Proc. U. S. Natl. Mus.,* **47**: 377. NEW SYNONYMY.
Type species: *Gorgora morga* Dyar, 1914. Original designation.

Chichimeca Hogue, 1963, *Contributions in Science,* **64**: 61. NEW SYNONYMY.
Type species: *Eulithosia thoracica* Hy. Edwards, 1884. Original designation.

The primary character defining the genus *Chrysoecia* is the presence of striations in the clear areas posterior to the apophyses anteriores in the eighth sternum of the female abdomen. In general the species have ovate forewings and never have forewings with produced apices. The clasper usually has a backward as well as a forward projection (text figure 80). The basal process of the sacculus is usually well developed and long, but is reduced in the species of the *thoracica* group. The uncus is short and swollen. The basal group of cornuti in the vesica tend to be on a sclerotized base although this base is reduced or absent in the *thoracica* group. The ovipositor lobes (text figures 80, 83, 85) are generally not as strongly sclerotized or well developed as in most other members of the subfamily. The apices of the lobes often are rounded, and in the *thoracica* group the ovipositor lobes are greatly reduced. The appendix bursae is well developed and rugose, and the corpus bursae is elongate, but without any angulation. The ostium is not contained in an indentation. The frontal process is variable in shape depending on the species group. Other characters are those found in the subfamily as a whole.

Several separate evolutionary lines are present in *Chrysoecia.* One line contains the three species *scira, gladiola,* and *requies.* Two of these species (*scira* and *gladiola*) occur in the United States. *Chrysoecia requies* Dyar was listed from the United States by Franclemont and Todd (1983). But, I have seen no specimens, and I suspect the record is based on a misidentification of *gladiola.* Therefore, the species is not treated but distinguishing characters are given under *gladiola.* The second phylogenetic line contains the single unusual species *Chrysoecia morga* (Dyar) (NEW COMBINATION). This species morphologically is clearly related to the *scira* group but has a distinctive forewing pattern. This unusual pattern led Hogue (1963) to use a separate generic name for the species. I do not believe this is justified given the unusual plasticity of the pattern in the Stiriinae. *Chrysoecia morga* is restricted to Mexico. The third lineage is the *atrolinea* species group containing the United States species *atrolinea* and at least two others from Mexico, *Chrysoecia dela* (Druce) and *hemicrocea* (Dyar). These three species have a distinctive yellow-orange forewing crossed by black lines. The male and female genitalia, however, show a close relationship with the *scira* group. The fourth major branch contains the single U.S. species *Chrysoecia thoracica. Chrysoecia thoracica* is one of a number of closely related species centered in Mexico. The specific limits within this complex of Mexican species are unclear and more material is needed before it can be adequately dealt with. This lineage of the genus is characterized by structural reductions. Specifically the clasper is a single, not double, projection, and the basal process of the sacculus is much smaller than in the other lineages of the genus. The ductus bursae and appendix bursae are unsclerotized (in contrast to the other branches in *Chrysoecia*) and the ovipositor lobes are greatly reduced in size.

Chrysoecia scira (Druce)
PL. 4, FIG. 6: TEXT FIG. 80 *b, c, e* (♂ gen.); TEXT FIG. 80 *a* (♀ gen.); TEXT FIG. 80 *d* (head); TEXT FIG. 81 (map) (RWH 9761).

Anthoecia scira Druce, 1889, *in* Godman and Salvin, *Biologia Centrali-Americana, Insecta, Lepidoptera-Heterocera,* **1**: 301, pl. 28, fig. 5.
Type locality: Mexico, Guerrero, Rincon. [BMNH]

Chamoclea benjamini Hill, 1924, *Bull. So. California Acad. Sci.,* **23**: 158.
Type locality: [USA], southern Arizona. [USNM]

Chrysoecia scira has a distinctively ovate forewing. The forewing subterminal area is cream white and contrasts with the purple-black upper half of the median area. The basal area and the lower half of the median area are both dark, brassy tan. The reniform mark is a thick, lunate white line and the upper half of the median area has distinct light-blue reflections if seen in the right light. The hindwing is dark brown. The closest relatives of *scira* are *gladiola* and *requies.* Both of them are smaller (compare figures 6 and 7 on plate 4). Both *gladiola* and *requies* have a cream-white patch between the reniform and orbicular marks. The outer rim of the frontal process is very weak, particularly its ventral margin. The prothoracic tibial claw is weak and sometimes hidden in the tibia's vestiture. The uncus (text figure 80 *b, c, e*) is short and greatly swollen. The apical projection of the clasper is thin and recurved. The basal group of cornuti in the vesica is partially fused into a ridge of sharp teeth. The apical group is relatively small, consisting of moderately strong cornuti. The ovipositor lobes (text figure 80 *a*) are not developed into long, strongly sclerotized projections and both sets of apophyses are weak. Wing length: mean = 12.75 mm, standard deviation = 0.81 mm, n = 10.

The larva and its foodplants are unknown.

Chrysoecia scira is known from Pima, Santa Cruz, and western Cochise Counties in southern Arizona. In Mexico it has been collected in the states of Veracruz, Morelos, and Guerrero. There is little or no variation in the species. Adults have been collected in July and August.

Chrysoecia gladiola (Barnes)
PL. 4, FIG. 7; PL. L, FIGS. 13, 14 (♂ gen.); PL. Q, FIG. 7 (♀ gen.); TEXT FIG. 82 (map) (RWH 9763).

Chamaeclea gladiola Barnes, 1907, *Can. Ent.,* **39**: 67.
Type locality: [USA], Arizona, Santa Catalina Mts. [USNM]

NOTE—*Chrysoecia gladiola* was described from one male and one female. The specimen bearing the label "type ♂" has been labeled and is hereby designated the LECTOTYPE.

Chrysoecia gladiola is closely related to *scira* and *requies.* It is smaller than *scira* (compare the figures in the color plates), and a large ovate to square patch of white is present between the reniform and the orbicular marks. This patch almost reaches the costa although the costa is dark brown. This patch is also present in *requies* but is much smaller and is either triangular or comma shaped. The forewing median area of *gladiola* is not separated into upper and lower regions of different color but is unicolorous, some-

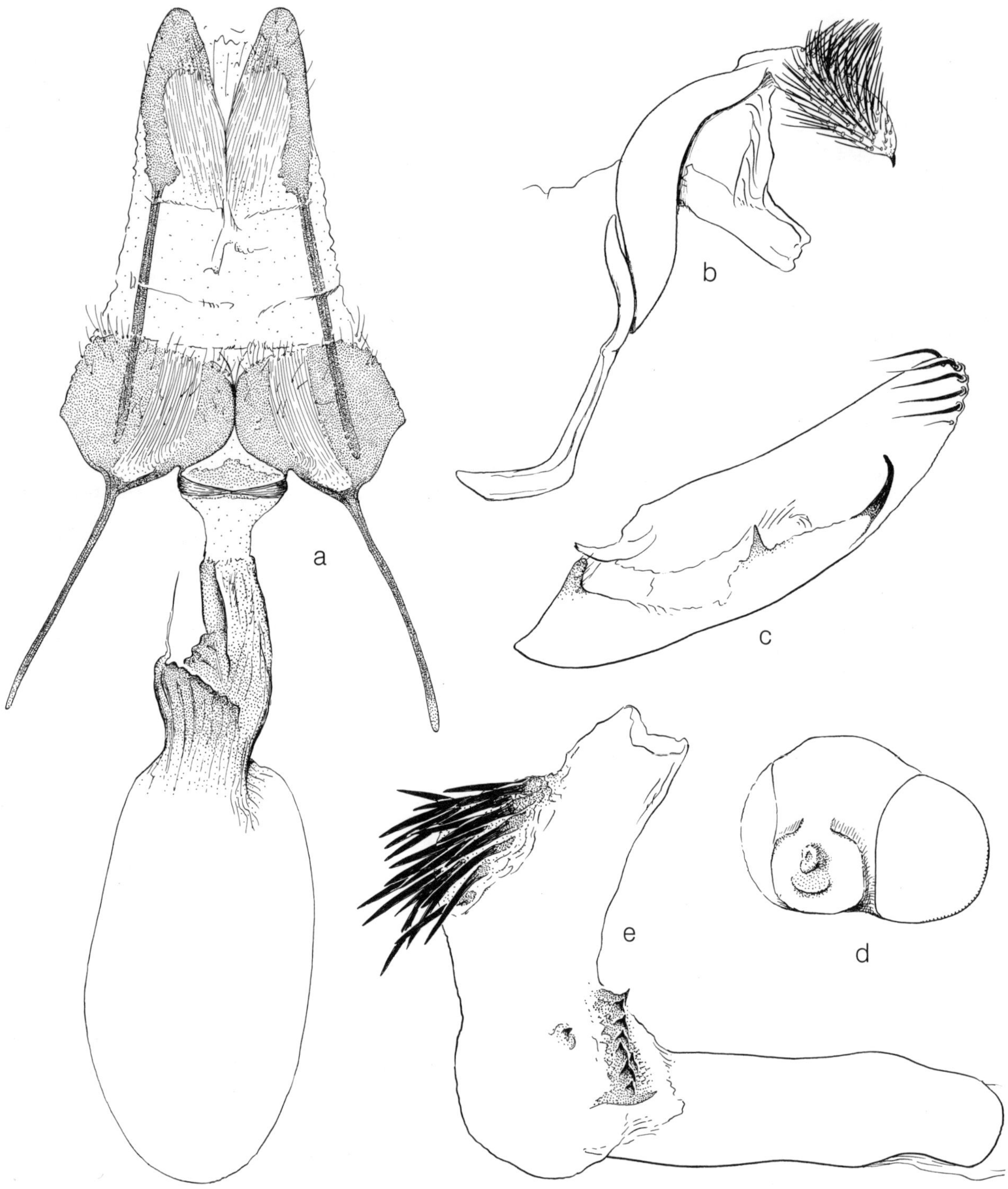

FIGURE 80: STRUCTURAL FEATURES OF *CHRYSOECIA SCIRA*
a. Female genitalia. *b.* Lateral view of male genitalia. *c.* Right valve of male genitalia. *d.* Head. *e.* Aedoeagus.

what contrasting with the cream-white basal area. The male genitalia are similar to those of *scira.* However, the uncus is not as short nor quite so swollen. The basal band of the vesica is not partially fused into a ridge but consists of approximately twice as many cornuti as in *scira.* The female genitalia are essentially as in *scira.* Wing length: mean = 12.23 mm, standard deviation = 0.19 mm, n = 8.

FIGURE 81: DISTRIBUTION OF SPECIMENS EXAMINED OF *CHRYSOECIA SCIRA*

FIGURE 82: DISTRIBUTION OF SPECIMENS EXAMINED OF *CHRYSOECIA GLADIOLA*

The larva and its foodplants are unknown.

In the United States *gladiola* is known only from the Arizona counties of Santa Cruz, Pima, and western Cochise. There is no important variation in the short series of specimens available. The only other known locality for *gladiola* is the Santo Domingo Valley in southern Baja California. *Chrysoecia gladiola* has not been collected in California. This large disjunction in range suggests the species may be widespread in northwestern Mexico, but no specimens have been collected from this region of the country. Adults have been collected in September.

Chrysoecia atrolinea (Barnes and McDunnough), REVISED STATUS

PL. 4, FIG. 8; TEXT FIG. 83 *b, c, e* (♂ gen.); TEXT FIG. 83 *a* (♀ gen.); TEXT FIG. 83 *d* (head); TEXT FIG. 84 (map) (RWH 9764, part).

Antaplaga atrolinea Barnes and McDunnough, 1912, *Contrib. Nat. Hist. Lep. N. Am.*, **1**(5): 25, pl. 1, fig. 23.

Type locality: [USA], Arizona, Cochise Co., Chiricahua Mts. [USNM]

NOTE—*Chrysoecia atrolinea* is based on 3 males and 1 female. The specimen bearing the label "Type ♂" has been labeled and is hereby designated as LECTOTYPE.

Chrysoecia atrolinea is yellow orange. The forewing maculation consists entirely of fine black lines. The forewing postmedial line is a double black line, which is entire near the costa and inner margin, but breaks into a series of black dots in the middle of the wing. The hindwing is brown to dark brown. The central areas of the ventral surface of the forewing and hindwing are dark brown surround by a yellow-orange border. The frontal process of the head is laterally elongate and the central tubercle is absent. The male genitalia are like those of *gladiola* except the outer process of the clasper is shorter and stubbier and the inner process larger. The basal process of the sacculus is larger than in either *scira* or *gladiola.* The basal group of cornuti in the vesica consists of only three large spines. The female genitalia are similar to those of *scira*; however, the sclerotization of the ductus bursae is slightly stronger and the ovipositor lobes slightly weaker. Wing length: mean = 14.09 mm, standard deviation = 0.68 mm, n = 10.

This species was incorrectly synonymized with *Chrysoecia dela* (Druce) by Franclemont and Todd (1983). There are at least two species of *Chrysoecia* related to *atrolinea* in south-central Mexico. *Chrysoecia dela* (Druce) is similar to *atrolinea* with three important superficial exceptions: 1) The forewing postmedial line is not broken up into a series of black points in the middle of the wing but rather is entire in this region and consists of two teeth on veins M_2 and M_3. 2) Usually there is a blackish suffusion in the cell of the forewing, and often in other parts of the median area, that is lacking in

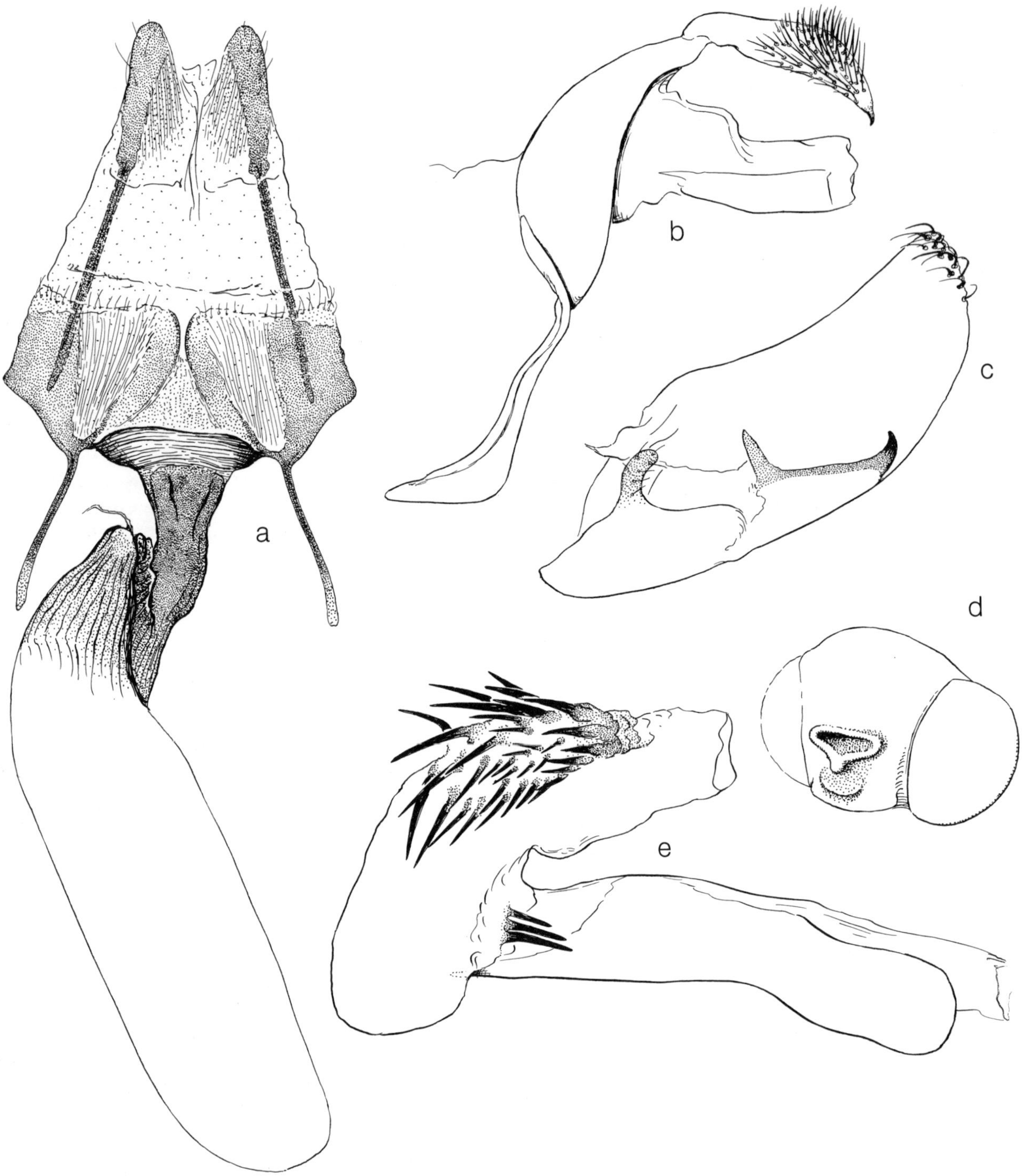

FIGURE 83: STRUCTURAL FEATURES OF *CHRYSOECIA ATROLINEA*
a. Female genitalia. *b.* Lateral view of male genitalia. *c.* Right valve of male genitalia. *d.* Head. *e.* Aedoeagus.

atrolinea. 3) The orbicular mark of *dela* is usually lighter than the surrounding ground color and is often closed on its costal side. The orbicular mark of *atrolinea* is not lighter than the surrounding ground color and is almost always open on its costal side. The posterior projection of the clasper is thin in *atrolinea* but larger and broader in *dela.* The second species in south central Mexico is *Chrysoecia hem-*

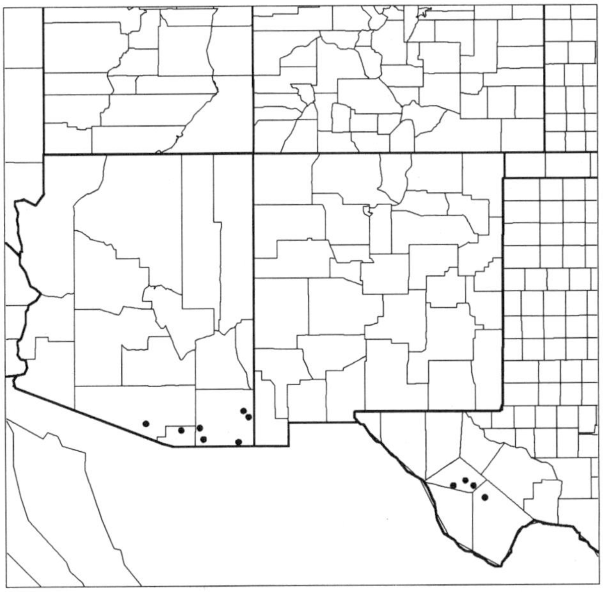

FIGURE 84: DISTRIBUTION OF SPECIMENS EXAMINED OF *CHRYSOECIA ATROLINEA*

icrocea (Dyar) (REVISED STATUS). The forewing of this species lacks all maculation except for vague hints of the reniform and orbicular marks. Further undescribed species may exist because the series of *dela* in the USNM contains considerable variation in size, color, and degree of suffusion in the median area of the cell of the forewing.

The larva and its foodplants are unknown.

Chrysoecia atrolinea has been collected from Pima, Santa Cruz, and Cochise Counties in southern Arizona, and Jeff Davis and Brewster Counties in western Texas. It has not been collected in New Mexico or the northern states of Mexico, but undoubtedly occurs there as well. Females tend to be slightly larger and darker than males. There is no apparent geographical variation. Adults have all been collected in late summer and early autumn.

Chrysoecia thoracica (Hy. Edwards)
PL. 4, FIGS. 9, 10; TEXT FIG. 85 *b, c, e* (♂ gen.); TEXT FIG. 85 *a* (♀ gen.); TEXT FIG. 85 *d* (head); TEXT FIG. 86 (map) (RWH 9774).

Eulithosia thoracica Hy. Edwards, 1884, *Papilio,* **4**: 44.
Type locality: [USA], Arizona. [USNM]

Chrysoecia thoracica is the U. S. representative of a large complex of species from central and southern Mexico. This genus contains *Chrysoecia anaemica* (Draudt) (*Antaplaga suffumosa* form *anaemica* Draudt), NEW COMBINATION, NEW STATUS; *Chrysoecia dulcita* (Schaus) (*Grotella dulcita* Schaus), NEW COMBINATION; *Chrysoecia muricolor* (Dyar) (*Antaplaga muricolor* Dyar), NEW COMBINATION; *Chrysoecia pseudovarra* (Dyar) (*Antaplaga pseudovarra* Dyar), NEW COMBINATION, *Chrysoecia salacon* (Druce) (*Grotella salacon* Druce), NEW COMBINATION; *Chrysoecia simplicia* (Dyar) (*Antaplaga simplicia* Dyar), NEW COMBINATION; *Chrysoecia spreta* (Draudt) (*Antaplaga spreta* Draudt), NEW COMBINATION; *Chrysoecia subfumosa* (Dyar) (*Antaplaga subfumosa* Dyar), NEW COMBINATION; and *Chrysoecia varrara* (Dyar) (*Antaplaga varrara* Dyar), NEW COMBINATION. The specific limits in this group of species are difficult to determine, and the genus needs its own systematic study. Its interest to us lies in its sole U. S. representative, *thoracica.*

Chrysoecia thoracica and its Mexican relatives represent a combination of modification and reduction of the basic *Chrysoecia* ground plan. The forewing is devoid of all maculation, the color varying from off-white to yellow or deep orange in the various species. The forewing of *thoracica* is an off-white with a slight gray tinge. The thorax has a strong orange tint. The male genitalia are simplified. The clasper of the valve has a single, small projection located at approximately ⅓ the length of the valve from the base. The clasper has two projections in all other U.S. species of *Chrysoecia.* The ovipositor lobes are weakly developed in common with other species of *Chrysoecia,* but more so than any other U.S. species in the Stiriinae. Both pairs of apophyses are long, and in one Mexican species of the group extraordinarily long. The ductus bursae and appendix bursae are not sclerotized unlike most other species in *Chrysoecia* and the Stiriinae in general. The corpus bursae is elongate. There appears to be a natural transition series beginning with the average length corpus bursae found in *scira* to the moderately long corpus bursae of *atrolinea* to the greatly elongate corpus bursae in *thoracica* and its relatives. The frontal process of the head is round with a large central projection. Wing length: mean = 12.89 mm, standard deviation = 1.11 mm, n = 10.

The closest relative of *thoracica* is *Chrysoecia subfumosa* (Dyar) from south-central Mexico. The two are superficially identical. However, the uncus of *subfumosa* is longer than in *thoracica* and not swollen. The apical patch of cornuti of the aedoeagus is slightly more diffuse in *subfumosa* than in *thoracica.*

The larva and its foodplants are unknown.

Chrysoecia thoracica is known only from Cochise and Santa Cruz Counties in southern Arizona. A specimen has also been taken near Buenaventura in northern Chihuahua, near the border with New Mexico. Specimens vary slightly in forewing color from cream white to white with a distinct gray tinge. Specimens also vary somewhat in size. Adults have been collected from July to early September.

GENUS

Basilodes Guenée

> *Basilodes* Guenée, 1852, *in* Boisduval and Guenée, *Histoire Naturelle des Insectes. Species Général des Lépidoptères,* **6**: 348.
> Type species: *Basilodes pepita* Guenée, 1852. Monotypy.

> *Deobriga* Walker, 1869, *Characters of Undescribed Lepidoptera Heterocera,* p. 41.
> Type species: *Deobriga chrysopasa* Walker, 1869. Monotypy.

The species of *Basilodes* are broad winged with a golden, metallic sheen to the forewing. The apex of the sacculus of the valve is produced into a long, apical spine, and a posterior projection of variable intensity is also present. The corona is present and moderately strong. The basal group of cornuti is absent, and the large apical group covers most of one side of the vesica. The uncus is relatively short and swollen in the middle. The ostium is the most deeply invaginated of all of the genera with this structure in the Stiriinae. The rim formed by the eighth sternum encircles the ostium. Minute spines are present within the ostium. The ovipositor lobes are very large. Clear areas are present to either side of the ostium, but these areas lack linear striae. The frontal process is circular and lacks a central tubercle. The other characters of *Basilodes* are typical of the subfamily; therefore, no formal description is given.

In the United States there are three species of *Basilodes*: *pepita, straminea,* and *chrysopis.* The Mexican species *inquinatus* was listed from the United States in Hodges et al. (1983) although I have not seen authentic specimens of it from outside of Mexico. The record is probably a misidentification of *straminea.* These two species are closely similar superficially. *Basilodes chrysopis* is the most easily recognized of the three species. The forewing is suffused with light gray giving the forewing a frosted appearance. The two remaining species are difficult to separate but may have disjunct ranges. *Basilodes pepita* and *straminea* are characterized by the golden, metallic sheen of the forewing. There are small, but not always consistent, differences in the forewing maculation. The postmedial line of *pepita* is straight from its angle near the apex of the wing to the inner margin except for one triangular indentation between veins CuA_2 and 1A. The postmedial line of *straminea* has a series of small scallops along its entire length with the indentation between CuA_2 and 1A not particularly stronger than any of the others. In *pepita* there is usually a brownish patch in the lower third of the basal area, although of variable development. In *straminea* this patch is reduced or absent. A single outwardly pointed tooth usually is present on the antemedial line of *pepita,* but generally there are two teeth in *straminea.* Finally, the forewing of *straminea* is a distinctly lighter, less brassy yellow than *pepita.* The proximal edge of the ostium is flat in *straminea* and rounded in *pepita.* The ductus bursae of *pepita* is approximately twice as long as in *straminea.* The appendix bursae is round in *pepita,* but convoluted with sclerotized ridges in *straminea.* The male genitalia of *straminea* are unknown. *Basilodes straminea* is known only from southern Arizona and southern New Mexico. In contrast *pepita* is not known from either region.

Basilodes pepita Guenée

PL. 4, FIG. 13; PL. L, FIG. 15; PL. M, FIG. 1 (♂ gen.); PL. Q, FIG. 8 (♀ gen.); TEXT FIG. 87 (map) (RWH 9781).

> *Basilodes pepita* Guenée, 1852, *in* Boisduval and Guenée, *Histoire Naturelle des Insectes. Species Général des Lépidoptères,* **6**: 348, p. 12, fig. 1.
> Type locality: [USA], Florida. [USNM]

> *Deobriga chrysopasa* Walker, 1869, *Characters of Undescribed Lepidoptera Heterocera,* p. 41.
> Type locality: no locality given. [BMNH]

Basilodes pepita is a distinctive species that should be immediately recognizable from the figure in plate 4, figure 13. The characters distinguishing it from *straminea* were given in the generic discussion. Its male genitalia cannot be compared with *straminea* because the male genitalia of *straminea* are unknown. There are two major differences between it and *chrysopis.* First, the two-pronged saccular pro-

FIGURE 85: STRUCTURAL FEATURES OF *CHRYSOECIA THORACICA*
a. Female genitalia. *b.* Lateral view of male genitalia. *c.* Right valve of male genitalia. *d.* Head. *e.* Aedoeagus.

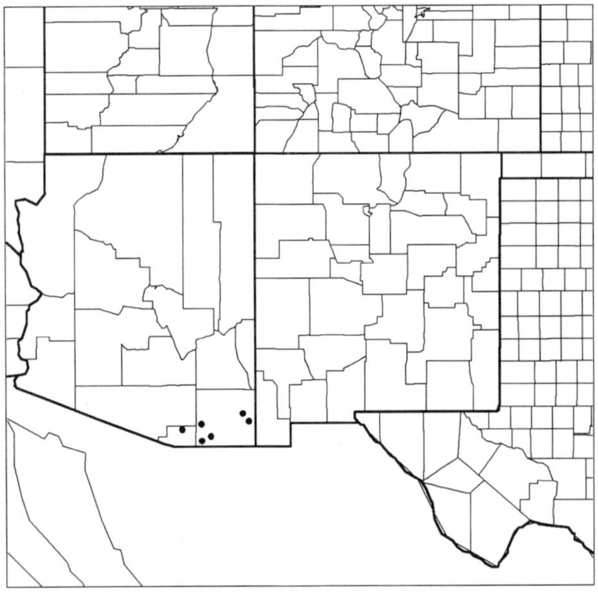

FIGURE 86: DISTRIBUTION OF SPECIMENS EXAMINED OF *CHRYSOECIA THORACICA*

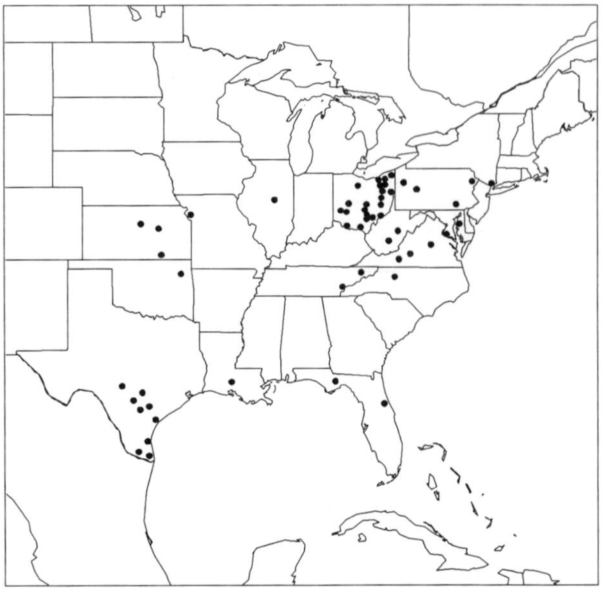

FIGURE 87: DISTRIBUTION OF SPECIMENS EXAMINED OF *BASILODES PEPITA*

cess in the valve is much stronger in *pepita* than in *chrysopis.* Second, a strong gradation in the size of the cornuti of the apical group in the vesica exists in *pepita* but is absent in *chrysopis.* The female genitalia of *pepita* and *chrysopis* are virtually indistinguishable. Wing length: mean = 19.42 mm, standard deviation = 1.58 mm, n = 10.

The larva has been reared from *Verbesina alternifolia* L. (Asteraceae) a perennial sunflower plant related to *Helianthus.* The larva is described in Crumb (1956). The coloration of the dorsum of the larvae is composed of transverse black and orange stripes. The setigerous tubercles on the mesothorax and metathorax are contained in a broad orange stripe margined on either side by black lines. A conspicuous subventral white stripe is present. The head is orange brown and slightly freckled with black.

Basilodes pepita has a wide range throughout the eastern and middle United States although it does not appear to be common anywhere. It occurs in the east from as far north as northeast Pennsylvania (Wayne County) and New York City, and southward to northern Florida. *Basilodes pepita* ranges westward to central Illinois in the north to central and eastern Texas in the south and appears to occur throughout the southeastern United States and the southern Great Plains states. The species is not geographically variable. The most significant individual variation is the extent of the brown patch in the lower third of basal area of the forewing. It may be strong or nearly absent. Adults have been collected from August to October.

Basilodes straminea Poole, NEW SPECIES
PL. 4, FIG. 14; PL. Q, FIG. 9 (♀ gen.); TEXT FIG. 88 (map).

Basilodes straminea Poole.
Type locality: USA, Arizona, Cochise Co., Huachuca Mts., Paradise. [USNM]

The characters separating *straminea* from *pepita* are discussed in the generic diagnosis. Only three females of *straminea* are known. This may be the species reported as *inquinatus* by Franclemont and Todd (1983). The forewing coloration is lighter yellow and less brassy than that of *pepita.* There are two other closely related Mexican species, *Basilodes inquinatus* Hogue and an apparently undescribed species from a series of specimens in the LACM collected at Todos Santos, Baja California del Sur. *Basilodes inquinatus* is found in the volcanic belt in and around Mexico City. *Basilodes inquinatus* is very similar to *pepita* and *straminea* but is generally a browner and duller species. This difference is particularly apparent in side by side series of the two species. The indentation of the postmedial line between veins CuA_2 and 1A in *pepita* is absent in *inquinatus.* Figures of the male and female genitalia of *inquinatus* can be found in Hogue (1963). The

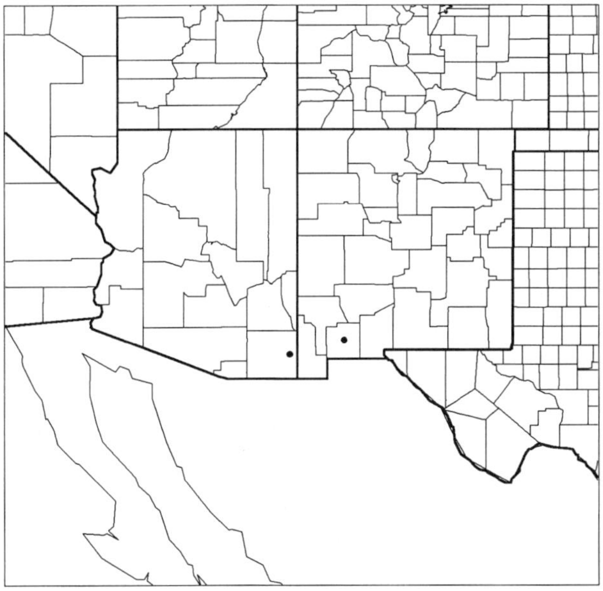

FIGURE 88: DISTRIBUTION OF SPECIMENS EXAMINED OF *BASILODES STRAMINEA*

apparently new species from southern Baja California has the maculation of *straminea* and *inquinatus*. It lacks the dark-brown suffusion of *inquinatus* but has the brown basal patch in the forewing that is absent in *straminea*. Wing length: mean = 19.67 mm, standard deviation = 1.02 mm, n = 5.

The larva and foodplants are unknown.

TYPES. Holotype: ♀. [USA], Ariz.[ona], Cochise Co., [Chiricahua Mountains], Paradise; female genitalia on slide USNM 43,773. USNM. Paratype: 1 ♀. Arizona, [Cochise Co.], Huachuca Mts. USNM. A single female from Deming, New Mexico has not been included in the type series.

Basilodes straminea is known only from the three localities listed above. Its range and the range of *pepita* are not known to overlap. Possible overlap with the two Mexican species cannot be determined because of the absence of specimens from northern Mexico. The single dated specimen was collected in early September.

Basilodes chrysopis Grote

PL. 4, FIGS. 11, 12; TEXT FIG. 89 *b, c, e* (♂ gen.); TEXT FIG. 89 *a*, (♀ gen.); TEXT FIG. 89 *d* (head); TEXT FIG. 90 (map) (RWH 9780).

Basilodes chrysopis Grote, 1881, *Papilio*, **1**: 154. Type locality: [USA], Arizona. [USNM]

Basilodes catharops Dyar, 1911, *Proc. Ent. Soc. Washington*, **13**: 64. Type locality: [USA], Texas, San Diego. [USNM]

NOTE—Dyar based *catharops* on three males. The specimen bearing Dyar's original type label has been labeled and is designated the LECTOTYPE.

Basilodes chrysopis looks like *pepita* and *straminea*, but the forewing is suffused with a covering of frosty white scales obscuring the metallic-bronze sheen of those two species. The median area is usually tinted with gray and, as a consequence, the orbicular and reniform marks are usually more contrasting than those of either *pepita* or *straminea*. *Basilodes chrysopis* lacks a dark-brown patch in the basal area. The forewing appears narrower than in either *pepita* or *straminea*. The frontal tubercle is larger in *pepita* than in *chrysopis*. The backward projecting spur of the sacculus is much larger in *pepita* than in *chrysopis*. The dorsal edge of the sacculus of *pepita* has many spiny knobs; these knobs are poorly developed or nearly absent in *chrysopis*. The cornuti of the large, apical patch of cornuti of the vesica are roughly equal in size in *chrysopis*, but in *pepita* there is a gradation from large cornuti dorsally (as viewed in the figure) to small cornuti ventrally. The female genitalia of the two species are similar, but the appendix bursae is slightly larger in *pepita* than in *chrysopis*. The corpus bursae of *straminea* is much longer than for *chrysopis* and the ductus bursae shorter and broader. Wing length: mean = 18.84 mm, standard deviation = 0.82 mm, n = 10.

The larvae has been described by Crumb (1956) from specimens reared from *Verbesina enceliodes* (Cav.) Benth. and Hook. (Asteraceae), the same genus of foodplants recorded for *pepita*. The larvae are similar to those of *pepita*. However a pair of distinctive, thin black lines is present in *chrysopis* below the line of the spiracles. These lines are absent in *pepita*, or if present, broad, indistinct, and blended with the general pattern.

Basilodes chrysopis has a moderately wide distribution in the western United States. It occurs in Colorado, mostly east of the Rocky Mountains, probably in the short grass prairie, and extending eastward into southwestern Kansas. It has been taken in northern Texas but is apparently commonest throughout southern Texas and westward through most of New Mexico and Arizona. It has not been collected in western Colorado, Utah, Nevada, or California, nor has the species been taken in Mexico, although it undoubtedly occurs in the bordering states. No apparent geographical variation exists in the species. Within population variation is present,

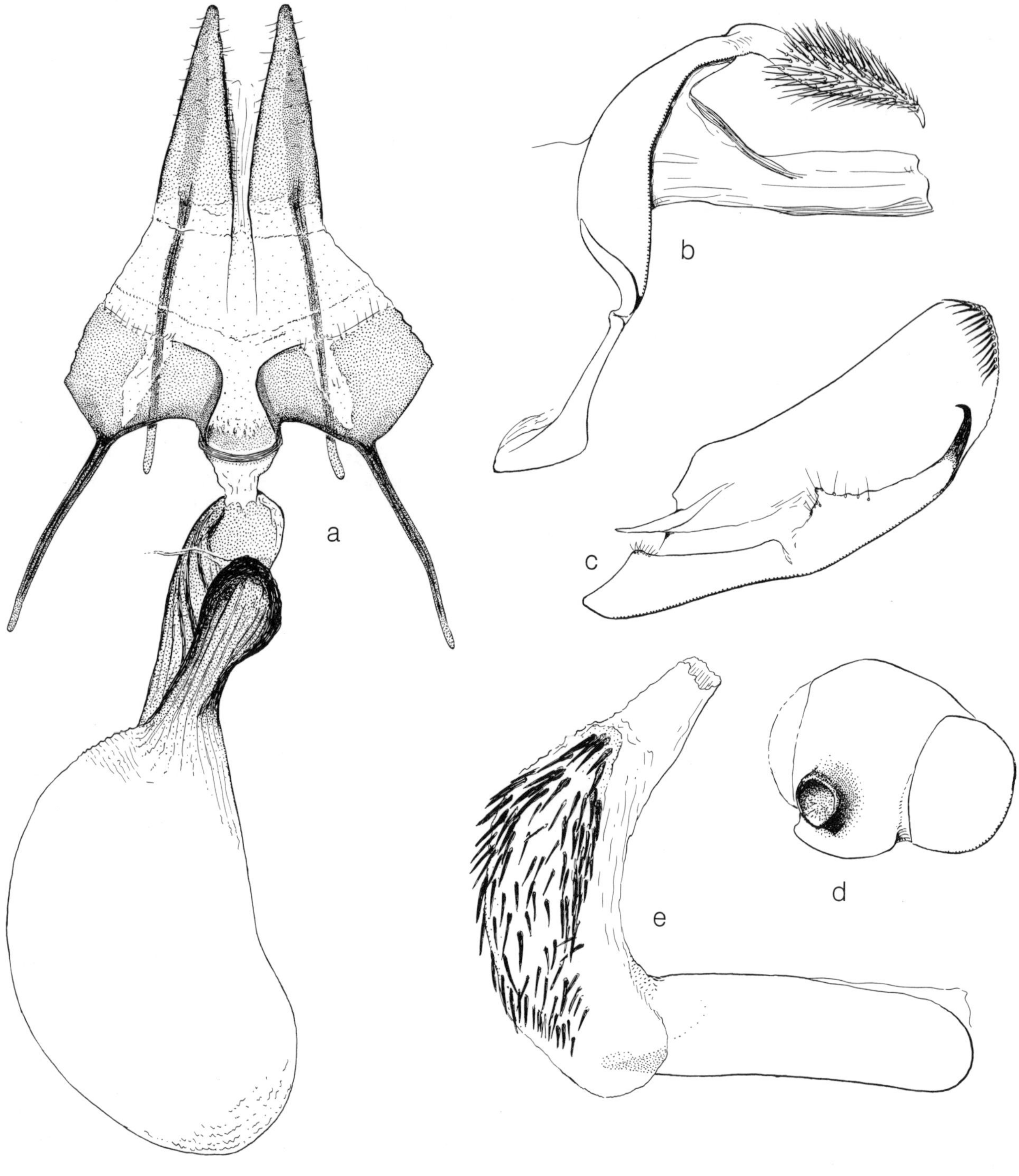

FIGURE 89: STRUCTURAL FEATURES OF *BASILODES CHRYSOPIS*
a. Female genitalia. *b.* Lateral view of male genitalia. *c.* Right valve of male genitalia. *d.* Head. *e.* Aedoeagus.

primarily in the intensity of the gray shading in the forewing median area. The gray shading is very strong in some specimens, and the orbicular and reniform marks contrast sharply with the rest of the median area. In other specimens the gray scaling may be almost absent. The amount of frosty white scaling is also variable. The markings are almost totally obscure in some individuals but well marked in others. The reniform mark usually has a black dot in it, but the dot is very weak or absent in a few in-

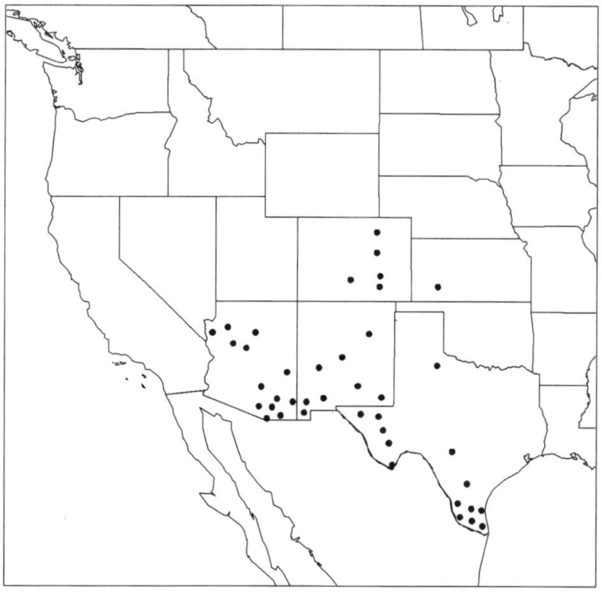

FIGURE 90: DISTRIBUTION OF SPECIMENS EXAMINED OF *BASILODES CHRYSOPIS*

dividuals. Adults have been collected from April to October. Although collected in all the intervening months, the distribution of dates suggests the possibility of two broods, one centered around May and June, and a second, larger brood in August and September.

GENUS
Hoplolythrodes Poole, NEW GENUS

Hoplolythrodes Poole.
Type species: *Lythrodes arivaca* Barnes, 1907.

Hoplolythrodes arivaca and its sister species from Mexico, *Hoplolythrodes prepontendyta* (Dyar) (*Antaplaga prepontendyta* Dyar), NEW COMBINATION, were considered by Hogue (1963) to be of indeterminate position. There is no doubt that they are species of the subfamily Stiriinae; however, their placement within the subfamily is uncertain. Females are not known for either *arivaca* or *prepontendyta.* The type of *arivaca* is a male, not a female as stated in the original description. The frontal process of the head is semicircular, ovate dorsally but flat ventrally. A large nose-shaped tubercle, flat on top, is present along the ventral margin. The foretibial claw is elongate. The uncus is short with a ventral bend and a strong apical tooth. The vesica is more globular than in other genera of the subfamily, and the basal group of cornuti is absent. The sacculus has both forward and rearward projecting processes. The corona is absent as is any indication of spines. The forewing pattern is unique in the subfamily, and the forewing is longer and narrower than in any other group in the Stiriinae. The female genitalia are unknown. Other characters are as in the subfamily. The relatively short uncus, the double process of the sacculus, and the absence of a basal group of cornuti in the vesica have led me to position the genus next to *Basilodes* in the phylogeny.

The larva and its foodplants are unknown.

Hoplolythrodes arivaca (Barnes), NEW COMBINATION

PL. 4, FIG. 15; PL. M, FIGS. 2, 3 (♂ gen.); TEXT FIG. 91 (map) (RWH 9770).

Lythrodes arivaca Barnes, 1907, *Can. Ent.,* **39**: 66.
Type locality: [USA], southern New Mexico, Deming. [USNM]

Hoplolythrodes arivaca is a distinctive species known only from the holotype from southern New Mexico. The forewing is elongate. The forewing ground color is white although the subterminal region is vaguely suffused with yellow brown. The median area is the most distinctive feature of the forewing. The antemedial line is a vague yellow-brown line curving from the costa inward toward the inner base of wing. No distinction in color exists between the basal area and the inner half of the median area. In contrast the median area between the strongly marked medial and postmedial lines is yellow brown. The forewing appears as if it has an extremely narrow, dark, yellow-brown median area. The orbicular mark is a small white circle outlined by a fine brown line, and the reniform mark is kidney shaped with a central dark-brown lunule. All the veins in the subterminal region are strongly accented with yellow brown. Wing length: mean = 15.45 mm, n = 1.

The closest relative of *arivaca* is probably *prepontendyta* Dyar from the state of Guerrero in Mexico. However, the forewing appears broader in *prepontendyta,* and the forewing is shiny white, not dull and suffused with yellow brown. The antemedial line is black and strong. Although the median area is narrow and distinct, the maculation leading to this condition is different between the two species. The veins in the subterminal area of *prepontendyta* are accented with black. Although both superficial appearance and the structure of the male genitalia indicate a sister species relationship between these two species, the different formation of the median area and the differences in wing shape leave room

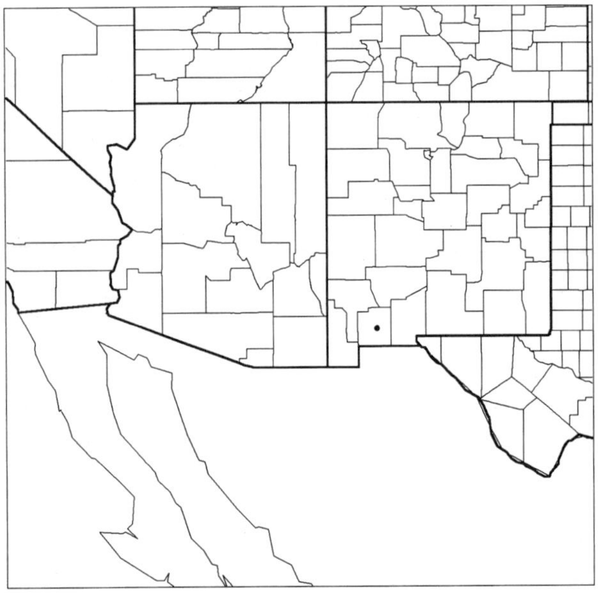

FIGURE 91: DISTRIBUTION OF SPECIMENS EXAMINED OF *HOPLOLYTHRODES ARIVACA*

for doubt. Females of both species are needed to resolve the problem.

The larva and its foodplants are unknown.

This species is known only from the holotype from Deming, in southwestern New Mexico. The holotypc was collected in September.

GENUS
Cirrhophanus Grote

> *Cirrhophanus* Grote, 1872, *Can. Ent.*, **4**: 187.
> Type species: *Cirrhophanus triangulifer* Grote, 1872. Monotypy.
>
> NOTE—The monotypic genus *Pumora* Dyar, 1918 was considered to be a junior synonym of *Cirrhophanus* Grote by Hogue (1963) and Poole (1989). Now, I consider *Pumora* to be a valid genus and restore it to generic rank, REVISED STATUS.

Cirrhophanus is a group of closely related species exemplified by the eastern North American species *triangulifer*. *Cirrhophanus* contains three United States species, *triangulifer, dyari* Cockerell, and *pretiosa* Morrison. Three additional species occur in Mexico; *Cirrhophanus dubifer* Dyar, *nigrifer* Dyar, and *magnifer* Dyar. The three U. S. species usually can be separated by locality. *Cirrhophanus triangulifer* occurs in the eastern United States west to the edge of the Great Plains. *Cirrhophanus pretiosa* occurs in the Great Plains, eastern Texas, and southern Arkansas. *Cirrhophanus pretiosa* is sympatric with *triangulifer* in the western parts of the range of *triangulifer* (e.g., localities in eastern Kansas). *Cirrhophanus dyari* occurs in southern Arizona, southern New Mexico, and western Texas. Its range is not known to overlap that of *pretiosa.*

The color of all species in *Cirrhophanus* is orange with a complex pattern of dark lines in the forewing. The frontal process of the head is triangular (text figure 93 *d*) with a concave dorsal lip, tapering ventrally to a knob. This shape is unlike the circular or hemispherical shape found in the other genera of the subfamily. The tibial claw is weak and in some slide mounts can be seen to be composed of at least two, partially fused, spines. The female has unsclerotized triangular to oval areas in the eighth sternum on either side of the ostium. The ostium is located in a depression in the eighth sternum although the depression is not as well developed as in *Basilodes.* Spines are present in the ostium. The apices of the ovipositor lobes are blunt and short relative to the typical condition in the Stiriinae. The corpus bursae is globular, not elongate. The ductus bursae is elongate. The appendix bursae is only weakly sclerotized and not well developed. The uncus is long without noticeable swollen regions. The clasper of the valve consists of a raised plate with a foreward projection and often a weak posterior projection. The basal process of the sacculus is long and well developed. The pattern of cornuti in the vesica consists of the usual apical group and basal row. Other characters are as in the subfamily description.

The genus *Cirrhophanus* of Hogue (1963) was more broadly defined and also contained the species of *Eulithosia.* The narrower definition of genera used in this revision has led me to separate the two groups of species into two genera. My reasons for this separation are given in the generic description of *Eulithosia.*

Cirrhophanus dyari Cockerell

> PL. 4, FIGS. 16, 17; PL. M, FIGS. 4, 5 (♂ gen.); PL. Q, FIG. 10 (♀ gen.); TEXT FIG. 92 (map) (RWH 9765).
>
> *Cirrophanus* [sic] *dyari* Cockerell, 1899, *Can. Ent.*, **31**: 288.
> Type locality: [USA], New Mexico, Organ Mts., Fillmore Canon. [USNM]

Cirrhophanus dyari is separable from *pretiosa* and *triangulifer* by the presence of a sharply defined, triangular indentation of the postmedial line near vein CuA_2. This indentation is not perfectly triangular in a few specimens, but the indentation is always clearly and sharply marked. This indentation is absent in *pretiosa* and *triangulifer,* and the area where it would be is usually obscurely marked. The male genitalia are virtually indistinguishable from those of *pretiosa.* The ovipositor lobes are moderately long and pointed at their apices, but the apices are blunted and square in both *pretiosa* and *triangulifer.* The two closest relatives of *dyari* are *dubifer* Dyar and *nigrifer* Dyar from Mexico. These two species (if they are not merely forms of each other) are larger than *dyari* and their coloration is slightly duller and not quite as yellow. Wing length: mean = 16.51 mm, standard deviation = 0.88 mm, n = 10.

The larva and its foodplants are unknown.

Cirrhophanus dyari ranges from the mountains of southern Arizona to western and central Texas. It has been taken as far east as Kerrville in Kerr County, Texas, but primarily has been collected in Jeff Davis County to the north of the Big Bend region. The type is from New Mexico, but *dyari* has not been collected there since then. In Arizona *dyari* has been collected in Cochise, Santa Cruz, and Pima Counties in southeastern Arizona and as far north as the White Mountains. There is a single specimen in the AMNH labeled "Poncha Springs, Colorado" [Chaffee County]. This locality is far outside the species' known range and the existence of *dyari* this far north needs to be confirmed. *Cirrhophanus dyari* has also been taken near Cuauhtemoc in the state of Chihuahua in northern Mexico. Intrapopulation variability consists primarily of the presence (and its strength) or absence of black shading in the hindwing. The hindwing is pure yellow orange in some specimens, but in others there is a black suffusion along the inner margin of the wing. The amount of black suffusion varies considerably but tends to be strongest in females. Adults have been collected in August and September.

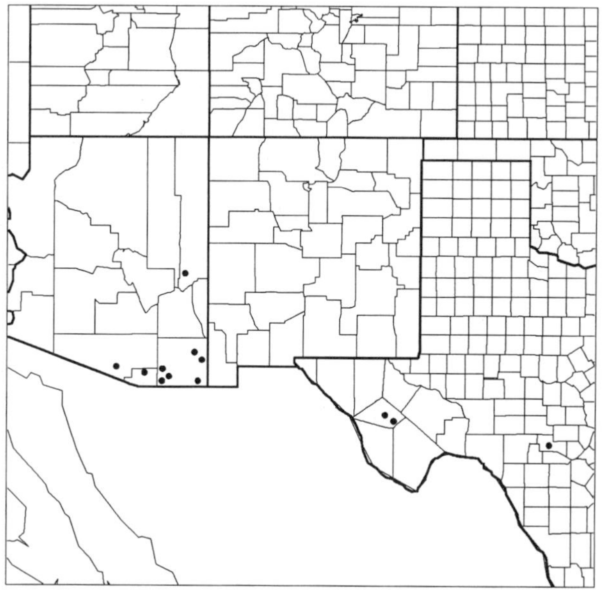

FIGURE 92: DISTRIBUTION OF SPECIMENS EXAMINED OF *CIRRHOPHANUS DYARI*

Cirrhophanus triangulifer Grote

PL. 4, FIGS. 19, 20; TEXT FIG. 93 *b, c, e* (♂ gen.); TEXT FIG. 93 *a* (♀ gen.); TEXT FIG. 93 *d* (head); TEXT FIG. 94 (map) (RWH 9766).

Cirrhophanus triangulifer Grote, 1872, *Can. Ent.,* **4**: 187.

Type locality: [USA], Missouri. [USNM]

Cirrhophanus triangulifer is one of the few eastern species of the Stiriinae. It looks like a larger, rounder version of *pretiosa.* The forewing length of *triangulifer* is 16–19 mm from base to apex, but only 13–15 mm in *pretiosa.* The hindwing and the ventral surface of the forewing of *triangulifer* normally have distinct postmedial lines. These postmedial lines are absent in *pretiosa.* The clasper of the male valve is elaborate in *triangulifer* (text figure 93 *b*) but simple in *pretiosa.* The top ⅕–¼ of the corpus bursae is sclerotized. The sclerotization of the corpus bursae is much more limited in *pretiosa.* Wing length: mean = 18.42 mm, standard deviation = 1.33 mm, n = 10.

The larva has been described and figured by Jones (1937) and described by Crumb (1956). The larva is striking and has a pronounced dorsal hump on A8 that is lacking in other species of the subfamily. This hump is plesiomorphic for the "trifid" noctuids. The pattern is complex but includes a white slanting stripe on each of the prolegs and a black band on the dorsum of each abdominal segment. Two larvae are illustrated in Jones (1937). The larvae feed on the flowers of their foodplants. Jones (1937) records *Bidens polylepis* Blake (Asteraceae) as a foodplant, and Crumb (1954) lists *Bidens trichosperma* (Michx.) Britton (Asteraceae).

Cirrhophanus triangulifer has been collected from southeastern New York and northern New Jersey south to northwestern Florida. It extends west to the Great Plains in Illinois, Kansas, and eastern Texas. It is sympatric with *pretiosa* along most of its western boundary, but the adults fly in the autumn. The adults of *pretiosa* fly in the spring and early summer. There is no discernible geographic variation in the species. Some intrapopulation variation

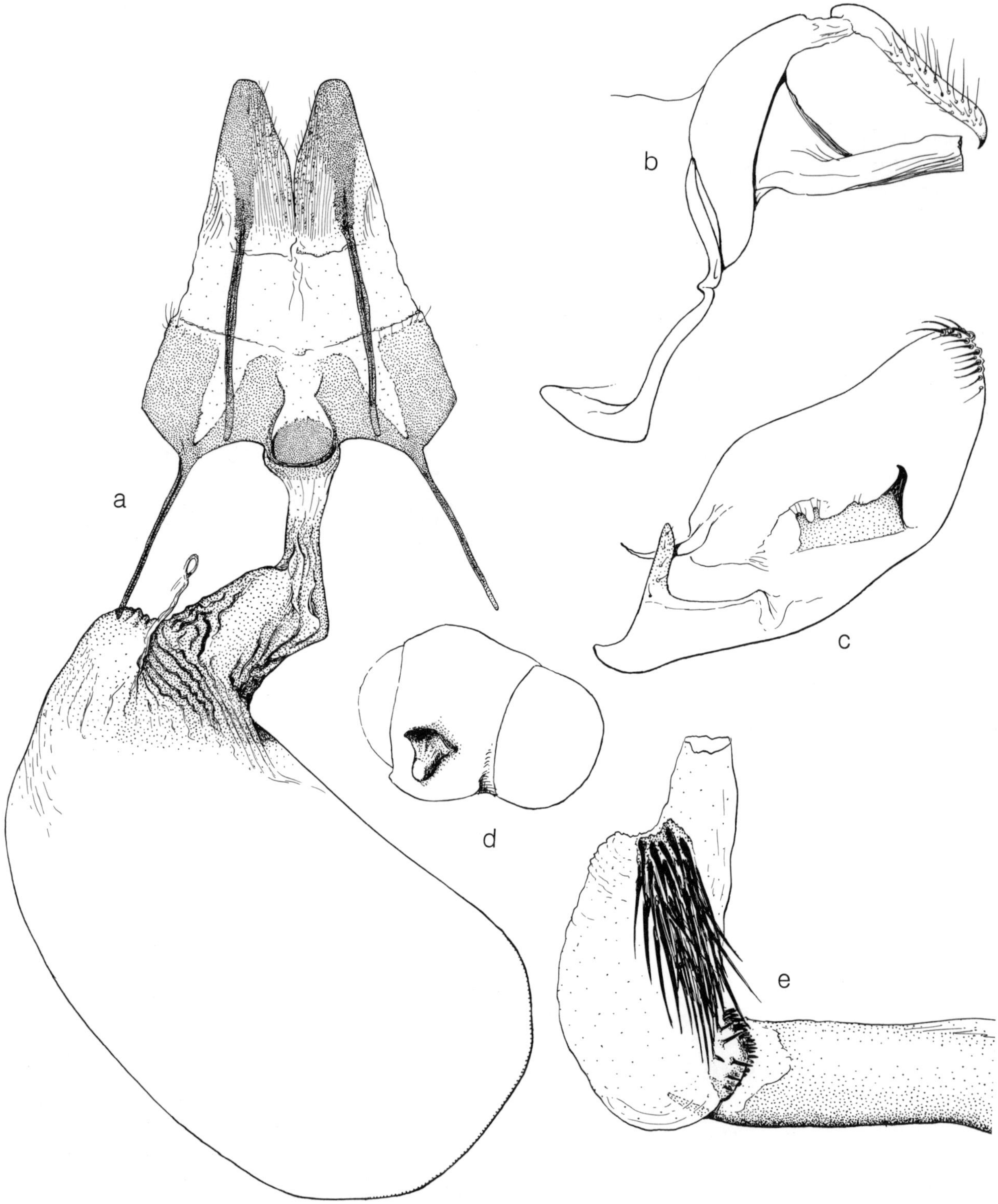

FIGURE 93: STRUCTURAL FEATURES OF *CIRRHOPHANUS TRIANGULIFER*
a. Female genitalia. *b.* Lateral view of male genitalia. *c.* Right valve of male genitalia. *d.* Head. *e.* Aedoeagus.

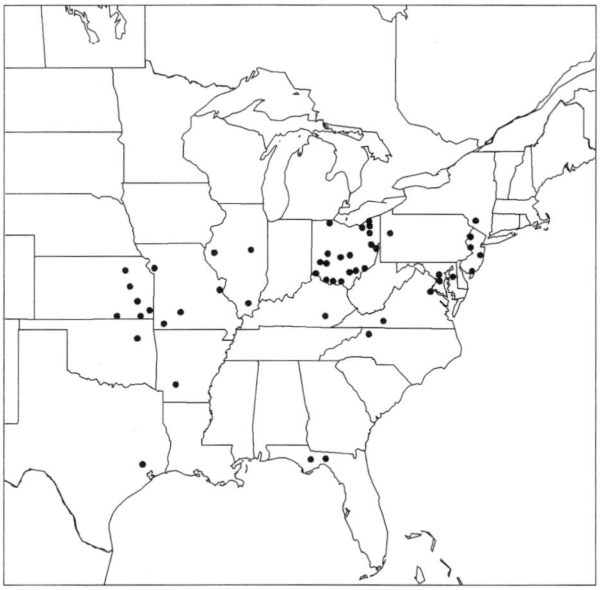

FIGURE 94: DISTRIBUTION OF SPECIMENS EXAMINED OF *CIRRHOPHANUS TRIANGULIFER*

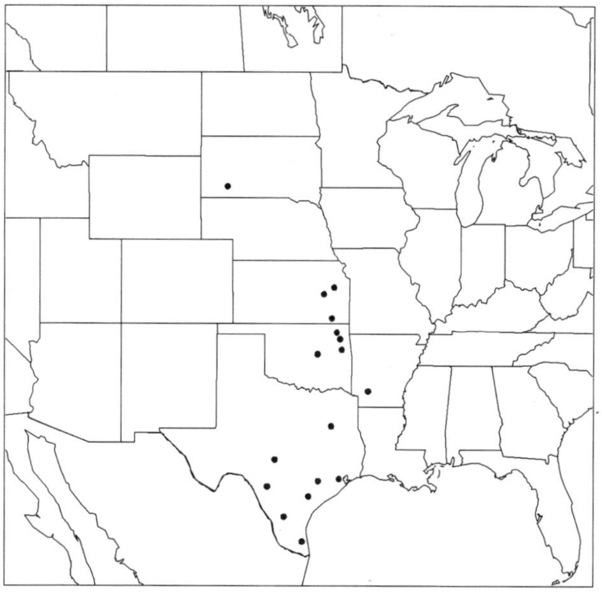

FIGURE 95: DISTRIBUTION OF SPECIMENS EXAMINED OF *CIRRHOPHANUS PRETIOSA*

in the intensity and darkness of the forewing markings exists. Females are slightly larger and darker on average than males. Considerable variation in the number and placement of knobs on the clasper of the male valve exists. Adults fly in August and September. Specimens are not abundant in collections, possibly because the adults are rather sedentary. Adults have been observed sitting on the flowers of their foodplants.

Cirrhophanus pretiosa (Morrison), REVISED STATUS

PL. 4, FIG. 18; PL. M, FIGS. 6, 7 (♂ gen.); PL. Q, FIG. 11 (♀ gen.); TEXT FIG. 95 (map) (RWH 9766, part).

Chariclea pretiosa Morrison, 1875, *Proc. Boston Soc. Nat. Hist.*, **18**: 122.
Type locality: [USA], Kansas, Leavenworth. [?]
NOTE—I have not seen the type of *pretiosa*. Morrison stated that the type was from the collection of Herrman Strecker, now in the FMNH. Nevertheless, the description refers to this species without any doubt.

Cirrhophanus pretiosa has been confused with *triangulifer*. It is smaller than *triangulifer* (forewing length 13–15 mm in *pretiosa* versus 16–19 mm in *triangulifer*). No postmedial line is present in the hindwing of *pretiosa*, but there usually is one in *triangulifer*. The ventral surface of the forewing of *triangulifer* normally has a distinct postmedial line but that of *pretiosa* does not. The male genitalia are distinct. The clasper is an elaborate affair in *triangulifer* but far simpler in *pretiosa*. The male genitalia of *pretiosa* are virtually identical with those of *dyari*. The ovipositor lobes of *dyari* are stronger and sharper than the blunt lobes of *pretiosa*. The ovipositor lobes of *triangulifer* are also blunted, but the sclerotization of the ductus bursae continues about 1/4 the way down the corpus bursae. The sclerotization of the ductus bursae extends only slightly into the corpus bursae in *pretiosa*. Wing length: mean = 14.22 mm, standard deviation = 0.65 mm, n = 10.

The larva and its foodplants are unknown.

Cirrhophanus pretiosa appears to have a distribution largely restricted to the central Great Plains. It has been collected in the eastern and central parts of Texas, Kansas, and Oklahoma. It has also been taken in southwestern South Dakota. The most eastern record is from southwestern Arkansas. In Kansas, Oklahoma, eastern Texas, and Arkansas it is sympatric with *triangulifer*. Little, if any, discernible variation exists except in the degree of black suffusion in the hindwing. Specimens have been collected in April–July. The adults of *triangulifer* fly in August and September. A temporal displacement of *pretiosa* and *triangulifer* in habitat or foodplant requirements may exist.

GENUS
Eulithosia Hy. Edwards

Eulithosia Hy. Edwards, 1884, *Papilio,* **4**: 43.
Type species: *Eulithosia composita* Hy. Edwards, 1884. Subsequent designation by Kirby, 1892, *Synonymic Catalogue of the Lepidoptera Heterocera,* **1**: 371.

Hoplolythra Hampson, 1910, *Catalogue of the Lepidoptera Phalaenae in the British Museum,* **9**: 214. NEW SYNONYMY.
Type species: *Lythrodes discistriga* Smith, 1903. Original designation.

The genus *Eulithosia* comprises a small group of species with similar wing shape and size, but divergent superficial coloration and maculation. The four North American species placed in *Eulithosia* (*plesioglauca, discistriga, composita,* and *papago*) were included in *Cirrhophanus* by Hogue (1963). I have segregated them into *Eulithosia* on the basis of a number of differences between the male and female genitalia of the two genera. In particular the ovipositor lobes of *Eulithosia* tend to be short and stubby with the apices splayed outwardly. The exception to this state is *Eulithosia papago.* The lobes are short and not well developed in *papago* and are more similar to those found in *Cirrhophanus* than *Eulithosia.* Wing shape is ovate elongate although this wing shape is certainly not unique within the subfamily. The frontal process is hemispherical with a flat ventral side and a medial protuberance on the ventral margin. In contrast the frontal process of *Cirrhophanus* is triangular. The prothoracic tibial claw is usually weak (except in *papago*). The male uncus is shorter than in the species of *Cirrhophanus,* and the uncus is swollen medially. The clasper is not contained on a platelike protuberance as in *Cirrhophanus,* and a posterior projection is absent. The basal process of the sacculus is weak to almost absent. Coronal spines are very weak or absent in three species but are strong in *papago.* The basal row of cornuti in the vesica is modified, either into a group of large knobs with short terminal spines (*discistriga* and *composita*) or fused together on a sclerotized base (*papago* and *plesioglauca*). Clear areas with small spines, but no linear striae, exist to either side of the female ostium in the eighth sternum. The ostium is contained within a depression of the eighth sternum. The corpus bursae is elongate. The exception to this condition is the peculiar, lateral, ovate corpus bursae and very long ductus bursae found in *composita.*

The U. S. species *papago* and its Mexican relative *miaiphona* are superficially similar to the species of *Cirrhophanus* both in color and the generally streaky pattern of the forewing. The forewing of *papago* lacks the complex pattern of the species in *Cirrhophanus* and in its place has merely darker veins, darker postmedial and antemedial lines, and a darker but obscure reniform mark. The internal morphology of these two species shares features with both *Eulithosia* and *Cirrhophanus.* The exact phylogenetic position of the two species is doubtful. They combine features found in both genera. The forewing maculation and the shape of the ovipositor lobes agree with *Cirrhophanus.* The vesica spining, frontal process, shape of the uncus, and the shape of the corpus bursae argue for a placement in *Eulithosia.* The two species are possibly evolutionarily intermediate between the two genera.

Eulithosia composita Hy. Edwards
PL. 4, FIGS. 21, 22; TEXT FIG. 96 *a, d, e* (♂ gen.); TEXT FIG. 96 *c* (♀ gen.); TEXT FIG. 96 *b* (head); TEXT FIG. 97 (map) (RWH 9773).

Eulithosia composita Hy. Edwards, 1884, *Papilio,* **4**: 44.
Type locality: [USA], Arizona. [USNM]

Eulithosia composita is easily recognized. It is small and is the only pure yellow-orange U. S. species of the subfamily. All maculation is absent. The hindwing is either yellow or gray black. Both the male and female genitalia are unique for the Stiriinae. The vesica is elongate with an apical group of fused cornuti. The basal group consists of a small number of nodules with short spikes as in *discistriga.* The female genitalia are unique and have an elongate ductus bursae and an ovate, not elongate, corpus bursae. Wing length: mean = 10.56 mm, standard deviation = 0.58 mm, n = 10.

The larva and its foodplants are unknown.

Eulithosia composita is known from southeastern Arizona and Presidio County in western Texas. The color of the hindwing varies from yellow orange to dark gray. Females are more likely to have totally dark hindwings than males. All dated specimens were collected in August.

Eulithosia discistriga (Smith), NEW COMBINATION
PL. 4, FIG. 23; TEXT FIG. 98 *b, c, e* (♂ gen.); TEXT FIG. 98 *a* (♀ gen.); TEXT FIG. 98 *d* (head); TEXT FIG. 99 (map) (RWH 9769).

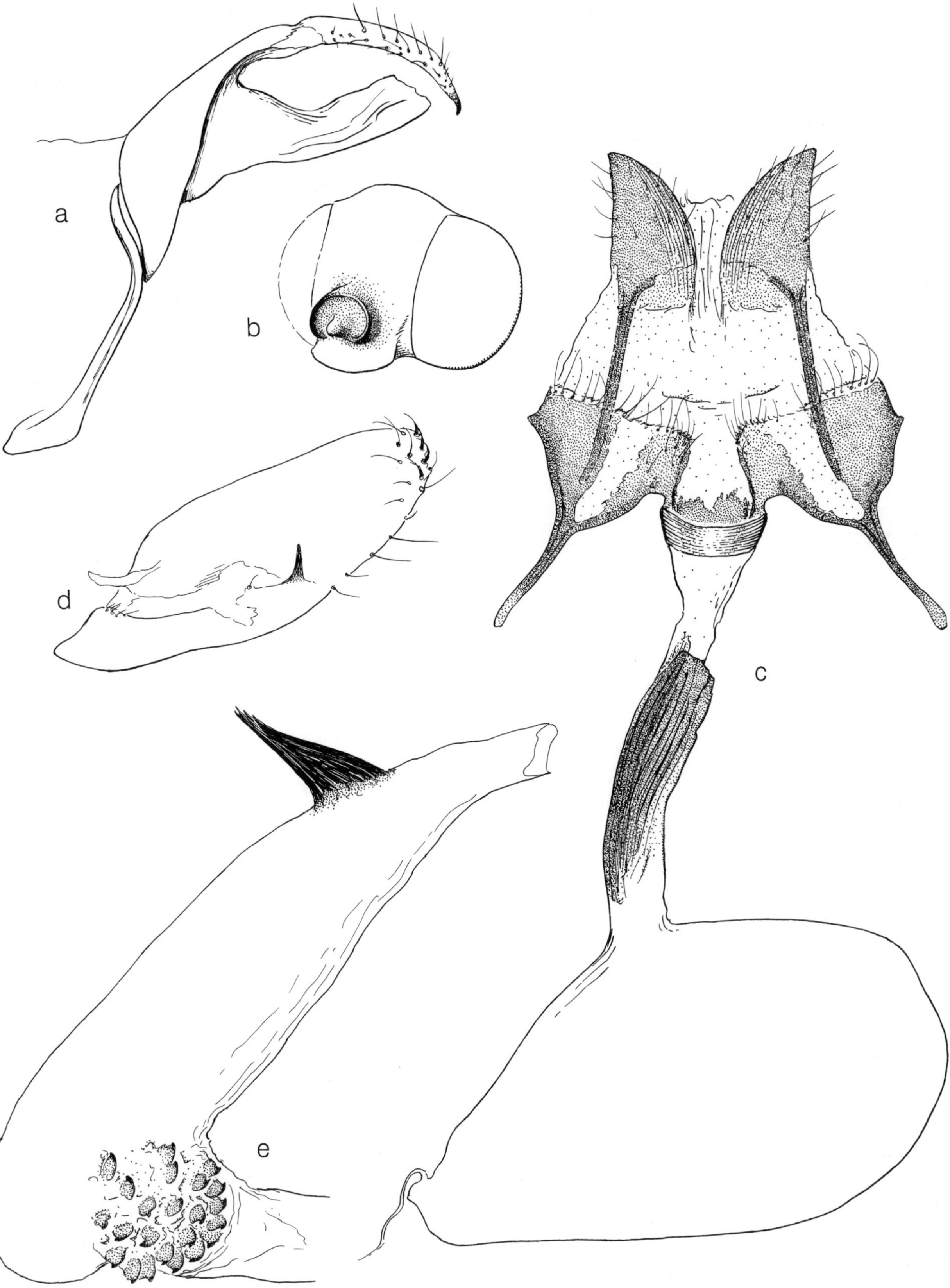

FIGURE 96: STRUCTURAL FEATURES OF *EULITHOSIA COMPOSITA*

a. Lateral view of male genitalia. *b.* Head. *c.* Female genitalia. *d.* Right valve of male genitalia. *e.* Aedoeagus.

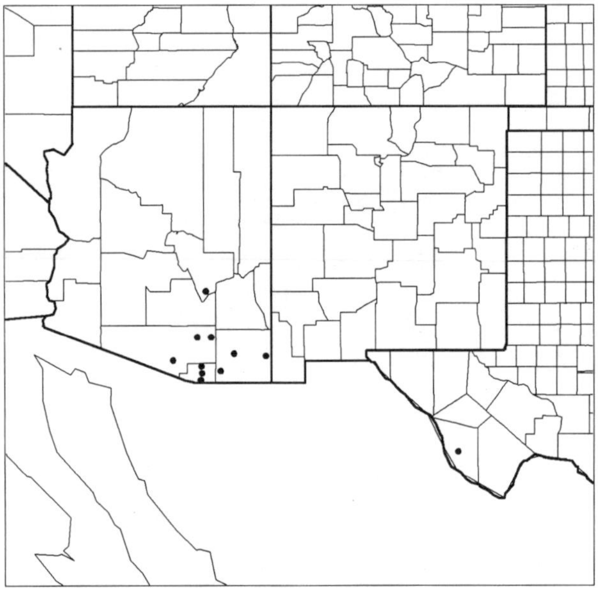

FIGURE 97: DISTRIBUTION OF SPECIMENS EXAMINED OF *EULITHOSIA COMPOSITA*

Lythrodes discistriga Smith, 1903, *Trans. Am. Ent. Soc.*, **29**: 207.
Type locality: [USA], California, Walters Station. [AMNH]; lectotype designated by Todd (1982).

Eulithosia discistriga is a pretty, but uncommon, little species of the southwestern United States. The forewing is yellow white and the hindwing is yellow. The forewing veins are accented with dark brown, and the accents widen into small triangles along the outer margin. The basal setal group of the vesica consists of a large rounded knob with small terminal spines, and the apical group of spines is reduced to a few enlarged setae. This condition is also found in *Eulithosia composita.* The ovipositor lobes are short and splayed outward. Wing length: mean = 13.27 mm, standard deviation = 0.37 mm, n = 10.

The larva and its foodplants are unknown.

Eulithosia discistriga has been collected in western Texas, southern and western Arizona, southern Nevada, southeastern California, and northern Baja California. Californian specimens seem to be limited to the eastern parts of San Bernardino, Riverside, and Imperial Counties. The adults appear to be two brooded with an early spring generation and a late summer one.

Eulithosia plesioglauca (Dyar), NEW COMBINATION

PL. 4, FIG. 24; TEXT FIG. 100 *b, d, e* (♂ gen.); TEXT FIG. 100 *a* (♀ gen.); TEXT FIG. 100 *c* (head); TEXT FIG. 101 (map) (RWH 9767).

Antaplaga plesioglauca Dyar, 1912, *Proc. U. S. Natl. Mus.*, **42**: 68.
Type locality: Mexico, Sierra de Guerrero. [USNM]

Chamoclea [sic] *comstocki* Hill, 1924, *Bull. So. California Acad. Sci.*, **23**: 184, pl. 3, fig. 6.
Type locality: [USA], southern Arizona, Baboquivari Mts. [USNM]

Eulithosia plesioglauca is a very distinctive species. The forewing has a pearly luster with a deep-violet suffusion of the inner half. The apex of the forewing has a slight, off-green metallic luster. The forewing is round and slightly elongate. Wing length: mean = 14.68 mm, standard deviation = 0.54 mm, n = 7.

The larva and its foodplants are unknown. Given the propensity of the adults in the subfamily to sit on the flowers of their foodplants, a composite with purple flowers would be a good first guess.

Eulithosia plesioglauca has been collected in Pima, Santa Cruz, and Cochise Counties in southeastern Arizona, mostly near the Mexican border. I have seen Mexican specimens from the states of Morelos, Guerrero, and Oaxaca at the Isthmus of Tehuantepec. Undoubtedly, it has a more extensive distribution. Mexican specimens tend to be slightly darker and slightly more violet than those from Arizona. Adults have been collected in August and September.

Eulithosia papago (Barnes), NEW COMBINATION

PL. 4, FIGS. 25, 26; TEXT FIG. 102 *b, c, e* (♂ gen.); TEXT FIG. 102 *a* (♀ gen.); TEXT FIG. 102 *d* (head); TEXT FIG. 103 (map) (RWH 9768).

Cirrophanus [sic] *papago* Barnes, 1907, *Can. Ent.*, **39**: 95.
Type locality: [USA], southern New Mexico, [Deming]. [USNM]

NOTE—Barnes states "type male and female." There is only a single female specimen labeled type in the USNM. This specimen is selected and has been labeled LECTOTYPE. Barnes records the type locality as "southern New Mexico," but the lectotype is labeled more specifically Deming, New Mexico.

The yellow-orange coloration of *papago* makes it look like a species of the genus *Cirrhophanus.* How-

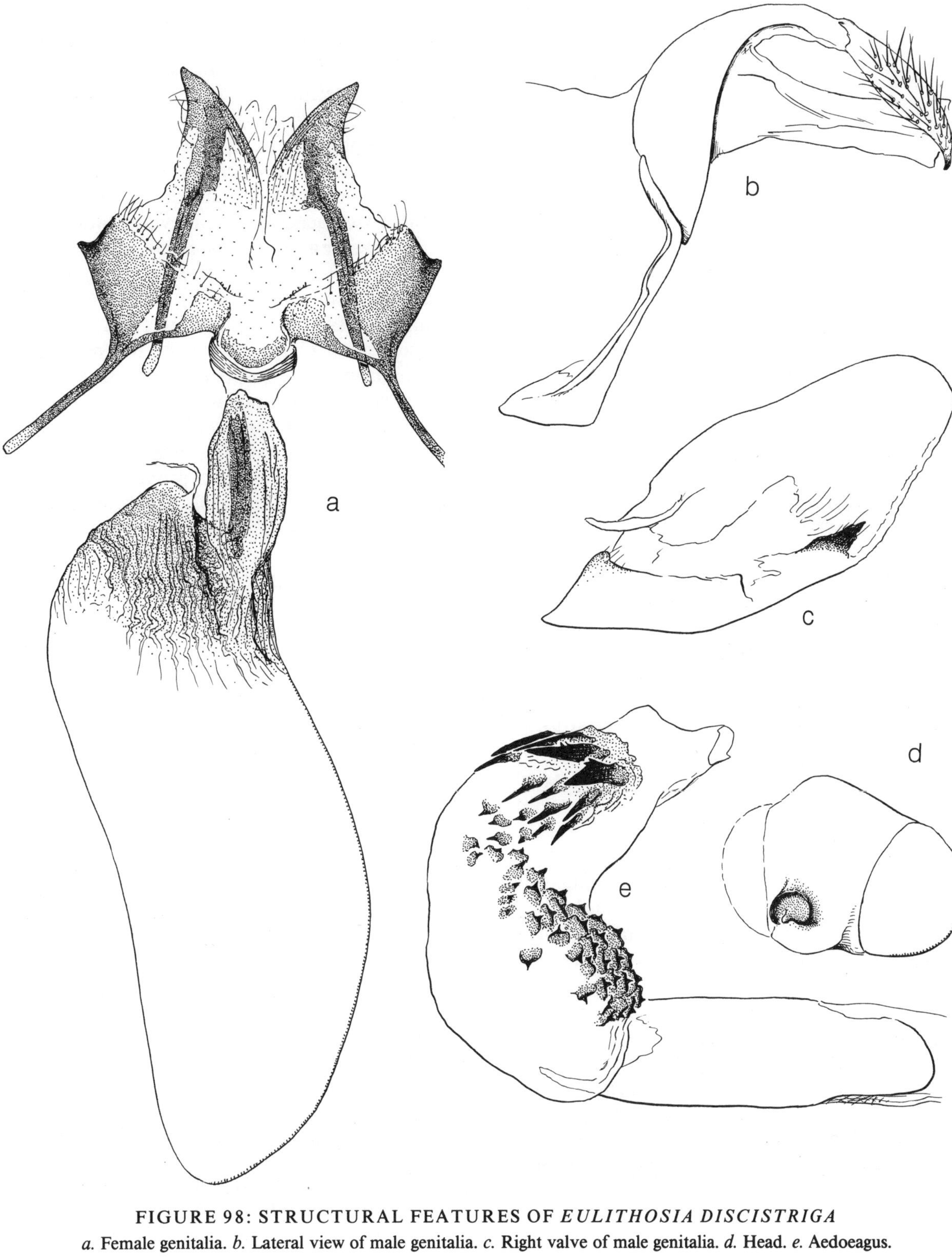

FIGURE 98: STRUCTURAL FEATURES OF *EULITHOSIA DISCISTRIGA*
a. Female genitalia. *b.* Lateral view of male genitalia. *c.* Right valve of male genitalia. *d.* Head. *e.* Aedoeagus.

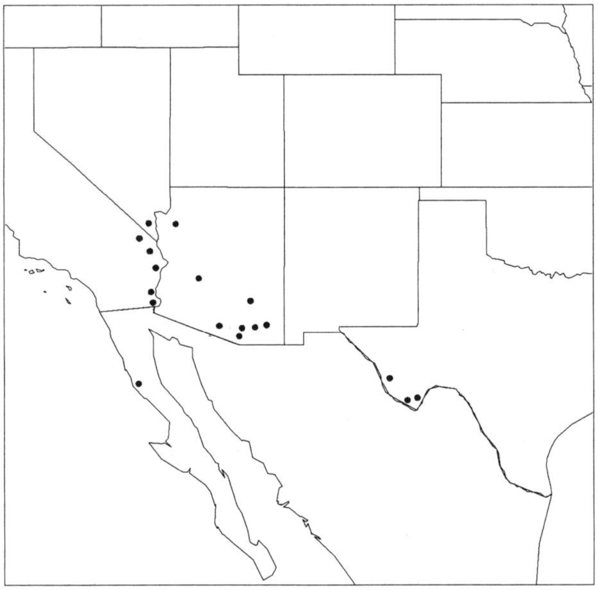

FIGURE 99: DISTRIBUTION OF SPECIMENS EXAMINED OF *EULITHOSIA DISCISTRIGA*

ever, the intricate forewing pattern of *triangulifer* has been replaced by a much simpler pattern of streaked veins, vague postmedial and antemedial lines, and a vague reniform mark. The hemispherical, not triangular, frontal process will also distinguish *papago* from any of the species in *Cirrhophanus. Eulithosia papago* has a close relative in south-central Mexico, *Eulithosia miaiphona* (Dyar) (*Cirrhophanus miaiphona* Dyar), NEW COMBINATION. Wing length: mean = 13.71 mm, standard deviation = 0.51 mm, n = 10.

The larva and its foodplants are unknown.

Eulithosia papago is known almost exclusively from the Davis Mountains of western Texas, mostly specimens collected by André Blanchard. Curiously neither Blanchard nor several other collectors have collected this species in the Big Bend region just to the south of the Davis Mountains. The type series came from Luna County in southwestern New Mexico, and there is a single specimen in the LACM from the White Mountains of east-central Arizona. The species has not been collected in northern Mexico. There is little, if any, variation in the species. Adults have been collected in July, August, and September.

GENUS

Angulostiria Poole, NEW GENUS

Angulostiria Poole.
Type species: *Basilodes chryseochilus* Dyar, 1909.

Angulostiria is proposed for *Basilodes chryseochilus* Dyar. This species was placed in *Cirrhophanus* by Hogue (1963). The single included species displays characters intermediate between those of *Cirrhophanus* and *Eulithosia* along with a short series of autapomorphic ones. These autapomorphies lead me to propose this new genus for *chryseochilus.* The apex of the forewing is pointed, not rounded as in the other two genera. The forewing pattern is basically the same as in *Cirrhophanus,* but the forewing is dull light brown with violet in the median and basal areas, not orange as in *Cirrhophanus.* The frontal process of the head is large, ovate to slightly triangular with a large, flat central tubercle, a shape different from the hemispherical process with a slender ventral tubercle found in *Eulithosia* or the small, compressed, triangular process with dorsal shelf found in *Cirrhophanus.* The clasper is not a raised plate and consists solely of a proximal (toward the base of the valve) raised knob. The basal process of the sacculus is small. The uncus is long and not swollen. The vesica has a basal diverticulum, and this diverticulum and its immediate vicinity are covered with a mat of small spinules. This is the only species of stiriine I have seen with a diverticulum. The basal group of cornuti is small and has shifted dorsally (as viewed in the figure). The ovipositor lobes are splayed outward as in *Eulithosia.* The ostium is in a depression. There are clear areas to either side of the ostium with small spines in the membrane, but striations are absent. The spinules in the ovipositor neck are not arranged in lines. The appendix bursae is absent, and the corpus bursae lacks sclerotized areas except for the striations running into it from the ductus bursae. The corpus bursae is roughly of the same dimensions as in *Cirrhophanus.* The remaining characters are as in the subfamily.

The larva and its foodplants are unknown.

Angulostiria chryseochilus (Dyar), NEW COMBINATION

PL. 4, FIG. 27, TEXT FIG. 104 *b, c, e* (♂ gen.); TEXT FIG. 104 *a* (♀ gen.); TEXT FIG. 104 *d* (head); TEXT FIG. 105 (map).

Basilodes chryseochilus Dyar, 1909, *Proc. Ent. Soc. Washington,* **11**: 25.
Type locality: Mexico, Puebla, Tehuacan. [USNM]

NOTE—*Basilodes chryseochilus* was described from two females. The specimen bearing the red USNM type label has been labeled and is hereby designated LECTOTYPE.

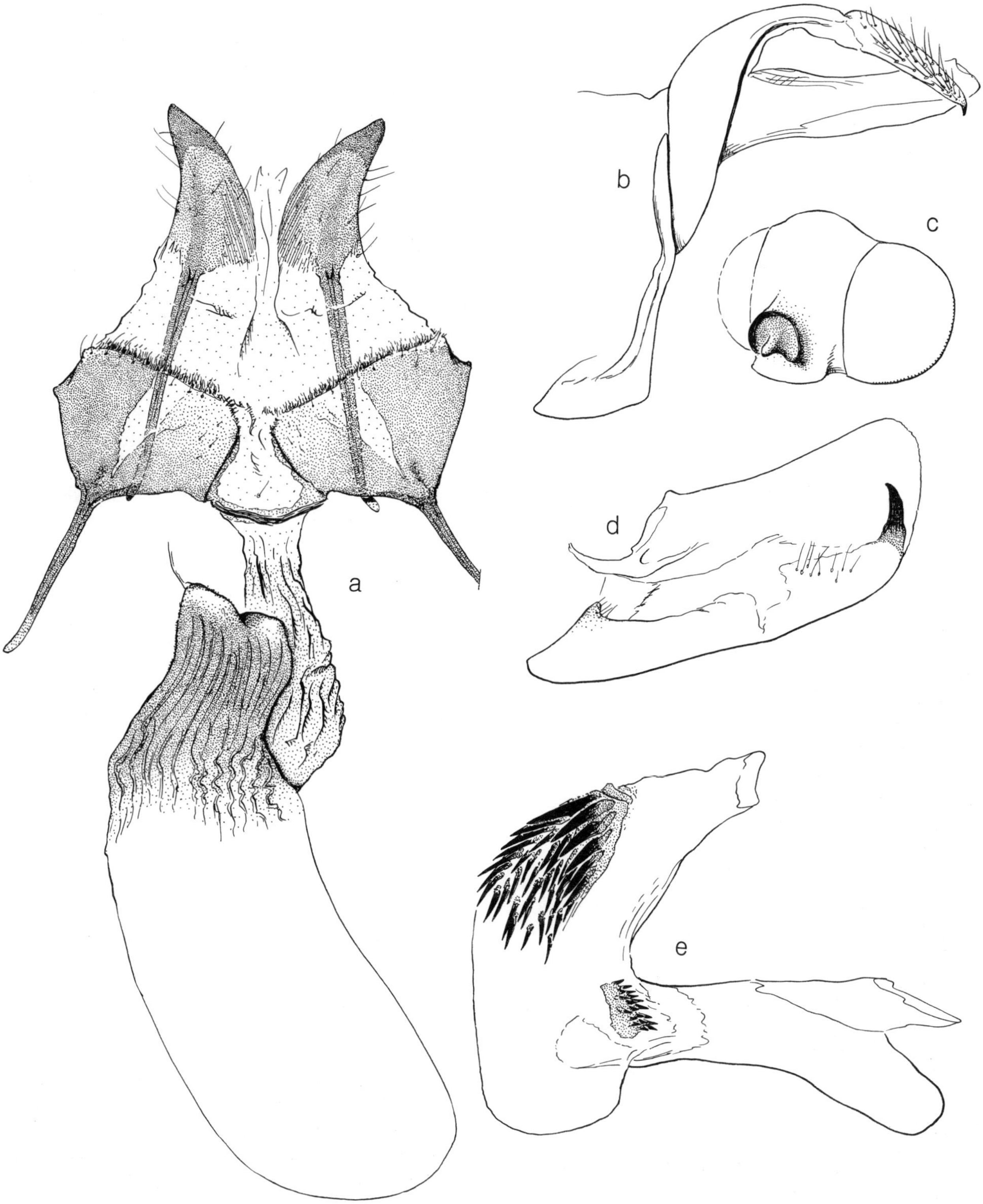

FIGURE 100: STRUCTURAL FEATURES OF *EULITHOSIA PLESIOGLAUCA*
a. Female genitalia *b.* Lateral view of male genitalia. *c.* Head. *d.* Right valve of male genitalia. *e.* Aedoeagus.

FIGURE 101: DISTRIBUTION OF SPECIMENS EXAMINED OF *EULITHOSIA PLESIOGLAUCA*

Angulostiria chryseochilus is reported for the first time from north of Mexico. The apices of the forewing are pointed, not rounded as in its closest relatives of the genera *Cirrhophanus* and *Eulithosia.* The ground color of the forewing is light brown, but the basal and median areas are strongly tinged with violet. The basal and antemedial lines are absent, and the veins in these two regions are accented with darker violet. The postmedial line is well defined and sets off the distinctly light-tan outer fourth of the forewing from the darker, violet inner three-fourths. Wing length: mean = 14.70 mm, standard deviation = 1.37 mm, n = 6.

Angulostiria chryseochilus has been collected by E. C. Knudson from Terrel County in western Texas. The species is otherwise known only from the type locality in the Mexican state of Pueblo. Mexican specimens appear to be slightly larger than the single Texas specimen I have seen. Adults have been collected in September.

GENUS
Stiria Grote

Stiria Grote, 1874, *Bull. Buffalo Soc. Nat. Sci.,* **2**: 32.
Type species: *Stiria rugifrons* Grote, 1874. Original designation.

The members of the genus *Stiria* are readily recognized by the yellow forewing with patches of brown along 1) the outer margin, 2) usually on the inner margin in the median area, and 3) usually at the base of the wing. The frontal process of the head is hemispherical, flat ventrally, with a small tubercle located at the midpoint of the ventral margin. The uncus is long, without swellings. A group of strong spines is present at the apex of the valve. The clasper consists of a single distal tooth although there may be small tubercles along the dorsal margin of the sacculus. The vesica contains the usual basal and apical groups of cornuti, but the apical group is very large and consists of very long cornuti. The ovipositor lobes are variable in length among species. The spinules in the ovipositor neck are the largest found in the Stiriinae. The ostium is located in a depression of the eighth sternum. Clear areas exist on both sides of the ostium, and the membranes in these areas have spinules, but no striations. The eighth sternum is very narrow. The ostium is broad, and the region contains a multitude of minute spinules. The ductus bursae is long, 5 × longer than wide. The appendix bursae is present, and there is sclerotization at the junction of the ductus bursae and the corpus bursae. The corpus bursae is ovate and about 3 × longer than wide.

The genus is most diverse in the southwestern United States and throughout Mexico, but one species occurs in eastern and central North America, i.e., *rugifrons.* Species' boundaries in *Stiria* are difficult to establish. The male genitalia are extremely variable within species, and the range of variability of each species is apparently greater than the differences among species. I have based my specific limits on superficial differences although even these distinguishing features can be subtle. Each species is superficially variable as well. This variability makes species definitions and delimitations difficult to determine. Consequently this revision may not be the last word on the subject but hopefully may be somewhat near the truth. The reader should consult the Stiriinae key to species and the individual species treatments for identification. The following summary may help.

1. Eastern North America and the Great Plains region.—The only species in this region is *rugifrons.*
2. Texas.—Four species occur in Texas. Only *rugifrons* is known to occur in the plains region of northern Texas although *intermixta* may eventually be found there. The brown patch in the median area along the inner margin is distinctly frosted with light-gray scales in *rugifrons.* This patch is usually more distinctly brown in *intermixta.* In south-central Texas only *intermixta* is known to occur. In western Texas *intermixta, sulphurea,* and *blanchardi* occur.

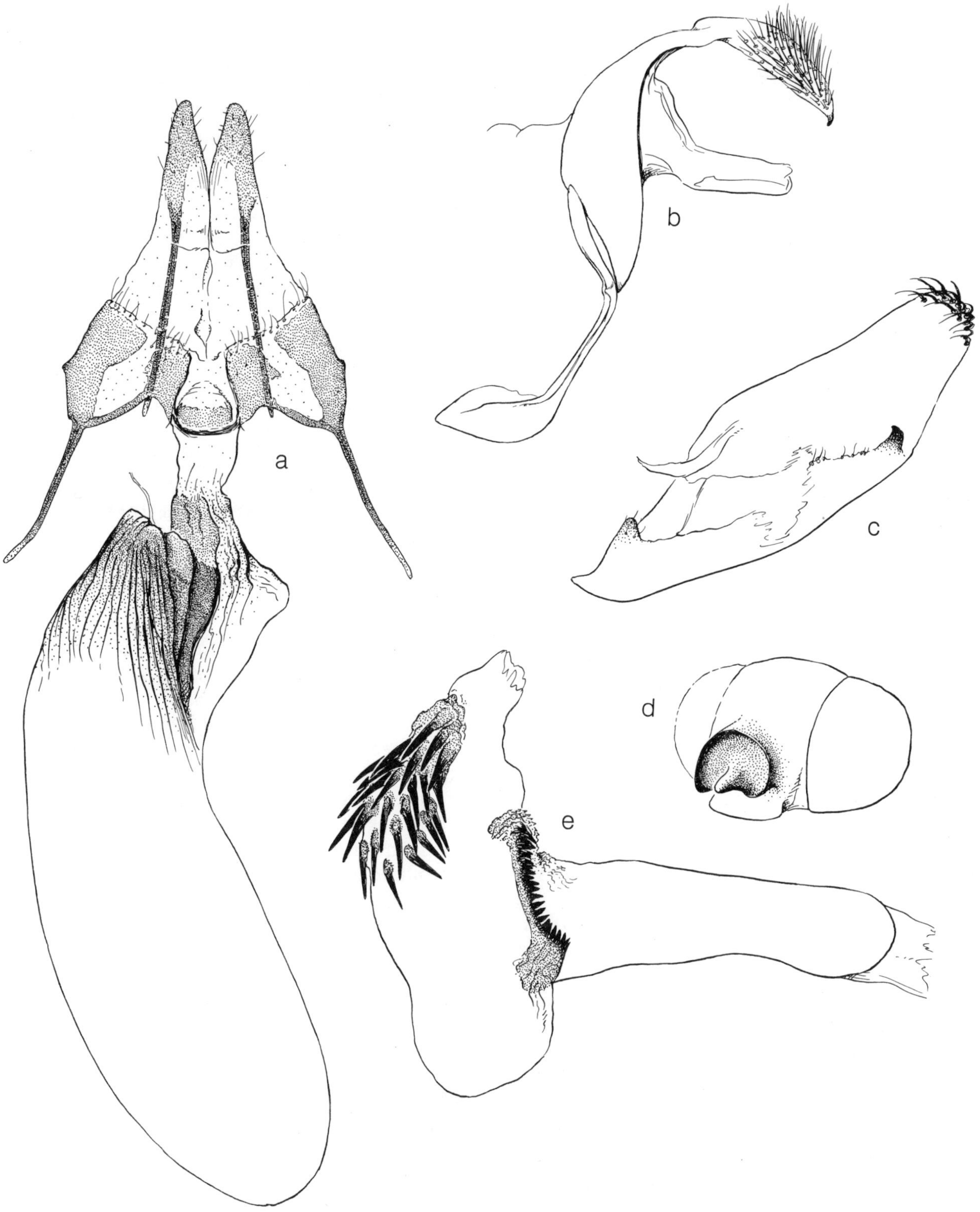

FIGURE 102: STRUCTURAL FEATURES OF *EULITHOSIA PAPAGO*
a. Female genitalia. *b.* Lateral view of male genitalia. *c.* Right valve of male genitalia. *d.* Head. *e.* Aedoeagus.

FIGURE 103: DISTRIBUTION OF SPECIMENS EXAMINED OF *EULITHOSIA PAPAGO*

Stiria blanchardi is the only species lacking a patch of brown scales in the median area at the inner margin. *Stiria sulphurea* is distinctly smaller than *intermixta* (14–17 mm forewing length versus 16–19 mm).

3. Colorado.—*Stiria intermixta* and *rugifrons* both live in Colorado. The region east of the Rocky Mountains appears to be the exclusive domain of *rugifrons* and the western slopes of *intermixta*. The superficial differences between the two species were discussed above.

4. New Mexico.—Three species live in New Mexico: *sulphurea, blanchardi,* and *intermixta.* The absence of the median patch of brown scales will immediately identify *blanchardi. Stiria sulphurea* is distinctly smaller and the maculation crisper than in *intermixta.*

5. Arizona.—Several species occur in Arizona: *intermixta, sulphurea, consuela, satana,* and *dyari. Stiria consuela* is the easiest one to identify. Its coloration and maculation are washed out, and the forewing is suffused with a vague white overlay. The median patch of brown scales on the forewing inner margin is semicircular not square or trapezoidal as in the other species. *Stiria intermixta* is larger than the remaining three species (16–19 mm forewing length versus 14–17 mm). Both *satana* and *sulphurea* have a distinct brown basal patch although the patch may be small in *sulphurea.* This forewing basal brown patch is missing in *dyari.* The hindwing of *dyari* tends to be pure white but is varyingly suffused with brown in *satana* and *sulphurea.* The hindwing is heavily suffused with brown in *satana,* and the forewing basal brown patch is large and confluent with the median patch. In contrast the hindwing of *sulphurea* is generally only lightly suffused with brown. The basal brown patch is small and triangular and never confluent with the median patch.

6. California.—Both *consuela* and *dyari* occur in southern California. The forewing of *dyari* is bright yellow and the maculation crisp and distinct; the hindwing is pure white. The forewing of *consuela* is whitish yellow, and the maculation is diffuse.

Stiria rugifrons Grote

PL. 4, FIGS. 28, 29; PL. M, FIGS. 8, 9 (♂ gen.); PL. Q, FIG. 12 (♀ gen.); TEXT FIG. 106 (map) (RWH 9785).

Stiria rugifrons Grote, 1874, *Bull. Buffalo Soc. Nat. Sci.,* **2**: 73.
Type locality: [USA], Kansas. [BMNH]

NOTE—Grote described *Stiria rugifrons* from a series of syntypes. The syntype bearing the label "*Stiria rugifrons* type Grote" in the BMNH has been labeled and is hereby designated the LECTOTYPE.

Stiria rugifrons and *intermixta* are extremely close morphologically and superficially. *Stiria intermixta* may be the southwestern U. S.-Mexican race of *rugifrons.* The ranges of variability in the male and female genitalia overlap and cannot be used to separate the two species consistently. The outer margin of the sacculus before the clasper is slightly knobby in both species with small setae on the knobs. These knobs are usually, but not consistently, larger in *intermixta* than in *rugifrons.* The forewing median area along the inner margin is usually wider and more rectangular in *rugifrons* than in *intermixta.* This median patch is usually distinctly suffused with light-gray scales in *rugifrons* although this suffusion is visible only in fresh specimens. The forewing median area of *intermixta* is more nearly square shaped and its color is browner than that of *rugifrons.* These superficial differences are not consistent. The best recognition character probably is range. *Stiria rugifrons* occurs primarily in eastern and central North America and *intermixta* in the southwestern United States and Mexico. Both species occur in Texas and Colorado, but their ranges are not known to overlap. In Texas *rugifrons* occurs in the northern Plains region of the state, and *intermixta* apparently is limited to western and south-central Texas. There is a third, closely related, but undescribed species, in central Mexico. This species is distinctly larger than

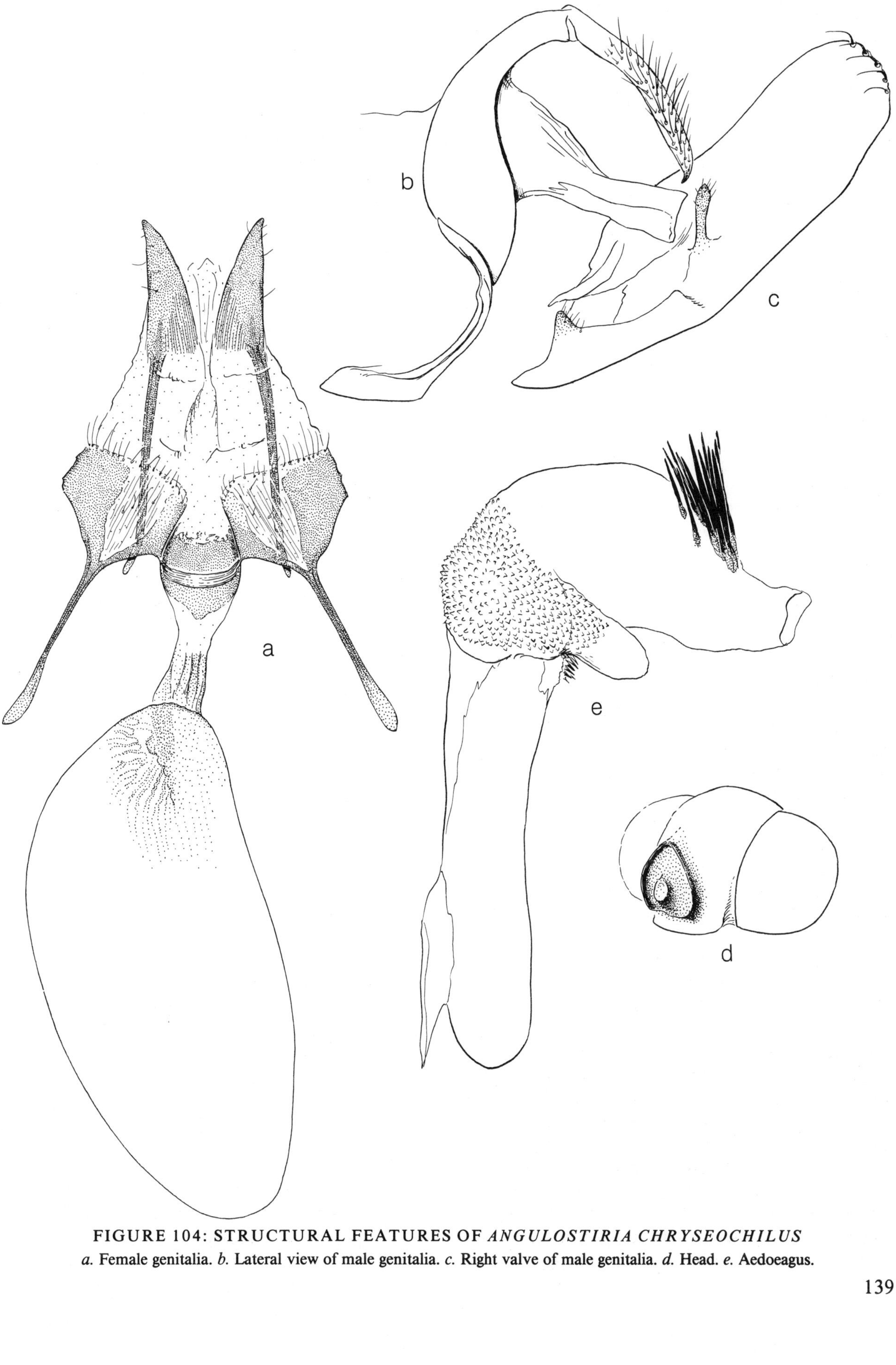

FIGURE 104: STRUCTURAL FEATURES OF *ANGULOSTIRIA CHRYSEOCHILUS*

a. Female genitalia. *b.* Lateral view of male genitalia. *c.* Right valve of male genitalia. *d.* Head. *e.* Aedoeagus.

FIGURE 105: DISTRIBUTION OF SPECIMENS EXAMINED OF *ANGULOSTIRIA CHRYSEOCHILUS*

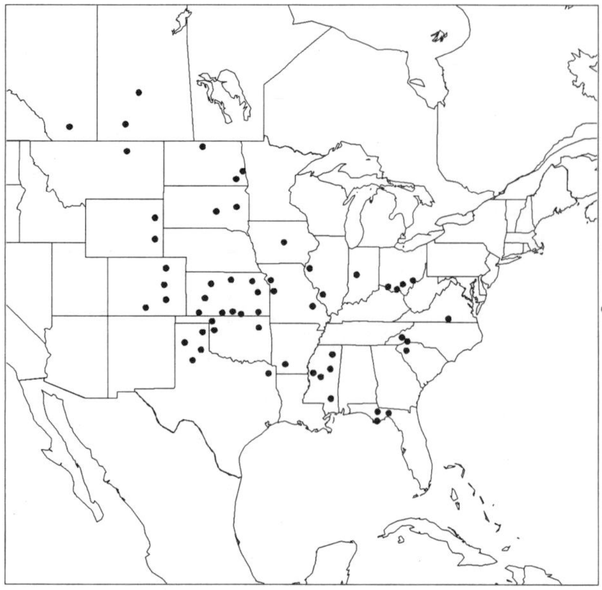

FIGURE 106: DISTRIBUTION OF SPECIMENS EXAMINED OF *STIRIA RUGIFRONS*

either *rugifrons* or *intermixta.* It is sympatric with *intermixta* in the Mexico City region. Wing length: mean = 19.45 mm, standard deviation = 1.39 mm, n = 10.

The larva has been reared in southern Ohio by Eric Metzler from *Helianthus microcephalus* Torr. & Gray (Asteraceae) who has kindly supplied me with color slides and notes. The foodplant is mostly found in disturbed habitats. The larva is characterized by a series of fleshy tubercles on the dorsum of each abdominal segment, the mesothorax, and the metathorax. Each abdominal segment has six tubercles. The two largest are located along the caudal margin of the abdominal segment. The lower half of each of these two tubercles is red and the upper half is black. A second pair of tubercles is located near the anterior margin of the abdominal segment. These tubercles are yellow orange or orange and smaller and closer together than the larger, caudal pair. Finally there is a third pair of tubercles each of which is located half way between the spiracle and one of the larger, caudal tubercles. The dorsal and subdorsal surfaces of the abdomen are yellow shading into green below the spiracles. There are black blotches in the membranes between the abdominal segments. The abdominal prolegs are yellow to yellow green. The head is black with a yellow patch on either side dorsal to the ocelli. The pattern viewed from a dorsal position is very disruptive and the larva is very cryptically patterned if the larva is sitting on the flowers of its foodplant.

Stiria rugifrons is one of the few species of the Stiriinae occurring in the eastern United States. It is apparently rare along the East Coast. I have seen specimens from northern Florida in the south as far north as Virginia and westward into southern Ohio, Indiana, and Illinois. *Stiria rugifrons* is more widespread and common in the Great Plains and southeast. It probably occurs throughout the southeastern states. In the Great Plains it ranges as far north as Saskatchewan and Alberta and as far south as northern Texas. The western boundary of its range reaches the foothills of the Rockies in Colorado and Wyoming. Although the hindwing is usually pure white, some specimens have a slight brown suffusion along the outer margin, particularly females. The adults fly in late summer through middle autumn. Eric Metzler (pers. communication) states that the larva is much commoner than the adult.

Stiria intermixta Dyar

PL. 4, FIGS. 30–32; PL. M, FIGS. 10, 11 (♂ gen.); PL. R, FIG. 1 (♀ gen.); TEXT FIG. 107 (map).

Stiria intermixta Dyar, 1918, *Proc. U. S. Natl. Mus.,* **54**: 349.

Type locality: Mexico, Veracruz, Zacualpan. [USNM]

Stiria intermixta is the sister species of *rugifrons* and may prove to be only a southwestern and Mexican race of that species. The characters distinguishing the two species are discussed under *rugifrons.* These characters are subtle and are not consistently different; however, the ranges of the two species are not known to overlap. Wing length: mean = 17.98 mm, standard deviation = 0.73 mm, n = 10.

The larva and its foodplants are unknown.

Stiria intermixta occurs from south-central and western Texas westward to central Arizona. The species occurs sparingly in Utah and Colorado and throughout Arizona and New Mexico. In Mexico it ranges south to the Mexico City region and the states of Morelos and Mexico. Although a variable species, some specimens are decidedly aberrant. A male from south Texas (Kerrville) lacks postmedial and antemedial lines and is slightly darker than the average specimen from the region. Two females from southern Arizona have dark-brown hindwings in contrast to the white hindwing of most specimens. The basal brown patch of the forewing varies slightly in size. No geographical differences between populations have been noted. The adults fly in August and September.

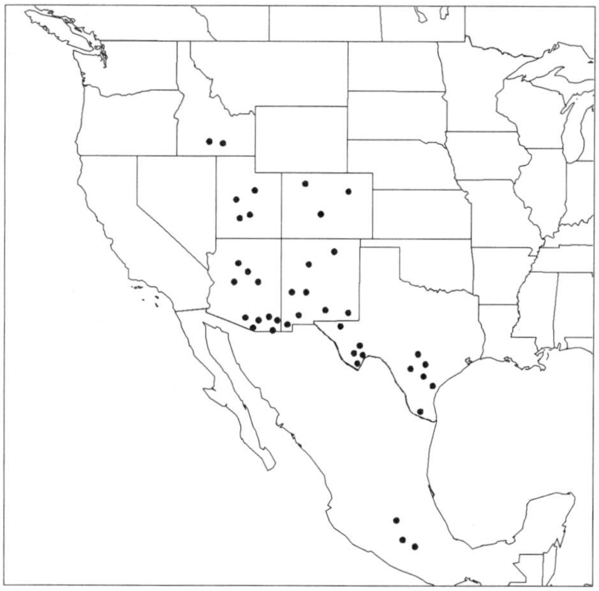

FIGURE 107: DISTRIBUTION OF SPECIMENS EXAMINED OF *STIRIA INTERMIXTA*

Stiria dyari Hill

PL. 4, FIGS. 33, 34; PL. M, FIGS. 12, 13 (♂ gen.); PL. R, FIG. 2 (♀ gen.); TEXT FIG. 108 (map) (RWH 9786).

Stiria dyari Hill, 1924, *Bull. So. California Acad. Sci.,* **23**: 183, pl. 3, fig. 3.
Type locality: [USA], California, San Diego. [USNM]

Stiria dyari lives in southern California, northern Baja California, and western Arizona. The species resembles *intermixta* and *rugifrons.* However the forewing basal brown patch found in *rugifrons* and *intermixta* is absent from almost all specimens of *dyari.* A few females have a slight hint of a basal brown patch. The male hindwing is entirely pure white. A hint of brown along the outer margin is usually present in *rugifrons* and *intermixta. Stiria dyari* has a slightly brighter, cleaner appearance than *rugifrons* or *intermixta* (compare the colored figures of the three species). The forewing costa often has a slight light-gray to white frosting of scales. This frosting is also found in *blanchardi.* I have not found any consistent differences between the male genitalia of *dyari* and *intermixta.* However, the setose knobs of the costal margin of the sacculus tend to be less well developed, on average, than in *rugifrons* and *intermixta.* Wing length: mean = 15.45 mm, standard deviation = 1.02 mm, n = 10.

The larva and its foodplants are unknown.

Stiria dyari has been collected in southern California in the Mohave Desert region and at several localities in northern Baja California. It occurs in extreme southern Nevada and western and southwestern Arizona as far east as the Baboquivari Mountains. Some variability exists in the strength of the forewing postmedial and antemedial lines. Specimens without the lines are commoner than those with them. The female hindwing is slightly tinged with brown. Specimens from Arizona tend to be smaller than those from California, and the forewing is not as bright. The forewing is slightly suffused with brown in Arizona specimens. Adults have been collected from April to June in contradiction to the autumn flying habits of other known species. A single specimen in the LACM from California has a collection date of October.

Stiria blanchardi (Hogue)

PL. 4, FIGS. 35, 36; PL. M, FIGS. 14, 15 (♂ gen.); PL. R, FIG. 3 (♀ gen.); TEXT FIG. 109 (map) (RWH 9783).

Basilodes blanchardi Hogue, 1965, *J. Research Lep.,* **4**: 275, fig. 1 (adult), fig. 2 (male genitalia), fig. 3 (female genitalia).
Type locality: [USA], New Mexico, Eddy County, Carlsbad Caverns. [LACM]

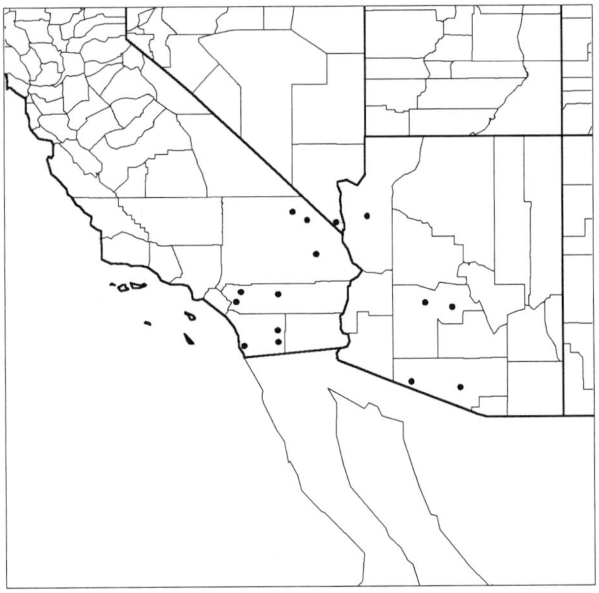

FIGURE 108: DISTRIBUTION OF SPECIMENS EXAMINED OF *STIRIA DYARI*

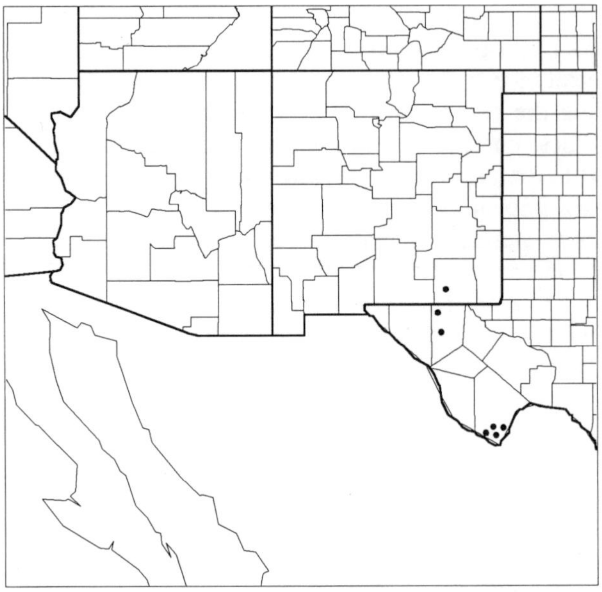

FIGURE 109: DISTRIBUTION OF SPECIMENS EXAMINED OF *STIRIA BLANCHARDI*

Stiria blanchardi is the easiest species of the genus to identify. The median brown patch along the inner margin of the forewing, present in all other members of *Stiria,* is absent. *Stiria blanchardi* and *dyari* share a slight white frosting of the costa of the forewing and a pure white male hindwing. The male genitalia of *blanchardi* and *dyari* are identical, and the two are probably sister species. Wing length: mean = 16.27 mm, standard deviation = 0.63 mm, n = 10.

The larva and its foodplants are unknown.

Stiria blanchardi is known from a narrow strip in western Texas (Brewster and Culberson Counties) and southeastern New Mexico (Eddy County). There is no apparent inter- or intra-population variation. Adults have been collected in late summer and early autumn.

Stiria sulphurea Neumoegen

PL. 4, FIG. 37; PL. N, FIGS. 1, 2 (♂ gen.); PL. R, FIG. 4 (♀ gen.); TEXT FIG. 110 (map) (RWH 9787).

Stiria sulphurea Neumoegen, 1882, *Papilio,* **2**: 135.
Type locality: [USA], Arizona, Prescott, [USNM]

NOTE—Neumoegen based *Stiria sulphurea* on one male and one female. The male syntype in the USNM has been labeled and is hereby designated the LECTOTYPE.

Stiria sulphurea is the smallest U. S. species of *Stiria.* Forewing length is 14–17 mm. Forewing length in *intermixta,* the species with which it occurs most commonly, is almost always over 17 mm. The male genitalia are also smaller on average than those of other species in the genus. Other than size there are no specific characters separating the male genitalia of *sulphurea* from those of other species. The markings and patches of the forewing are usually crisper than in *intermixta,* and the yellow coloration of the forewing is slightly darker. A triangular, basal brown patch is present on the forewing. Both male and female hindwings are suffused with brown, another useful feature separating *sulphurea* from *intermixta.* Wing length: mean = 16.16 mm, standard deviation = 0.75 mm, n = 10.

The sister species of *sulphurea* appears to be *Stiria sisaya* Dyar from south-central Mexico. The two species are superficially similar. The forewing median patch is usually mushroom shaped in *sisaya* but quadrate in *sulphurea.* The forewing postmedial and antemedial lines are consistently present in *sisaya,* but often vague or absent in *sulphurea.* The brown suffusion of the hindwing is considerably darker in *sisaya* than in *sulphurea.*

The larva and its foodplants are unknown.

Stiria sulphurea ranges from western Texas, through southern and western New Mexico, and most of Arizona. It appears to be common in the mountains of southern Arizona. The forewing post-

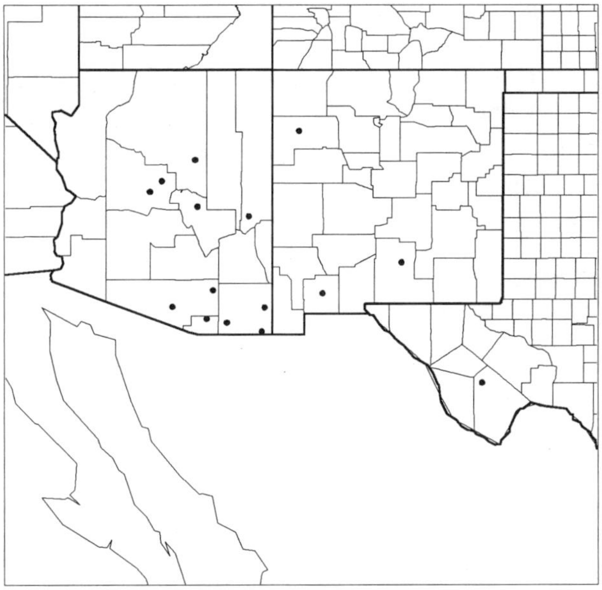

FIGURE 110: DISTRIBUTION OF SPECIMENS EXAMINED OF *STIRIA SULPHUREA*

FIGURE 111: DISTRIBUTION OF SPECIMENS EXAMINED OF *STIRIA SATANA*

medial line varies from clearly present to totally absent. The patch of brown in the subterminal area also varies in strength from present to nearly absent. No clear geographical trend in this variation is obvious. All adults have been collected in late summer and early autumn.

Stiria satana Poole, NEW SPECIES
PL. 4, FIG. 38; PL. N, FIGS. 3, 4 (♂ gen.); TEXT FIG. 111 (map).

Stiria satana Poole.
Type locality: [USA], Arizona, Pima Co., Baboquivari Mts. [USNM]

Stiria satana looks like a small version of *intermixta.* However the basal brown patch in the forewing is large and confluent with the median patch. The basal patch is not as large and never confluent with the median patch in *intermixta.* The hindwing is heavily suffused with brown, a feature distinguishing it from both *intermixta* and *dyari.* The basal process of the male sacculus is larger and more elongate than in any other species in the genus (plate N, figure 3). The clasper is thicker than in other species, and it has a strong apical hook. Wing length: mean = 15.82 mm, standard deviation = 2.31 mm, n = 2.

The sister species of *satana* appears to be *Stiria ischune* Dyar from south-central Mexico. *Stiria ischune* is larger than *satana* by about 2 mm in forewing length. The clasper of *satana* is smaller than that of *ischune* and is located near the inner margin of the valve. The clasper of *ischune* is strong and stubby and not located near the inner margin of the valve. The photographs of the adults and male genitalia of *satana* will characterize this species.

The larva and its foodplants are unknown.

TYPES. Holotype: ♂ [USA], Ariz[ona], Pima Co., Baboquivari Mts., 1–15 Sept. 1923, O. C. Poling. USNM. A single male specimen labeled "Arizona" is not included in the type series.

This species is so far known from two males from southern Arizona, one labeled simply "Arizona." The single dated specimen was collected in September.

Stiria consuela Strecker
PL. 4, FIGS. 39, 40; PL. N, FIGS. 5, 6 (♂ gen.); PL. R, FIG. 5 (♀ gen.); TEXT FIG. 112 (map) (RWH 9784).

Stiria consuela Strecker, 1900, *Lepidoptera, Rhopaloceres and Heteroceres, Supplement* **3**: 34.
Type locality: [USA], Arizona. [FMNH]

Stiria consuela is a distinctive species. The forewing has a whitish, washed out appearance. The brown patch along the outer margin of the forewing is uniquely absent in *consuela.* The median brown patch along the inner margin of the forewing is semicircular, not square or rectangular as in other species of the genus with a median patch. *Stiria consuela* does not have any close relatives in the United States

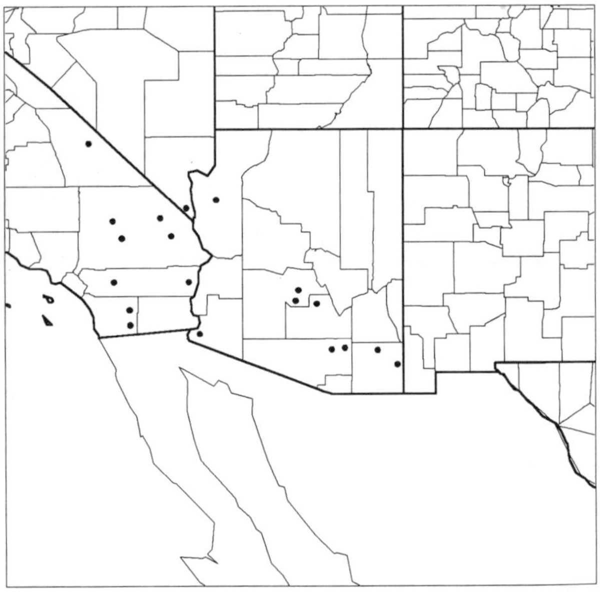

FIGURE 112: DISTRIBUTION OF SPECIMENS EXAMINED OF *STIRIA CONSUELA*

or Mexico. Wing length: mean = 15.95 mm, standard deviation = 1.04 mm, n = 10.

Stiria consuela has been collected in southern California, southern Nevada, and western and southern Arizona. A great deal of variation is present within populations. Thc amount of yellow in the forewing varies from obvious almost to absent. Generally, the yellower the individual the stronger the maculation although maculation is never very strong. Some specimens are almost white with even the median patch on the forewing nearly absent. The male genitalia are particularly variable, in both the strength of the development of the clasper and the degree of dispersion of the distal patch of cornuti in the vesica. There is no clear geographical variation in the species.

The larva and its foodplants are unknown.

Adults have been collected primarily in the spring, but two specimens from Arizona were taken in August suggesting the possibility of two broods in southern Arizona.

GENUS

Chalcopasta Hampson

Chalcopasta Hampson, 1908, *Catalogue of the Lepidoptera Phalaenae in the British Museum,* **7**: 4.

Type species: *Basilodes territans* Hy. Edwards, 1884. Subsequent designation by Hampson, 1910, *op. cit.,* **9**: 218.

Rodriguesia Dyar, 1912, *Proc. Ent. Soc. Washington,* **14**: 61.

Type species: *Deva ornata* Ottolengui, 1898. Monotypy.

The genus *Chalcopasta* is primarily defined by a single supposed synapomorphy, the presence of extensive metallic-green areas on the forewing. Areas of metallic-green scales are not limited to *Chalcopasta* however. For example, metallic-green areas occur in *Cuahtemoca,* a genus more closely related to *Plagiomimicus* on the basis of other structural characters. Metallic-green scaling is also found in the genus *Neumoegenia,* a genus of uncertain affinities included under the *incertae sedus* heading following the Stiriinae. Metallic green is even found in the forewing of the plusiine *Diachrysia balluca.* This metallic-green scaling is probably some modification of the metallic-bronze scaling found in *Basilodes* and *Angulostiria,* indicating a possibly closer relationship between *Chalcopasta* and these two genera, as postulated by Hogue (1963), than is evident in the arrangement used here. There are two additional possible shared derived characters defining the genus *Chalcopasta.* The basal patch of cornuti of the vesica is located on the right side of the vesica (in the orientation of the figures) and away from the junction of the vesica and the shaft of the aedoeagus. The basal patch of cornuti is located at the junction of the aedoeagus shaft and the vesica in almost all other genera of the Stiriinae. I am interpreting this loose group of cornuti to be equivalent to the basal patch (row) but am not entirely sure that it is. Second, the appendix bursae (the portion of the bursa bearing the ductus seminalis) is located in the left, caudal region of the bursa (in the orientation of the figures) in most genera of the Stiriinae. In contrast the appendix bursae is on the right, caudal side in *Chalcopasta* (except in the Mexican species *ellica* Dyar) (e.g., see text figure 113 *a*). The change in position suggests a 180 degree twist in either the ductus bursae or the caudal region of the bursa. The frontal process is round with a central tubercle. The uncus is long, and swellings are absent. The clasper is variable, and a digituslike structure is present in the species of the *acema* group. The ovipositor lobes are small (relative to those of most other genera of the subfamily) and their apices are slightly splayed outward as in the genus *Eulithosia.* Otherwise the characters of *Chalcopasta* are typical for the subfamily.

The genus is composed of three species groups. Only two of them are represented in the United

States, the *territans* and *acema* groups. The forewings of the species in the *territans* group are longer and more nearly rectangular than the shorter, almost ovate forewings of the species in the *acema* group. The species of the *acema* group are unique for the subfamily in having a digituslike structure in the valve.

territans GROUP

The *territans* group contains three U. S. species, *territans* (Hy. Edwards), *howardi* (Hy. Edwards), and *fulgens* Barnes and McDunnough, and a few other species in Mexico. The species are among the most distinctive in the Noctuidae with their metallic-green and yellow-green forewing patches. The center of the group's distribution appears to be Mexico; only *fulgens* occurs exclusively in the United States. The two remaining species are widely distributed in the drier regions of Mexico and reach the northern limits of their range in the southwestern United States.

The three species are relatively easy to separate in the United States (in Mexico the situation is more complex). *Chalcopasta fulgens* is the most distinctive with a whitish frosting over the forewing. The metallic-green portions of the forewing are yellower and not as reflective as in the other two species. A large brown area at the base of the forewing is present in *Chalcopasta howardi.* The brown patch is widest at the inner margin and narrows toward the costa. A patch of green is often isolated within the brown, triangular area. This triangular area is absent in *territans* or represented, at most, by a very small brown patch.

There are two specimens in the USNM of the Mexican species *Chalcopasta sinuata* Hampson labeled "Arizona." These labels appear to be the same bogus "Arizona" labels found on some other Mexican specimens in North American collections (e.g., see the discussion under the genus *Neumoegenia* in the *incertae sedus* heading following the Stiriinae). I have not seen any valid U. S. specimens of *sinuata* and am nearly certain that the species does not occur north of Mexico. *Chalcopasta sinuata* looks like a slightly larger, slightly darker version of *howardi.* It is the only species of the group with uniformly brown hindwings; all other species have white hindwings or white with brown outer margins. The forewings of *sinuata* also have a slightly rounder appearance than do any of the other three species. The vesica is distinctive and has only a small apical group of long setae; all other cornuti are short and stubby. The species has been collected in the states of Veracruz, Puebla, and Nayarit in Mexico.

Chalcopasta territans (Hy. Edwards)

PL. 4, FIG. 41; TEXT FIG. 113 *b, c, e* (♂ gen.); TEXT FIG. 113 *a* (♀ gen.); TEXT FIG. 113 *d* (head); TEXT FIG. 114 (map) (RWH 9775).

Basilodes territans H. Edwards, 1884, *Papilio,* **4**: 45.
Type locality: [USA], Arizona. [USNM]

NOTE—Edwards described this species from "3 males, 1 female, coll. Neumoegen." Only one male is labeled type in the USNM, the eventual resting place of the Neumoegen collection. This specimen bearing Edwards' type label and the USNM type number 33951 is designated and has been labeled LECTOTYPE.

Plusia arianda Druce, 1889, *in* Godman and Salvin, *Biologia Centrali Americana, Insecta, Lepidoptera-Heterocera,* **1**: 329, pl. 30, fig. 11. NEW SYNONYMY.
Type locality: Mexico, northern Sonora. [BMNH]

NOTE—This name was based on two syntypes, both now in the BMNH, even though one was indicated as in "Mus. Dognin," a collection now in the USNM. The female type from northern Sonora is in far better condition than the Arizona specimen and is designated and has been labeled LECTOTYPE.

Plusia arizona French, 1889, *Can. Ent.,* **21**: 161.
Type locality: [USA], Arizona. [probably lost]

Chalcopasta anopis Dyar, 1918, *Proc. U. S. Natl. Mus.,* **54**: 348. NEW SYNONYMY.
Type locality; Mexico, Morelos, Cuernavaca. [USNM]

Chalcopasta chalcophanis Dyar, 1918, *Proc. U. S. Natl. Mus.,* **54**: 348. NEW SYNONYMY.
Type locality: Mexico, Mexico City. [USNM]

Chalcopasta territans is the only U. S. species of the *territans* group without a brown, triangular basal region in the forewing. A small, light-brown area is sometimes present at the base of the inner margin, but when present it is insignificant. The spination of the vesica is very heavy. In contrast only about half as many cornuti are present in *howardi.* The secondary group of cornuti (interpreted here as the basal row and located on the lower right part of the vesica in the figures) is fused into a spiny node in *howardi,* but in *territans* the spines are separate. The setal pattern of the vesica is similar in *howardi* and *fulgens*; however, there are more cornuti and the

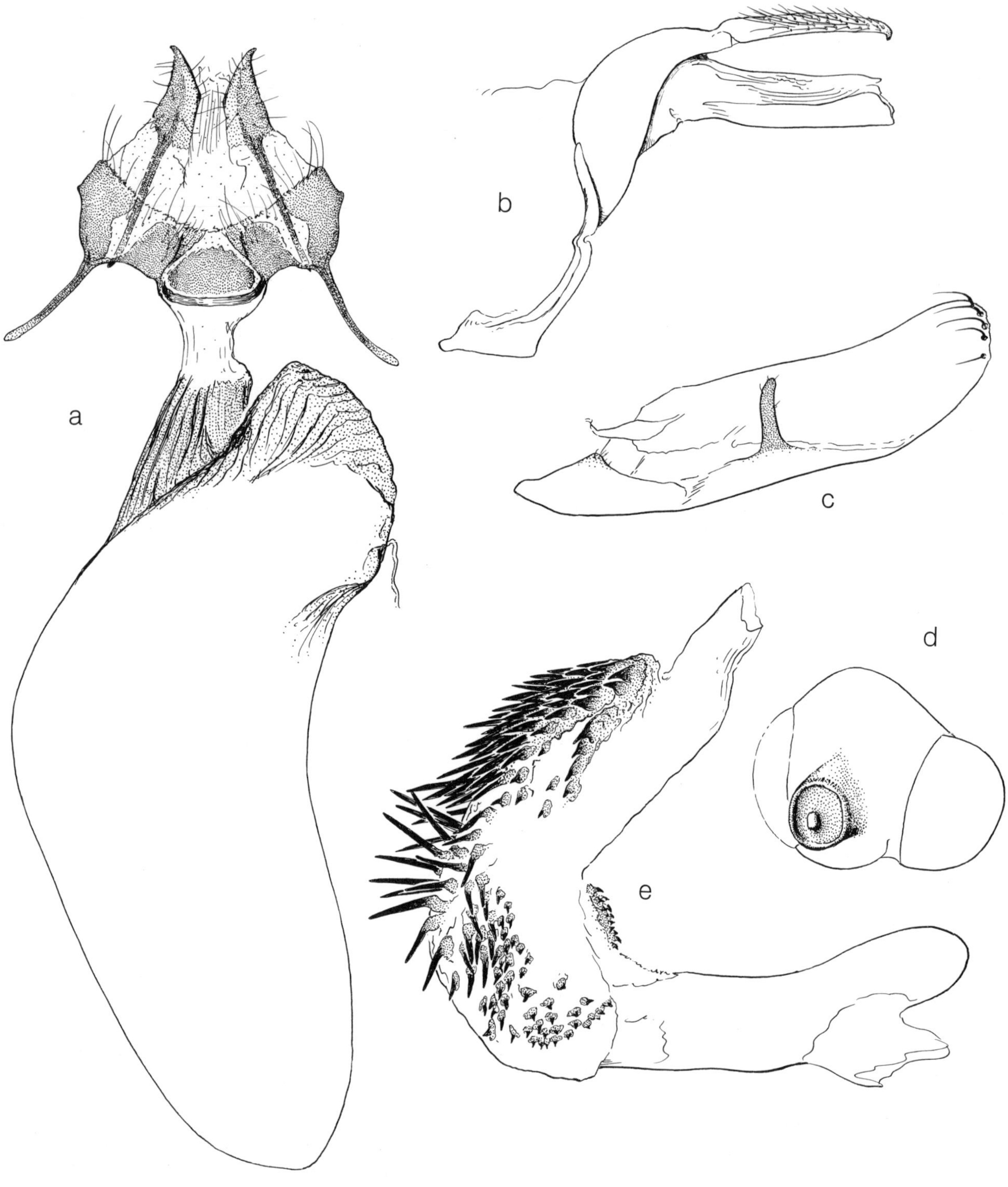

FIGURE 113: STRUCTURAL FEATURES OF *CHALCOPASTA TERRITANS*
a. Female genitalia. *b.* Lateral view of male genitalia. *c.* Right valve of male genitalia. *d.* Head. *e.* Aedoeagus.

basal setal knob is more denticulate in *fulgens* than *howardi.* The female genitalia are similar in all three species. The ovipositor lobes are relatively short and the tips recurved in *territans* but are longer and more nearly straight in the other two species. Wing length: mean = 16.56 mm, standard deviation = 1.01 mm, n = 10.

The larva and its foodplants are unknown.

FIGURE 114: DISTRIBUTION OF SPECIMENS EXAMINED OF *CHALCOPASTA TERRITANS*

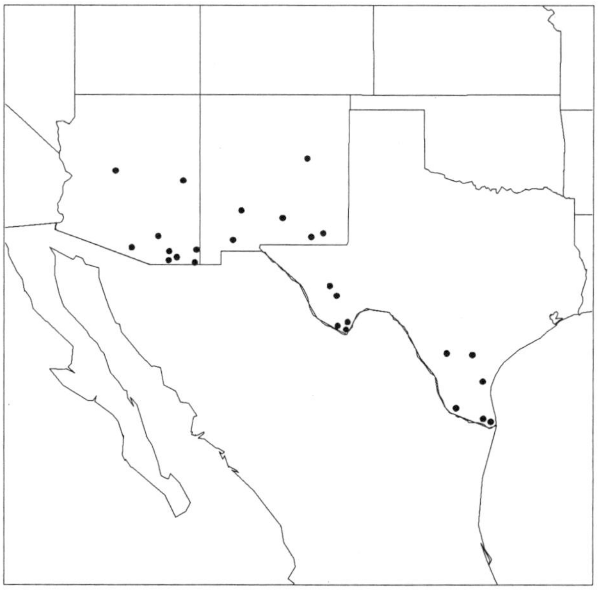

FIGURE 115: DISTRIBUTION OF SPECIMENS EXAMINED OF *CHALCOPASTA HOWARDI*

Chalcopasta territans has been collected in Cochise, Pima, and Santa Cruz Counties in southern Arizona. It has also been collected commonly in the trans-volcanic belt in the general vicinity of Mexico City and in northern Sonora. It undoubtedly occurs more widely in Mexico, but western Mexico has not been well collected. Most adults were captured in August and September.

Chalcopasta howardi (Hy. Edwards)
PL. 4, FIGS. 42, 43; PL. N, FIGS. 7, 8 (♂ gen.); PL. R, FIG. 6 (♀ gen.); TEXT FIG. 115 (map) (RWH 9776).

Plusia howardi Hy. Edwards, 1877, *Pacific Coast Lepidopterist,* no. 25, p. 1.
Type locality: [USA], Arizona, Prescott. [AMNH]

Deva ornata Ottolengui, 1898, *Can. Ent.,* **30**: 106, pl. 5, fig. 1.
Type locality: [USA], New Mexico, Hot Springs. [USNM]

Chalcopasta houardi Hampson, 1910, *Catalogue of the Lepidoptera Phalaenae in the British Museum,* **9**: 219.
Type locality: [USA], Arizona, Prescott. [AMNH]

NOTE—*Chalcopasta houardi* is an unjustified emendation of *howardi* and has the same type and type locality.

Chalcopasta howardi is the first of two species with a triangular brown area at the base of the forewing. It looks like *territans* superficially, but the basal brown patch is absent in *territans* or, at most, very small. The closest relative of *howardi* appears to be *fulgens.* The forewing of *fulgens* is suffused with white giving it a soft appearance and making the metallic green yellower and duller than in *howardi.* Wing length: mean = 16.31 mm, standard deviation = 0.68 mm, n = 10.

The larva and its foodplants are unknown.

Chalcopasta howardi has been collected in most of southern and central Arizona, but not from the western part of the state or counties north of the Mogollon Rim. It has been collected from only the southern counties of New Mexico, throughout the southern part of Texas, eastward as far as Brownsville, and only in the state of San Luis Potosi, Mexico. Adults have been captured in April and May and July through October.

Chalopasta fulgens Barnes and McDunnough
PL. 4, FIG. 44; PL. N, FIGS. 9, 10 (♂ gen.); PL. R, FIG. 7 (♀ gen.); TEXT FIG. 116 (map) (RWH 9777).

Chalcopasta fulgens Barnes and McDunnough, 1912, *Contrib. Nat. Hist. Lep. N. Am.,* **1**(4): pl. 25, fig. 3 [1 July]; **1**(5): 25 [10 July].
Type locality: [USA], Arizona, Redington. [USNM]

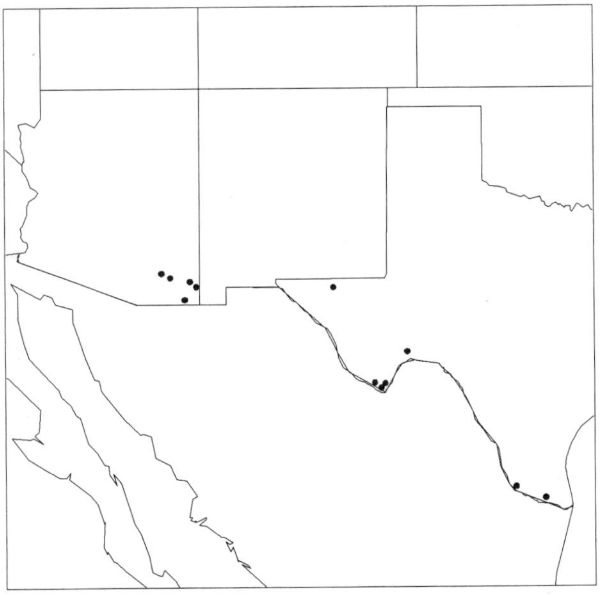

FIGURE 116: DISTRIBUTION OF SPECIMENS EXAMINED OF *CHALCOPASTA FULGENS*

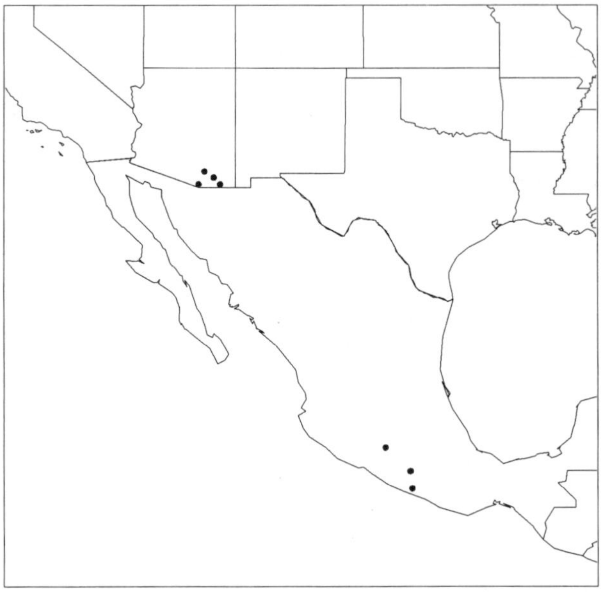

FIGURE 117: DISTRIBUTION OF SPECIMENS EXAMINED OF *CHALCOPASTA ACEMA*

NOTE—The earlier figure of the "type male" is an indication in the sense of the code. Therefore the name must date from the earlier reference. The specimen labeled "type ♂" is selected and has been labeled LECTOTYPE.

Chalcopasta fulgens looks like *howardi* but appears as if the forewing were covered with speckles of white paint. Consequently the metallic green of the forewing is slightly yellower and duller than in the other two species of the group. The vesica is similar to that of *howardi,* but a larger number of cornuti is present in *fulgens,* and the knob at the base of the vesica formed by the fusion of the basal setal group is more denticulate. Wing length: mean = 15.96 mm, standard deviation = 0.54 mm, n = 10.

The larva and its foodplants are unknown.

Chalcopasta fulgens has essentially the same distribution as *howardi.* The species is known from southeastern Arizona. It has not been recorded from southern New Mexico, but undoubtedly occurs there as well. The species occurs in the western and southern counties of Texas and has been collected near the southeastern tip of the state. I have not seen any specimens from Mexico. Adults fly in the autumn from August through October.

acema GROUP

The *acema* group consists of three species, *acema* (Druce), *chalcotoxum* (Dyar), and *acantha* (Druce). All three species are Mexican and only *acema* reaches the United States. These species resemble the *territans* group because of the metallic-green and brown coloration of the forewing. However, they are slightly smaller, and the brown areas of the forewing are much darker. The male genitalia are structurally unique. The inner margin of the valve is produced into a digituslike spine (plate N, figure 11). The dark part of the eighth sternum of the male abdomen is closed, not U-shaped, and the distal edge is slightly sclerotized.

Chalcopasta acema (Druce)

PL. 4, FIG. 45; PL. N, FIGS. 11, 12 (♂ gen.); PL. R, FIG. 8 (♀ gen.); TEXT FIG. 117 (map), (RWH 9778).

Plusia acema Druce, 1889, *in* Godman and Salvin, *Biologia Centrali-Americana, Insecta, Lepidoptera-Heterocera,* **1**: 330, pl. 30, fig. 14. Type locality: Mexico, Guerrero, Tierra Colorada. [BMNH]

Chalcopasta acema resembles the species of the *territans* group but is smaller, and the brown areas of the forewing are darker and more extensive. Its closest relative is the Mexican species *Chalcopasta chalcotoxum* (Dyar). *Chalcopasta chalcotoxum* is not known from the United States but could conceivably be found in the U. S. some day. The two species are very close, particularly females; however, the forewing of *acema* appears rounder, and the green band on the forewing is broader and has more regular edges. The cornuti of the apex of the vesica are larger and more splayed in *acema* than in *chalcotoxum*. Wing length: mean = 13.43 mm, standard deviation = 0.89 mm, n = 7.

The larva and its foodplants are unknown.

Chalcopasta acema has been collected only in southern Arizona (Cochise and Santa Cruz Counties) in the United States. It is known from Guerrero and Michoacan in Mexico, but probably occurs throughout most the western dry areas of the country. Adults have been collected in July and August.

INCERTAE SEDIS

I have included under the heading *incertae sedis* a small number of genera whose phylogenetic positions I have not been able to determine. Some may be related to the Stiriinae, some may not. All bear some superficial resemblance to the Stiriinae and are placed here until their phylogenetic positions can be determined more exactly.

GENUS

Neumoegenia Grote

Neumoegenia Grote, 1882, *Papilio,* **2**: 132.
Type species: *Neumoegenia poetica* Grote, 1882. Monotypy.

Kallitrichia Ottolengui, 1898, *Can. Ent.,* **30**: 107.
Type species: *Kallitrichia pendula* Ottolengui, 1898. Subsequent designation by Hampson, 1910, *Catalogue of the Lepidoptera Phalaenae in the British Museum,* **9**: 224.

Trichocala Dyar, [1903] 1902, *Bull. U. S. Natl. Mus.,* **52**: 196.
Type species: *Kallitrichia pendula* Ottolengui, 1898. Subsequent designation for *Kallitrichia.*
NOTE—*Trichocala* is an unnecessary objective replacement name for *Kallitrichia* Ottolengui 1898.

The genus *Neumoegenia* is a small group of distinctive species. All are characterized by a basic white coloration overlain with extensive areas of metallic-green scales. A single species, *Neumoegenia poetica* Grote, occurs in the United States. *Neumoegenia poetica* superficially resembles an aberrant species of *Chalcopasta* because of the extensive areas of metallic green on the forewing and the presence of a basal brown patch. However, neither the male or female genitalia show any relationship to the species in the Stiriinae. A frontal ring is present but is represented only by a dorsal arc. A prothoracic tibial claw is almost always absent although a deciduous claw exists in some specimens, mostly newly emerged ones. This deciduous claw was noted by J. G. Franclemont (personal communication). Although a calosity is present at the apex of the tibia, I have found a claw in very few of the several hundred specimens in the USNM collection of this species.

Unfortunately, three of the type specimens in the genus are based on mislabeled specimens including the type of *Kallitrichia sagittalba* Ottolengui, which has been listed as a synonym of *poetica.* All of these mislabeled specimens have a typeset label reading "Arizona." These specimens are Mexican in origin. The labels are identical and probably all came from a single source. The types of *Kallitrichia albavena* Ottolengui, *Kallitrichia pendula* Ottolengui, and *Kallitrichia sagittalba* Ottolengui all bear these "Arizona" labels and all are mislabeled Mexican specimens. Similarly labeled specimens clearly of Mexican origin exist in other genera. For example, there are two specimens of the Mexican species *Chalcopasta sinuata* Hampson in the USNM bearing these false "Arizona" labels. Because of the confusion caused by these incorrect locality labels, a summary of the species in *Neumoegenia* follows.

Neumoegenia poetica Grote (southern Arizona).
Neumoegenia smithi Druce, 1889 (Mexico), REVISED STATUS.
 Kallitrichia sagittalba Ottolengui, 1898 (type mislabeled Arizona), NEW SYNONYMY.
Neumoegenia bellamusa Dyar, 1898 (*Neumoegenia albavena* subsp.) (Mexico), NEW STATUS.
Neumoegenia albavena (Ottolengui), 1898 (*Kallitrichia albavena*) (Mexico) (type mislabeled Arizona).
Neumoegenia coronides Druce, 1889 (*Plusia coronides*) (Mexico).
 Kallitrichia pendula Ottolengui, 1898 (type mislabeled Arizona).

The effect of this correction is to remove two of the species listed in Hodges et al. (1983) from the faunal lists of the United States, i.e., *Neumoegenia coronides* (Druce) (RWH 9738) and *Neumoegenia albavena* (Ottolengui) (RWH 9739), leaving only *poetica* to represent the genus north of the Mexican border.

Two of these species, the U. S. *poetica* and the Mexican *smithi,* are closely similar superficially. Although I have not seen *smithi* from the United States or *poetica* from south of the Mexican border, *poetica* undoutedly occurs in northern Mexico and conceivably *smithi* may some day be found in the southern United States. The white mark in the metallic-green part of the forewing is larger in *smithi* than in *poetica. Neumoegenia smithi* tends to be slightly larger than *poetica*; however, some geographic variation in size exists in *smithi,* and its overall larger size is not a totally reliable character. Several good differences between *poetica* and *smithi* are present in the male genitalia. A group of rounded cornuti with their bases fused is present near the base of the vesica in *poetica*; the vesica lacks cornuti of any type in *smithi.* The ductus bursae is about 3 × longer than wide in *poetica,* but is about as long as wide in *smithi.* The ostium has a ventrally projecting lip in *poetica,* but this lip is absent in *smithi.*

The formal description follows. Head: antenna of male and female simple, scaled above, naked, but finely ciliate below; no lashes from base of antenna or rear margin of eye; ocellus present; eye large, without hair; occiput of head closely covered with flat, white scales; no tufts of scales between antennal bases; front of head with large circular ring, but ring represented only by a dorsal raised arc, rest of ring absent; center of front slightly swollen; clypeus produced into shelf; palpus unmodified, first and second segments with ventral fringe of hairlike scales, strongest at base of first segment; third segment weak, buried in scales of second segment; proboscis unmodified. Thorax: wing venation identical to that of the Stiriinae; no sexual modifications of forewing or hindwing; wing shape ovate; hindwing with M_2 strong; patagium and front half of thorax with closely appressed white scales; weak tufts of gray-tinged hairlike scales from tegula; metathoracic tuft present, comprised of broad, flattened, glossy gray scales. Prothoracic leg: femur and tibia without tufts or ridges of scales, covered with closely appressed scales; femur 3 × longer than tibia; tibia usually without apparent tibial claw, but deciduous tibial claw present in some specimens, particularly freshly emerged ones, and represented by calosity in almost all museum specimens (J. G. Franclemont, personal communication); tibial claw thin and strongly curved; first tarsal segment approximately equal in length to remaining four; first four tarsal segments with three vague rows of spines, but spines somewhat hidden in scales on ventral surface of segment; fifth segment without ventral rows of setae but with a few scattered long lateral setae; tarsal claw without hook. Mesothoracic and metathoracic leg: length of tibia relative to femur about 0.75 in mesothoracic leg and 1.3 in metathoracic leg; tibia with some longer hairlike scales forming ill-defined fringe; tarsal segments as in prothoracic leg, but fifth segment with three rows of spines, although sometimes poorly defined. External tympanic structure: hood strong, located dorsal, not caudal, of spiracle; alula weak but well defined by large, flattened scales; no bullae present in intersegmental membrane. Internal tympanic structure: not examined. Male abdomen: segments without dorsal tufts, covered with closely appressed flat scales; basal hair pencils and Stobbe's glands absent; eighth tergum with V-shaped sclerotization; eighth sternum with U-shaped sclerotization and strong central row of long hairlike scales, possibly pheremonal in function. Male genitalia: plate N, figures 13, 14; valve long, rectangular with strong apical corona; corona followed by field of long, fine hairs; clasper strong, medially placed, slightly curved and with apex flattened; basal process of sacculus present, about 1.5 × longer than wide; uncus with strong medial swelling followed by a strong apical hook; tegumen unmodified; free pleurite present, straplike; saccus of vinculum rounded at apex; juxta consisting of two opposed, but separate, triangular plates; transtilla present, but weak; vesica (in type species *poetica*) globular, without cornuti except for basal group of stubby cornuti with rounded apices and fused together at their bases. Female genitalia: plate R, figure 9; ovipositor lobes unmodified; eighth sternum with a strong, concave bulla to either side of ostium and before the apophyses anteriores; ostium strongly sclerotized with a ventrally projecting lip (in *poetica*); ductus bursae (in *poetica*) strong, sclerotized, about twice as long as wide and with internal plicae; corpus bursae elongate-ovate without signa; appendix bursae present as a separate membranous lobe near junction of ductus bursae and corpus bursae; seventh sternum and tergum of abdomen not modified or especially sclerotized.

There is considerable variation in the spining patterns of the vesica among species. In particular, the Mexican species *albavena* and *coronides* have a spining pattern considerably different than that of *poetica* or *smithi*; however, *Neumoegenia bellamusa* has an intermediate pattern, and I have retained the current synonymy of *Kallitrichia* with *Neumoegenia.*

Neumoegenia poetica Grote

PL. 4, FIG. 46; PL. N, FIGS. 13, 14 (♂ gen.); PL. R, FIG. 9 (♀ gen.); TEXT FIG. 118 (map) (RWH 9737).

Neumoegenia poetica Grote, 1882, *Papilio,* **2**: 132.
Type locality: USA, Arizona. [USNM]

Neumoegenia poetica is a small, but distinctive, species from southern Arizona. The forewing is ovate and mostly covered with metallic-green scales. The dorsal one-fourth of the wing along the costa is white and contrasts strongly with the rest of the wing. A basal brown patch is present near the base and inner margin, and a roughly triangular white patch surrounded by metallic green exists in the center of the wing. This triangular patch is elongated along the veins at a position corresponding to the lower part of the reniform mark if it were present. All other maculation is absent. The hindwing is tinged with light dull gray. Wing length: mean = 12.16 mm, standard deviation = 0.40 mm, n = 10.

The larva has not been described. John G. Franclemont, however, has reared the species from *Bidens leptocephala* Sherff (Asteraceae) and has kindly provided me with color slides of the larvae. The larva is gray with a slight pink tinge from the dorsum to below the level of the spiracles. The ventral region is darker and browner. Dorsal and subdorsal pink lines are accented with brown on either side. A strong, contrasting yellow, subspiracular line accented with black is present. The pinacula are yellow and surrounded by a black ring. Dorsal chevrons of dark gray are present on abdominal segments one, two, seven, and eight. The eighth segment is humped. The head is not visible in the photographs.

Neumoegenia poetica is known only from Pinal, Pima, Santa Cruz, and Cochise Counties in southern Arizona although it is apparently common within its range. It probably occurs in western New Mexico and northern Mexico, but I have not seen any specimens from these localities. All of the Mexican specimens I have seen that might have been this species are the closely related sister species *smithi.* Some variation exists in *poetica,* particularly in the development of the forewing basal brown patch and the size of the median triangular white spot. Adults have been collected in July and August.

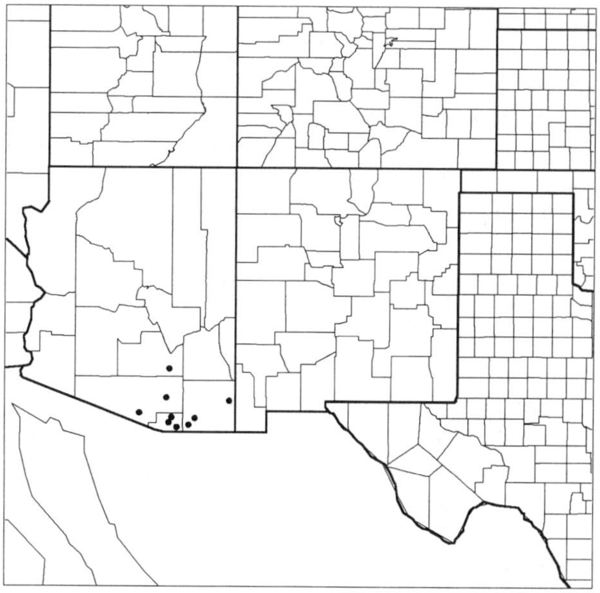

FIGURE 118: DISTRIBUTION OF SPECIMENS EXAMINED OF *NEUMOEGENIA POETICA*

GENUS
Argentostiria Poole, NEW GENUS

Argentostiria Poole.
Type species: *Antaplaga koebelei* Riley, 1893.

Argentostiria is monotypic containing only the type species *koebelei.* This species was placed in the stiriine genus *Chalcopasta* in Hodges et al. (1983) but with a notation that it was probably misplaced. *Argentostiria koebelei* is distinctive with a shiny white ground color and an overlying pattern of tan to yellow-tan bands. The phylogenetic placement of this genus is a mystery. *Argentostiria koebelei* has a strong superficial resemblance to the Stiriinae. I suspect its real affinities lie in the subfamily Psaphidinae, possibly within the tribe Triocnemidini. The strongest argument supporting this placement is the presence of a strong prothoracic tibial claw with an associated plate. This associated plate is found in many, if not most, species of the Psaphidinae, particularly in the tribes Psaphidini and Triocnemidini. The stiriine tibial claw is located in a socket and never occurs with an associated plate. A single mass of cornuti is present in the vesica and the bases of these cornuti combine to form a rugose region. No basal group of cornuti, as exists in most of the Stiriinae, is present. Some species of the Stiriinae, however, lack a basal spine group and have a structure similar to that found in this genus. The uncus is long and narrow. A long, narrow uncus is commonly found in the Psaphidinae but never in the Stiriinae. A frontal ring is found in both the Stiriinae and the Psaphidinae. The strongly sclerotized ovipositor lobes of *Argentostiria* are most characteristic of the Stiriinae and are rarely found in the Psaphidinae except in the tribe Nocloini. Strongly sclerotized ovipositor lobes, however, are commonly found in flower feeding noctuid species and could easily be an independent development with little phylogenetic information. The ductus seminalis arises laterally on the corpus bursae in *Argentostiria,* a condition commonly seen in the Triocnemidini and Nocloini but not the Stiriinae. The ductus bursae arises from the apex of the appendix bursae in the Stiriinae. A strong piece of evidence against placing *Argentostiria* in the Triocnemidini is the absence of an incomplete outer wall of the abdominal spiracle, one of the structures used to define the Triocnemidini.

The morphological information indicates a closer relationship between *Argentostiria* and the Triocnemidini than with the Stiriinae. The evidence, however, is equivocal. The larva, when it is discovered, may help to clarify the situation. I have placed *Argentostiria* nearer the Stiriinae, for the moment, because it superfically appears to belong to this group of the Noctuidae.

The formal description follows. Head: antenna simple, scaled dorsally, finely ciliate ventrally; lashes absent; eye normal, without hairs; vestiture of head of broad scales, without distinct tufts or ridges; frontal process present, rectangular; lower margin of frontal process confluent with clypeus, upper margin interrupted in middle; central, laterally compressed tubercle present, dorsal point of tubercle elongate and blending with dorsal opening of ring; palpus with short third segment, second segment with lower fringe of broad scales; palpus closely apressed against head; proboscis normal. Thorax: wing venation as in the Stiriinae; dorsal vestiture of broad scales, without tufts or ridges except for usual metascutal tuft. Prothoracic leg: femur about 6 × longer than wide, closely scaled, without hair, scale ridges or tufts; tibia short, about twice as long as wide; tibial claw present, strong, continued as a plate on upper posterior margin of tibia; tibial spines absent; first tarsal segment about 2.5 × as long as second; tarsal segments with three irregular rows of spines. Mesothoracic and metathoracic legs: femur and tibia about 5 × longer than wide, closely scaled, without ridges or tufts of scales or hairs; otherwise as in prothoracic leg. Internal tympanic region: not examined. External tympanic region: hood very strong; first intersegmental region caudal to hood without bullae or depressions; first abdominal tergum narrow with cephalic projection; alula strong. Male abdomen: dorsum without tufts of scales; sclerites strongly sclerotized; sclerotized pattern of eighth sternum U-shaped; eighth tergum V-shaped with small quadrate projections at cephalad margin; basal hair pencils and Stobbe's glands absent; shape of sternum 1+2 more suggestive of clade without potential hair pencils than clade potentially with hair pencils. Male genitalia: plate N, figure 15; plate O, figure 1; uncus elongate, not swollen, covered with small hairs; free pleurite present, only narrowly articulated with tegumen; valve simple, rectangular, broad; ventral margin of valve a narrow ridge; clasper absent; corona absent; basal process of sacculus absent; juxta a simple trapezoidal plate; saccus of vinculum ovate; vesica elongate, arising at right-angle to body of aedoeagus; a single large mass of cornuti with bases of cornuti fused into a rippled mass; no basal spine group as in Stiriinae. Female genitalia: plate R, figure 10; ovipositor lobes very long, strongly sclerotized, scimitar shaped with pointed apices and curved ventrally; lobes with small hairs; apophyses posteriores short relative to length of ovipositor lobes but very broad; ovipositor neck heavily covered with small hairs, slightly leathery; eighth sternum narrow, strongly sclerotized cephalad, less so caudally; ostium a simple, membranous region, with ventral lip of ostium continuous with membrane of seventh abdominal sternum (as in Psaphidinae); ductus bursae 4–5 × longer than wide with sclerotized pleats over lower ⅔ of its length; appendix bursae region large, about 1.5 × as long as wide; ductus seminalis arising laterally at level with junction of corpus bursae with ductus bursae; corpus bursae without signum, about twice as long as wide. Larva: Foodplants and larva unknown.

Argentostiria koebelei (Riley), NEW COMBINATION

PL. 4, FIG. 52; PL. N, FIG. 15; PL. O, FIG. 1 (♂ gen.); PL. R, FIG. 10 (♀ gen.); TEXT FIG. 119 (map) (RWH 9779).

Antaplaga koebelei Riley, 1893, *Insect Life,* **5**: 333, fig. 5.
Type locality: [USA], California, [Inyo County], Argus Mountains. [USNM]

NOTE—*Antaplaga koebelei* was based on three syntypes. The syntype bearing the USNM type number 280 has been labeled and is hereby designated as LECTOTYPE.

Chalcopasta coebelei Hampson, 1910, *Catalogue of the Lepidoptera Phalaenae in the British Museum,* **9**: 222.
Type locality: [USA], California, [Inyo County], Argus Mountains. [USNM]

NOTE—*Chalcopasta coebelei* is an unjustified emendation of *Antaplaga koebelei* Riley, 1893 and has the same type and type locality.

Argentostiria koebelei is another highly distinctive species. The forewing is shiny, almost metallic, white, with a strong orange-tan postmedial line angled outward at the reniform mark and connected by a thin band to a yellowish terminal region. The reniform mark is a small black dot. A strong orange-brown antemedial line is present. The postmedial line is yellowish in the center and darker along either edge. The apex of the wing is pointed. The hindwing is white and slightly tinted with orange. The prothoracic tibia has a strong claw, and the claw is associated with a plate along the distal edge of the tibia. Wing length: mean = 12.27 mm, standard deviation = 0.53 mm, n = 10.

The larva and its foodplants are unknown.

Argentostiria koebelei is primarily a denizen of the Mohave Desert region and has been found thoughout most of southern California, southern Nevada, and western Arizona. A single specimen has also been collected in northwestern Nevada. Little or no variation exists in the species. Adults have been collected from late March to mid-June.

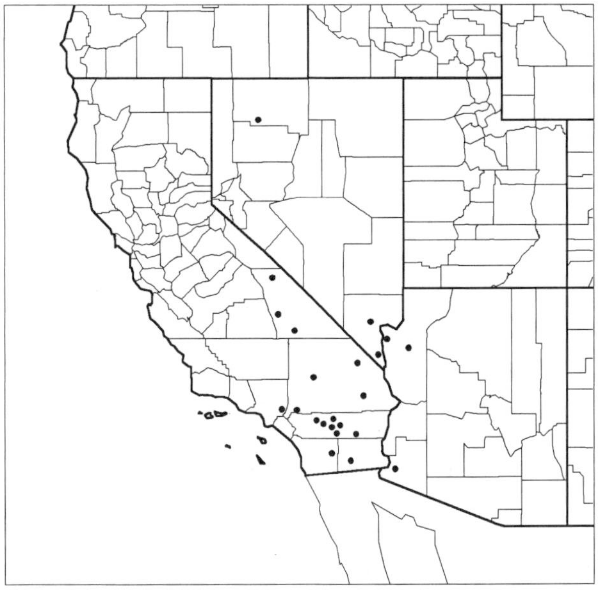

FIGURE 119: DISTRIBUTION OF SPECIMENS EXAMINED OF *ARGENTOSTIRIA KOEBELEI*

GENUS

Chamaeclea Grote

> *Chamaeclea* Grote, 1883, *Can. Ent.*, **15**: 76.
> Type species: *Chariclea pernana* Grote, 1883.
> Original designation.

Chamaeclea is a small genus with only two species. They are small with a yellow-cream ground color and strong purple or violet tints along the antemedial and postmedial lines. The correct phylogenetic placement of this genus is undetermined. The genus is placed near the Stiriinae because of its superficial appearance; however, it is clearly not a stiriine on structural evidence. The male genitalia are not typical of any noctuid group with which I am familiar and appear to be simplified as well. The ovipositor lobes are elongate and strongly sclerotized, but this is a recurrent modification found in flower-feeding noctuids, presumably an adaptation for laying eggs within flower buds. The interior of the lumen of the corpus bursae has some peculiar, small toothlike projections. The larva is unknown.

The formal description follows. Head: antenna simple, scaled above, finely ciliate below; lashes from base of antenna absent; eye normal, without hair; vestiture of occiput and front of ovate scales without distinct tufts or ridges; frontal process present, circular, small, located near dorsal margin of front, interrupted ventrally and with a dorsoventral central ridge, ridge running from center of process to dorsal margin of ring; palpus slightly reduced in size, thin; second segment with only a weak ventral fringe of scales, third segment about 1.5 × as long as wide; proboscis normal. Thorax: vestiture of strongly toothed scales, hairs absent; scales of posterior end of tegula rough, curved, giving rise to a spreading tuft; weak metascutal tuft present. Wing venation: as in the Stiriinae. Prothoracic leg: femur about 5 × longer than wide; vestiture of femur of ovate scales with only a very weak ventral fringe of scales; tibia about 3 × longer than wide; vestiture of tibia of longer scales than on femur; tibial claw absent; tibial spines absent; first tarsal segment as long as remaining four; tarsal spining of first tarsal segment irregular, confined to apical fifth of segment; tarsal spining on remaining tarsal segments irregular, spines mostly hairlike, although lateral spines stronger than inner spines. Mesothoracic and metathoracic legs: femur and tibia approximately equal in length; tarsal spining stronger than on prothoracic leg, strongly irregular and with no indication of three rows. Internal tympanic region: not examined. External tympanic structure: alula absent; hood absent; intersegmental membrane swollen ventral to spiracle, swelling curving upward around position of spiracle in a fat, C-shaped bulla. Male abdomen: dorsum without tufts; vestiture of scales, hair absent; basal hair pencils and Stobbe's glands absent; structure of sternum 1+2 most like species without potential for hair pencils; eighth sternum U-shaped with a weak,

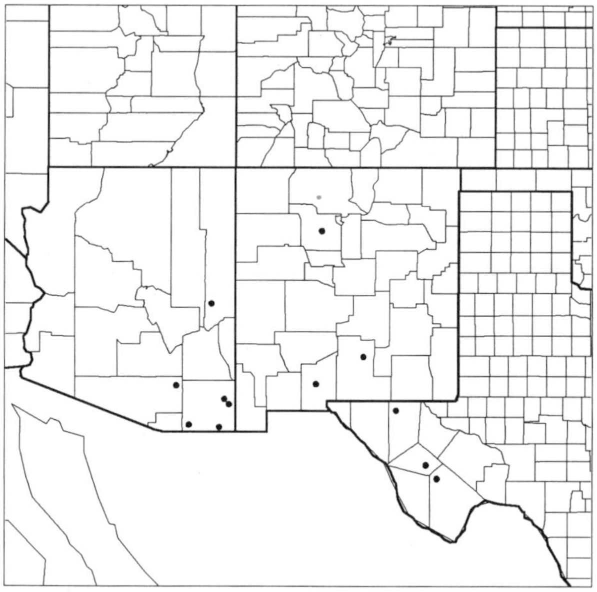

FIGURE 120: DISTRIBUTION OF SPECIMENS EXAMINED OF *CHAMAECLEA PERNANA*

but concentrated, hair ridge; eighth tergum without definite pattern of sclerotization. Male genitalia: plate O, figures 2, 3; uncus long, without swelling or long hairs; weak apical tooth present; free pleurite present, unmodified; valve simple; corona absent; clasper or clasper region an elongate ovate plate with an irregular dorsal margin, shape depending on species; basal process of sacculus absent; juxta diamond shaped, with a narrow, dorsal V-shaped indentation; saccus of vinculum triangular; vesica globular with a distal diverticulum covered with coarse hairs; a scobinate bar present at base of vesica. Female genitalia: plate R, figure 11; ovipositor lobes elongate, strongly sclerotized; apophyses posteriores and anteriores both long and strongly developed; ostium membranous; ductus bursae tubular and strongly sclerotized at its caudal end; appendix bursae a strong lobe, located to left of ductus bursae in ventral orientation of genitalia; corpus bursae very long, about 8 × longer than wide; interior of lumen of cephalic half of corpus bursae with small teeth. Larva and foodplants unknown.

Chamaeclea pernana (Grote)
PL. 4, FIGS. 47, 48; PL. O, FIGS. 2, 3 (♂ gen.); PL. R, FIG. 11 (♀ gen.); TEXT FIG. 120 (map), (RWH 9789).

Chariclea pernana Grote, 1881, *Papilio*, **1**: 155.
Type locality: [USA], Arizona. [USNM]

NOTE—*Chariclea pernana* was based on an unspecified number of specimens from the Neumoegen collection. There are three syntypes in the USNM, two males and one female. The type labels on all three specimens are different in shape and hand writing. The female syntype has the head of an unrelated species glued to it. One of the male syntypes bears the same [Arizona] label that has proven so problematical in *Neumoegenia*. The male syntype bearing the locality label "Ariz." has been labeled and is here designated as LECTOTYPE.

Chamaeclea pernana is a small species ranging from southern Arizona to western Texas. The forewing ground color is yellow cream with strong purple tints along the antemedial and postmedial lines and a slightly duller brown-purple tint in the region of the subterminal line. The purple tint is strongest along the lower half of the antemedial line and defines a strong contrast between the light basal area and the inner half of the median area. The antemedial line runs outward from the costa, is produced into two large teeth, and then runs inward at a strong slant to the inner margin. The lower of the two teeth does not reach the postmedial line. The other species in the genus, *basiochrea*, is superficially similar; however, the lower tooth of the antemedial line appears to join an inward pointing tooth of the postmedial line in *basiochrea*. *Chamaeclea basiochrea* is a slightly darker species than *pernana*. The best differentiating character between the two species is found in the hindwing. The hindwing of *pernana* is uniformly dull, light brown. The base of the hindwing is shaded with dull orange in *basiochrea*, and it contrasts with the darker brown outer half although no clear demarcation between the inner and outer color exists. The primary difference in the male genitalia is in the shape of the clasper. The clasper of *pernana* is a simple, rounded plate. The apical part of the clasper is produced into a small hook in *basiochrea*. There are major differences between the female genitalia of the two species. The ostium and the caudal ¼ of the ductus bursae of *basiochrea* form a Y-shaped structure with a strongly sclerotized tubular base. The Y-shape is created by a deep cephalic extension of the membrane in the ostium region. This Y-shaped structure is absent in *pernana*, and the ostium is unmodified and opens directly to a short tubular, sclerotized caudal section of the ductus bursae. The ductus bursae is approximately 3 × longer than wide in *pernana* but at least 5 × longer than wide in *basiochrea*. A very narrow constriction of the corpus bursae is present in *basiochrea* following the intersection of the ductus and corpus bursae. Although this constriction is present in *pernana,* it is not nearly as strongly pronounced. Wing length: mean = 11.96 mm, standard deviation = 0.73 mm, n = 10.

The larva and its foodplants are unknown. The elongate, sclerotized ovipositor lobes are characteristic of a flower-feeding species, and the violet tint of the forewing suggests that a purple composite would be a good first guess at a foodplant.

Chamaeclea pernana has a moderately wide distribution in the southwestern United States. It has been collected in southeastern Arizona, extending as far north as the White Mountains. The species has been taken widely in New Mexico, and reaches its eastern limit in the western counties of Texas. I have not seen *pernana* from Mexico, but it undoubtedly occurs in the northern states adjacent to the known U. S. range of the species. *Chamaeclea pernana* is not variable, although females are slightly larger, on average, than males. Adults have been collected in July and August.

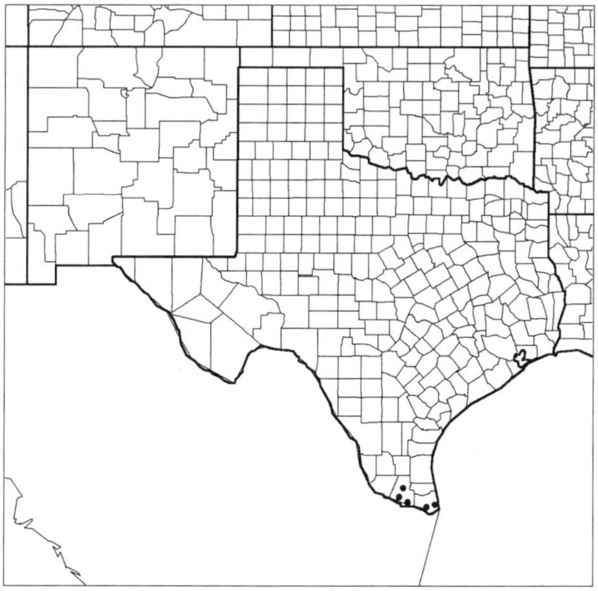

FIGURE 121: DISTRIBUTION OF SPECIMENS EXAMINED OF *CHAMAECLEA BASIOCHREA*

Chamaeclea basiochrea Barnes and McDunnough

PL. 4, FIGS. 49, 50; PL. O, FIGS. 4, 5 (♂ gen.); PL. R, FIG. 12 (♀ gen.); TEXT FIG. 121 (map) (RWH 9790).

Chamaeclea basiochrea Barnes and McDunnough, 1916, *Contrib. Nat. Hist. Lep. N. Am.,* **3**(1): 12, pl. 3, fig. 22.
Type locality: [USA], Texas, Brownsville. [USNM]

NOTE—*Chamaeclea basiochrea* was based on one male and six females. The syntype labeled "type ♂" is not in good condition. Therefore the syntype labeled "type ♀" has been labeled and is here designated as LECTOTYPE.

Chamaeclea mapensa Dyar, 1919, *Insecutor Inscitiae Menstruus,* **7**: 80. NEW SYNONYMY.
Type locality: Mexico, Veracruz, Panuco. [USNM]

NOTE—The type locality is in extreme northern Veracruz near Tampico.

The characters separating *basiochrea* from *pernana* have been discussed under *pernana.* Wing length: mean = 11.67 mm, standard deviation = 0.88 mm, n = 10.

The larva and its foodplants are unknown.

Chamaeclea basiochrea has a very restricted distribution in the United States. It has been collected only in Cameron and Hidalgo Counties in extreme southern Texas. The only known Mexican specimens are from Panuco in northern Veracruz. *Chamaeclea basiochra* probably occurs throughout eastern Mexico, possibly along the coastal region. The distinctness of the orange base in the hindwing is variable and in some females may be almost entirely swamped by dark brown. Males tend to be slightly smaller, on average, than females. Adults have been collected in late September, October, and early November.

GENUS
Bistica Dyar

Bistica Dyar, 1912, *Proc. U. S. Natl. Mus.,* **42**: 67.
Type species: *Bistica noela* Dyar, 1912. Original designation.

Bistica is a small genus of two species. The type species, *Bistica noela,* is known from the Santa Rita Mountains in southern Arizona south to Guatemala. The second species is undescribed and so far represented by a small series of specimens from the Mexico City region of Mexico. The two species are very similar. *Bistica noela* is small with metallic-yellow forewings. The apex of the wing is falcate. A tibial claw is absent but there is a large, round frontal process with a small central process. The main morphological features characterizing the genus are found in the male and female genitalia. The vesica is tubular, not globular as in the Stiriinae, and not oriented at right angles to the shaft of the aedoeagus.

The cornuti in the vesica consist of two converging basal rows. The valve is relatively simple and lacks a corona. The clasper is apical and large. The base of the sacculus is enlarged into a projecting shelf. The saccus is elongate and projects backward from the base of the valves (as seen in a lateral view of the male genitalia). The juxta is a large ovate to rectangular plate with a lower, raised portion and a dorsal (toward the uncus) free plate in the region of the transtilla. The ovipositor neck and apophyses posteriores are extremely long. The ovipositor lobes are also elongate and strongly sclerotized. Both conditions are typical of species ovipositing in flowers although the biology of *noela* is unknown. The ostium is shielded by a strongly sclerotized, connate plate projecting caudally as a rectangular plate. The corpus bursae is elongate, 6–7 × longer than its widest part. The larva and its foodplants are unknown. The phylogenetic affinities of this genus are unknown.

The technical description follows. Head: antenna of male and female simple, scaled dorsally, finely ciliate below; apex of head covered with flat scales, not arranged in tufts or rows; front of head with a large circular ring, with a small ovate central process; eye naked, unlashed, normal in size; ocellus present; palpus roughly scaled, third segment very short, nearly hidden by projecting scales; proboscis normal. Thorax: dorsum clothed with elongate, flat, dentate scales without intermixed hairs; no tufts or ridges of scales present; wing venation as in the Stiriinae; dorsum of wing clothed with flat, closely appressed, ovate, dentate scales. Prothoracic leg: coxa very large, approximately 4 × longer than wide; femur about 6 × longer than wide, slightly longer than coxa, covered with ovate scales, but no hairs; tibia about 4 × longer than wide, roughly scaled; tibial claw absent; first tarsal segment very long, approximately ⅔ as long as remaining four segments; tarsal segments four and five very short, about as long as wide; tarsal spines very weak, completely covered by large scales on ventral surface of tarsal surfaces. Mesothoracic and metathoracic legs: femur very elongate, thin; tibiae also elongate, swollen toward apex; first tarsal segment approximately ½ as long as remaining four segments; tarsal spines weak, concealed by ventral scaling of tarsal segments. Internal tympanic structure: not examined. External tympanic region: hood absent, no bullae in intersegmental membrane of first abdominal segment; spiracle not covered; alula absent; tympanic membrane without nodular sclerite. Male abdomen: vestiture of closely apressed scales tufting absent; basal hair pencils and Stobbe's glands absent; eighth sternum with vague U-shaped sclerotization; eighth tergum with two weak rectangular bars of sclerotization. Male genitalia: plate O, figures 6, 7; uncus about 5 × longer than wide, with a weak ventral carina, covered with rough hairs, slightly swollen toward tip, and with a hook at tip; valve roughly rectangular with round apex and without corona; clasper apical, elongate, projecting beyond margin of valve; base of sacculus produced into a large, free plate; saccus elongate, projecting backward from vinculum at right angles; juxta a large rectangular plate with medioventral raised ridge and dorsal (near transtilla) free projecting plate; vesica tubular with two basal converging bands of spines. Female genitalia: plate S, figure 1; ovipositor neck and apophyses posteriores very elongate; ovipositor lobes elongate, but not large, with a few spines; basal half (cephalic) of ovipositor neck appearing glandular; apophyses anteriores elongate; ostium surround by cylindrical, sclerotized structure, projecting caudally in a bilobed plate; ductus bursae about 4 × longer than wide, caudal third sclerotized and with complex sclerotized, dentate bulla in cephalic third; corpus bursae elongate, about 6 × longer than wide, without signum but finely spiculate internally. Larva and larval foodplants unknown.

Bistica noela (Druce)

PL. 4, FIG. 51; PL. O, FIGS. 6, 7 (♂ gen.); PL. S, FIG. 1 (♀ gen.); TEXT FIG. 122 (map) (RWH 9736).

Mychonia noela Druce, 1892, *in* Godman and Salvin, *Biologia Centrali-Americana. Lepidoptera-Heterocera,* **2**: 69, pl. 47, fig. 28.
Type locality: Guatemala, [Baja Verapaz], San Geronimo [San Jeronimo]. BMNH

Bistica noela Dyar, 1912, *Proc. U. S. Natl. Mus.,* **42**: 67.
Type locality: Mexico, [Puebla], Tehuacan. [USNM]

NOTE—*Bistica noela* Dyar, 1912 is a junior secondary homonym of *Mychonia noela* Druce 1892.

Bistica noela is a distinctive little species with a dull-yellow forewing and a strong metallic sheen. The apex of the forewing is falcate. The reniform mark is a brown line, the line sometimes connecting two white spots. The inner ¾ of the median area are distinctly lighter and less metallic than the outer ¼, the basal area, and the subterminal area. A vague brown patch is present between the top of the reniform mark and the origin of the postmedial line. The metallic-yellow tints tend to be strongest in the subterminal and terminal areas. The hindwing is dull yellow and strongly infused with dull brown. Wing length: mean = 12.36 mm, standard deviation = 0.51 m, n = 2.

The larva and its foodplants are unknown.

Bistica noela has a wide, but spotty distribution. It is known in the United States from a small series of specimens from the Santa Rita Mountains of southern Arizona collected by J. G. Franclemont. *Bistica noela* is also known from the state of Puebla in Mexico and the state of Baja Verapaz in Guatemala. Guatemala and U. S. specimens are virtually identical, but the inner ¾ of the median area of the forewing in specimens from Puebla (Tehuacan) is not nearly as light as material from the U. S. Specimens are too few to make any comments about individual variability. The Arizona and Puebla specimens were collected in September.

FIGURE 122: DISTRIBUTION OF SPECIMENS EXAMINED OF *BISTICA NOELA*

GENUS
Hemioslaria Barnes and Benjamin

> *Hemioslaria* Barnes and Benjamin, 1924, *Contrib. Nat. Hist. Lep. N. Am.,* **5**: 167.
> Type species: *Hemioslaria pima* Barnes and Benjamin, 1924. Original designation.

The genus *Hemioslaria* contains the single species *Hemisolaria pima.* The moth looks like a gigantic version of the genus *Grotella,* but the two genera are probably not related. The genus is defined by a number of synapomorphies. The forewings are ovate, not rectangular. The most striking generic characters are found in the prothoracic leg. The first tarsal segment is scaled dorsally, naked ventrally, and it has two, very large recurved spines along the lateral margin. All five tarsal segments are covered ventrally with long hairlike spines without any apparent arrangement. The last three tarsal segments are naked dorsally and covered dorsally with the same hairlike setae. The external tympanic region has a number of peculiarities. In particular the hood is located dorsal to the spiracle, not caudally, and forms an umbrella over it. A rectangular bulla is present ventral to the spiracle, but no bulla exists on the caudal side of the spiracle. This structure of the external tympanic region may be an extreme development of the Stiriinae type. The hood in the Stiriinae, although located basically in a caudal position relative to the spiracle, has shifted upward and forward slightly.

The affinities of *Hemioslaria* are unknown, although it shares the hairlike setae of the tarsal segment with the genus *Heminocloa.* The long hairlike setae and their lack of arrangement of the legs is a feature shared with the tribe Psaphidini of the Psaphidinae (discussed shortly in this work). The shortened prothoracic tarsal segments indicate a possible connection with the Nocloini of the Psaphidinae. The overall habitus and, perhaps, the tympanic structure suggest a connection with the Stiriinae. *Hemioslaria* may represent a relict of some ancestral group that gave rise to the three evolutionary lines mentioned above.

Head: antenna of males and female simple, finely ciliate ventrally; segments large giving base of antenna a fat look; front of head with circular frontal process, process with raised rim, evaginated in middle of bottom margin with evagination continued as a ridge through center of frontal process; dorsum of front clothed with scales, not arranged in discrete tufts; eyes unmodified; palpus unmodified, second segment triangular; proboscis unmodified. Prothoracic leg: femur closely scaled, without tufts or long hairs; tibia about 3 × longer than wide, epiphysis very large, tip projecting beyond end of tibia; first two tarsal segments clothed with scales dorsally, last three segments naked dorsally, with long hairlike setae ventrally; lateral margin of first tarsal segment with two very large, recurved claws; ventral side of all five tarsal segments with a covering of long hairlike setae, setae not arranged in definite rows; tarsal

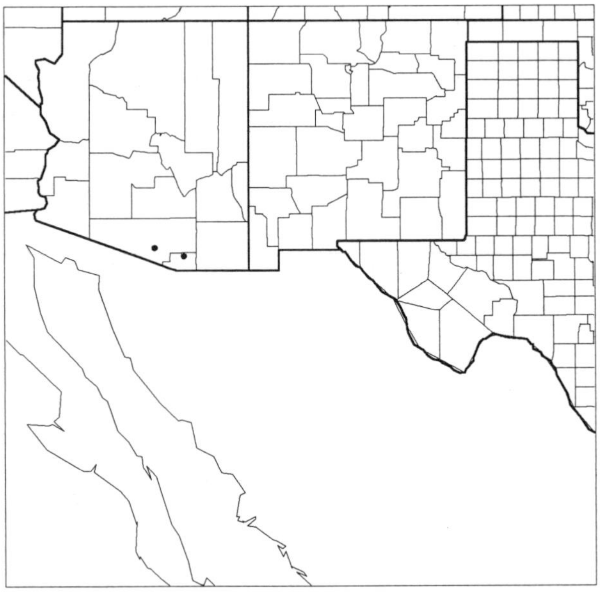

FIGURE 123: DISTRIBUTION OF SPECIMENS EXAMINED OF *HEMIOSLARIA PIMA*

claws without basal teeth; first four tarsal segments about as long as wide, fifth about 3 × longer than wide giving tarsal segments a short stocky appearance (as in the Nocloini). Mesothoracic leg: femur closely scaled; tibia long, triangular at apex, with vague tuft of scales at proximal end; first tarsal segment clothed with scales dorsally and ventrally, with three slightly irregular rows of spines; second segment naked ventrally with two lateral rows of spines, and with long hairlike setae medially; last three segments naked ventrally with only long hairlike setae; tarsal segments not shortened, at least twice as long as wide. Metathoracic leg: as in mesothoracic leg except for usual two pairs of spurs; tibia slightly more triangular. Thorax: vestiture mostly of long scales, but with a few hairlike scales; no tufts of any kind; forewing scales short, with small teeth, closely apressed to wing; forewing venation as in Stiriinae; hindwing with M_2 strong; cross veins mdc and ldc approximately equal in length. Internal tympanum: not examined. External tympanic region: alula strong; hood small, located dorsal to spiracle and forming a dorsal umbrella over spiracle; a dorsoventral elongate bulla in first abdominal segment located ventral to spiracle; no special bulla on caudal side of spiracle. Male abdomen: basal hair pencils and Stobbe's gland absent; no dorsal tufts; eighth sternum and tergum without any differential sclerotization except for caudal and cephalic margins. Male genitalia: plate O, figures 8, 9; uncus elongate, slightly swollen in apical half, with apical tooth; uncus without hairs or setae; free pleurite present, elongate, but weak, not clearly separated from vinculum; apex of saccus rounded; juxta a single trapezoidal plate; valve simple with caudally pointed clasper; apex of clasper rounded, projecting beyond costal margin of valve; apex of valve without spines; aedoeagal shaft about 4 × longer than wide; vesica globular with three terminal projections, one much larger than other two, all covered with fine hairs. Female genitalia: plate S, figure 2; ovipositor lobes basically unmodified but with setae at caudal end on sclerotized cones; ostium unmodified, but with spines from caudal margin of eighth sternum continuing down lateral sides of ostium; a double sclerotized ridge between the ovipositor lobes ventrally; ductus bursae unsclerotized, about 3 × longer than wide; corpus bursae very long, about 6 × longer than wide; caudal fourth of corpus bursae and appendix bursae rugose and slightly sclerotized; appendix bursae present, rectangular. Larva and foodplants: unknown.

Hemioslaria pima Barnes and Benjamin
PL. 4, FIG. 54; PL. O, FIGS. 8, 9 (♂ gen.); PL. S, FIG. 2 (♀ gen.); TEXT FIG. 123 (map) (RWH 9824).

Hemioslaria pima Barnes and Benjamin, 1924, *Contrib. Nat. Hist. Lep. N. Am.*, **5**(3): 167.
Type locality: [USA], Arizona, Baboquivari Mts. [USNM]

Hemioslaria pima is a pretty and easily recognized species from southeastern Arizona. The forewing is ovate and almost entirely glossy white. The only maculation consists of fine black lines along the costa and inner margin marking the antemedial, postmedial, and basal lines. The hindwing is dull, dark gray. The species superficially resembles a gigantic version of some of the species in the genus *Grotella.* Superficial appearance, the two large claws on the first prothoracic tarsal segment, the long irregularly spaced hairlike setae on the prothoracic tarsal segments, and the round frontal process with its central ridge will readily separate *Hemioslaria pima* from any other noctuid in North America. Wing length: mean = 17.22 mm, standard deviation = 0.37 mm, n = 7.

Hemioslaria pima has been collected only in the Santa Rita and Baboquivari Mountains of southern Arizona. Adults have been collected in August and September.

GENUS
Heminocloa Barnes and Benjamin

Heminocloa Barnes and Benjamin, 1924, *Contrib. Nat. Hist. Lep. N. Am.,* **5**(3): 165.
Type species: *Basilodes mirabilis* Neumoegen, 1884. Original designation.

Heminocloa and *Hemioslaria* are undoubtedly closely related despite the superficial differences between them. Both genera share a nearly identical external tympanic region structure (i.e., the dorsal umbrellalike hood and ventrocephalic bulla) and the presence of long, irregularly arranged, hairlike setae on the prothoracic tarsal segments (with the exceptions listed in the generic descriptions). *Heminocloa mirabilis* from southern Arizona is the only species of the genus.

The technical description follows. Head: antenna of both male and female simple, finely ciliate, basal segments fatter than normal (as in *Hemioslaria*); eye unmodified; lashes absent from both base of antenna and rear margin of eye; front with ovate frontal process, longer laterally than dorsoventrally, with raised outer rim; an inner tubercle present, located near dorsal margin of process; palpus unmodified; proboscis unmodified; scales on dorsum of head not arranged in either tufts or rows. Thorax: dorsal vestiture primarily of short, weakly dentate scales; patagia broad giving dorsum of thorax a flattened look; a darker green metathoracic tuft. Wing venation: as in *Hemioslaria.* Prothoracic leg: femur with concave cephalic margin with lateral rows of scales forming trough for reception of tibia; tibia with strong epiphysis, reaching distal tip of tibia, but not exceeding it as in *Hemioslaria*; tibia without spines or apical claw; tarsal segments not shortened as in *Hemioslaria*; tarsal segments with scales dorsally, naked ventrally; all tarsal segments covered ventrally with long, hairlike setae; fourth and fifth segments with long hairlike setae dorsally as well; first tarsal segment with one spine at proximal end larger and thicker than other setae, but not nearly as enlarged as the two setae in *Hemioslaria.* Mesothoracic leg: femur with weak ventral fringe of long hairs; tibiae without spines, with usual apical pair of spurs; first tarsal segment long, 5 × longer than second; ventral spines on first and second in four irregular rows, hairlike setae absent, apical spines larger than spines on remaining part of segment; ventral spines on second segment also in four irregular rows; ventral surface of last three segments without strong spines, only long hairlike setae; long hairlike setae present on dorsal surface of fifth tarsal segment. Metathoracic leg: as in mesothoracic leg, but tibia with usual two pairs of spurs. Internal tympanum: not examined. External tympanic region: almost identical with *Hemioslaria*; a slight swelling of intersegmental membrane present toward caudal margin of first abdominal segment. Male abdomen: vestiture of closely appressed, hairlike scales; basal hair pencils and Stobbe's gland absent, sternum 1+2 elongate, as long as wide; eighth sternum with cephalic margin sclerotized, but otherwise sternum and eighth tergum without differential sclerotization; hair pencils or modified scales absent from eighth sternum. Male genitalia: plate O, figures 10, 11; uncus elongate, not swollen anywhere, without hairs or setae; uncus with weak apical tooth; free pleurite present, strong, unmodified; juxta a trapezoidal plate; saccus of vinculum rounded at apex; valve simple with ventroapical clasper, clasper pointed at its apex; sacculus with a slight nob near middle; apex of valve without spines or corona; shaft of aedoeagus about 3 × as long as wide; vesica globular with three diverticula, one diverticulum 3 × larger than the other two; large diverticulum with many small cornuti, two smaller diverticula naked; diffuse batch of small cornuti from base of vesica to base of large diverticulum present. Female genitalia: plate S, figure 3; ovipositor lobes unmodified, although slightly pointed caudally; eighth sternum unmodified, with caudal row of spines; caudal margin of ostium membranous, continuous with sclerotized caudal third of ductus bursae; caudal third of ductus bursae sclerotized, with internal plicae and spinules; ductus bursae about 5 × longer than wide, caudal ⅓ sclerotized, cephalic ⅔ membranous; corpus bursae very elongate, without signum and C-shaped; appendix bursae a large lobe located near junction of corpus bursae and ductus bursae. Larva and foodplants: unknown.

Heminocloa mirabilis (Neumoegen)

PL. 4, FIG. 55; PL. O, FIGS. 10, 11 (♂ gen.); PL. S, FIG. 3 (♀ gen.); TEXT FIG. 124 (map) (RWH 9735).

Basilodes mirabilis Neumoegen, 1884, *Papilio,* **4**: 94.
Type locality: [USA], Arizona. [USNM]

Heminocloa mirabilis is an easily recognized species from southern Arizona. The forewing is light olive green, shading almost to white along the costal margin. The upper third of the basal area is dark olive green as is a subapical patch. A brown-green patch is present following the subterminal line near the

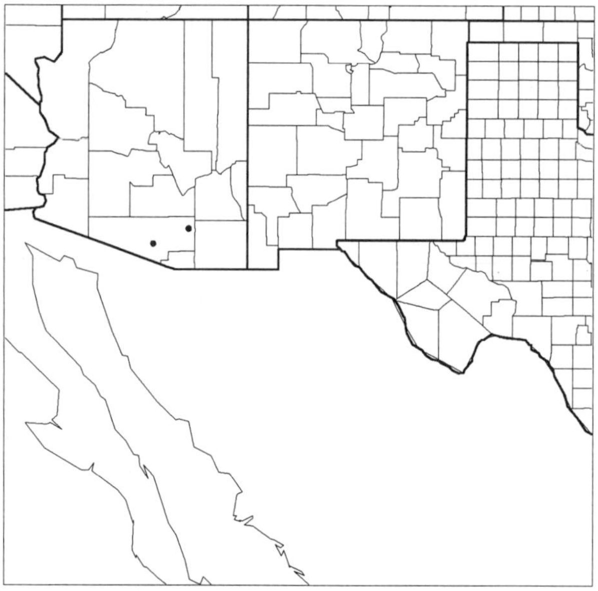

FIGURE 124: DISTRIBUTION OF SPECIMENS EXAMINED OF *HEMINOCLOA MIRABILIS*

tornus. The antemedial and postmedial lines are present and white. Dark-green shading exists along the inner side of the postmedial line. A frontal process is present and is laterally ovate with a dorsal inner tubercle. Wing length: mean = 15.82 mm, standard deviation = 0.95 mm, n = 6.

This species is known from the Santa Catalina and Baboquivari Mountains of Pima County, Arizona. No significant variation is present in the small number of specimens seen. Adults have been collected in August and September.

GENUS
Thurberiphaga Dyar

> *Thurberiphaga* Dyar, [1920] 1919, *Insecutor Inscitiae Menstruus,* 7: 188.
> Type species: *Thurberiphaga catalina* Dyar, [1920]. Monotypy.

Thurberiphaga contains the single species *diffusa.* It is characterized by a distinctive external appearance and an absence of internal synapomorphies to either categorize it as a group or to link it with other groups of the family. The species most distinctive feature is the larva, a borer in tree cotton (Malvaceae). *Thurberiphaga diffusa* superficially resembles some species in the Stiriinae, and none of its characters definitely argue against this placement. However no characters particularly support the placement of *Thurberiphaga* in the Stiriinae either, hence its placement among the *incertae sedis* genera. Crumb (1956) was unable to place this species on larval characters. The larva falls into the "larva liberae" branch of the "trifid" noctuid tree. The technical description follows.

Head: male antenna weakly bipectinate, pectinations weaker on one side than the other; female antenna serrate to very weakly bipectinate; eye unmodified, naked; lashes absent from base of antenna and rear margin of eye; scales along dorsum of head not arranged in rows or tufts; frontal process present, laterally ovate with raised rim, invaginated along ventral margin (upside down heart shape), and without central tubercle; palpus somewhat weaker than normal, but not significantly so; proboscis unmodified. Thorax: dorsal vestiture of mixed scales and hairlike scales; no tufts present; tegulae and patagia give thorax a square appearance; scales of wings long, dentate. Wing venation: as in the Stiriinae, but R_3 and R_4 connate about ½ (not ⅓) the distance from end of accessory cell to outer margin; R_{3+4} and R_5 connate just past apex of accessory cell, not at apex. Prothoracic leg: femur and tibia unspined; epiphysis unmodified; tibia unspined and without an apical claw; tarsal segments except fifth with three ventral rows of spines; rows of spines on first tarsal segment confined to distal half, absent from proximal half; tarsal claws with apical tooth. Mesothoracic leg: as in prothoracic leg except for usual differences. Metathoracic leg: as in mesothoracic leg except for usual differences; tarsal claw with particularly strong tooth. Internal tympanic structure: not examined. External tympanic structure: almost identical to that found in the Stiriinae; hood slightly shifted dorsally, somewhat intermediate between Stiriinae and the following two genera. Male abdomen: vestiture of closely appressed, long scales; basal hair pencils and Stobbe's gland absent; sternum 1+2 as long as wide; eighth sternum and tergum without differential sclerotization except for darker caudal and cephalic margins; no hair pencils or modified scales associated with eighth sternum. Male genitalia: plate O, figures 12, 13; uncus long, not swollen, with a weak apical tooth, and without hairs or setae; tegumen unmodified; free pleurite present, unmodified, broadly articulated with tegumen; juxta a trapezoidal plate with a V-shaped caudal indentation; transtilla thin near junction with valve, but expanded near its midpoint; anellus heavily dentate; valve elongate, thin; clasper reduced to a series of small, spinose bumps; corona and all large setae absent from apex of valve; saccus rounded at apex; aedoeagus shaft about 3 × longer than wide, angulate; vesica globular, without major cornuti, but with a multitude of diverticula; fine spicules present irregularly over vesica. Female genitalia: plate S, figure 4; ovipositor lobes basically unmodified, although spines well developed; a weak, almost absent, double dentate ridge present between ovipositor lobes ventrally; eighth sternum covered with long spines, spines not confined to caudal margin of sternum; ostium ovate, simple, with internal, short, stubby spines; ductus bursae weak, short, triangular; corpus bursae rectangular, laterally elongate, not the usual caudocephalic orientation; appendix bursae absent. Larva: internal borer in *Gossypium thurberi,* described by Crumb (1956).

Thurberiphaga diffusa (Barnes)
PL. 4, FIG. 53; PL. O, FIGS. 12, 13 (♂ gen.); PL. S, FIG. 4 (♀ gen.); TEXT FIG. 125 (map) (RWH 9817).

Alaria diffusa Barnes, 1904, *Can. Ent.,* **36**: 238.
Type-locality: [USA], Arizona, Huachuca Mts. [USNM]

NOTE—*Alaria diffusa* is based on one male and one female. The male syntype has been labeled and is here designated as LECTOTYPE.

Thurberiphaga catalina Dyar, [1920] 1919, *Insecutor Inscitiae Menstruus,* **7**: 188.
Type locality: [USA], Arizona, Santa Catalina Mts., Sabino Canyon. [USNM]

NOTE—*Thurberiphaga catalina* is based on five syntypes. Only one of these syntypes bears a Dyar handwritten type label. This syntype has been labeled and is here designated as LECTOTYPE.

Thurberiphaga diffusa is yellow with reddish-pink markings along the outer margin of the forewing and in the general region of the postmedial line. All other maculation is indistinct. A frontal process is present and resembles an inverted heart. Wing length: mean = 15.54 mm, standard deviation = 0.65 mm, n = 10.

The larva has been described by Crumb (1956) from "*Thurberia*" (*Gossypium thurberi* Todfaro) (Malvaceae), a wild species of cotton sometimes known as tree cotton. The larva are borers in the stems of the plant. Stem borers are unknown in this part of the noctuid phylogenetic tree, although they occur elsewhere in the family. The larvae are described as pink, tapering caudally, and covered with coarse granules. No specimens have been reared from cultivated cotton.

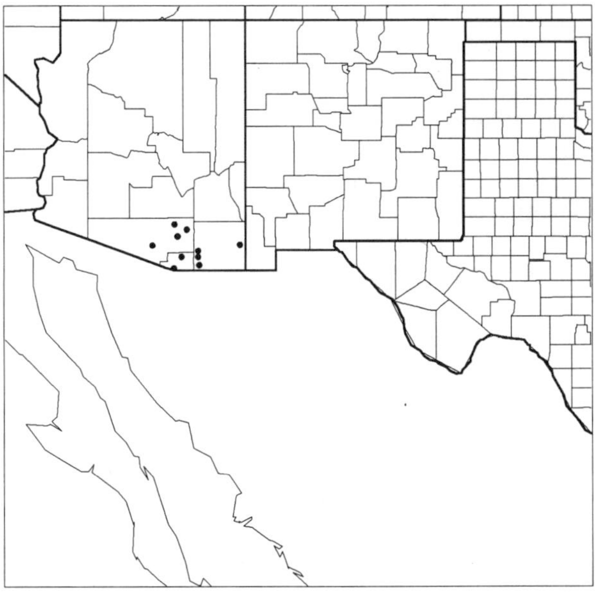

FIGURE 125: DISTRIBUTION OF SPECIMENS EXAMINED OF *THURBERIPHAGA DIFFUSA*

Thurberiphaga diffusa is known from Cochise, Santa Cruz, and Pima Counties in southern Arizona. There is a single specimen in the USNM labeled "Hot Springs, Green River, Washington," but this specimen is undoubtedly mislabeled. Some individual variation in size and the amount of pink on the forewing is present. Adults are fairly common in collections and have been collected from late July to mid-September.

SUBFAMILY
Psaphidinae Grote

Dicopinae Grote, 1883, *Proc. Amer. Philosophical Soc.,* **21**: 142.
Type genus: *Dicopis* Grote, 1874.

Psaphidiini Grote, 1895, *Abhandl. Naturwissenschaftlichen Vereins Bremen,* **14**: 70.
Type genus: *Psaphida* Walker, 1865.

NOTE—The tribal name Psaphidiini was based on the misspelling *Psaphidia* of *Psaphida* Walker and is correctly emended to Psaphidini.

NOTE—Grote (1895) proposed the family-group name Psaphidiini [sic] (based on the valid genus *Psaphida* Walker, 1865) as a replacement name for Dicopinae Grote, 1883 because he treated *Dicopis* Grote, 1874 as a junior synonym of *Psaphida.* Subsequently, Psaphidinae has been used as the valid family-group name for the taxon. According to Article 40(b) of the *International Code of Zoological*

Nomenclature Psaphidinae should be maintained as the valid family-group name.

The subfamily Psaphidinae (as defined in this fascicle) includes six tribes; the Psaphidini, Nocloini, Feraliini, Triocnemidini, Azeniini, and Grotellini. The tribes Psaphidini, Feraliini, Nocloini, and Grotellini were recognized as monophyletic groups in Hodges et al. (1983), but placed in different subfamilies. The new tribes Triocnemidini and Azeniini were briefly characterized in the introduction to this work. This amalgamation of tribes represents, to some degree, a marriage of convenience and could (and probably will) be modified in the future. The Psaphidini, Nocloini, and Triocnemidini are almost certainly sister groups on the basis of the shared, derived characters discussed below. I am far less certain that the Feraliini, Azeniini, and Grotellini are related to the first three tribes or to each other. Therefore the synapomorphies discussed in the following short justification of the subfamily Psaphidinae apply safely only to the Psaphidini, Nocloini, and usually the Triocnemidini. A complete discussion of the tribes Azeniini and Grotellini is deferred to a later fascicle. Knowledge of larval structures might help to resolve relationships, but I have not found larvae of any species in the Nocloini, Triocnemidini, Azeniini, or Grotellini. 1) A character uniting the Psaphidini, Nocloini, and Feraliini is the heavy and irregular spacing of the spines on the ventral surface of the tarsal segments. In most trifid noctuids the tarsal spines are arranged in three, sometimes four, rows. These rows may be regular or slightly irregular but are always clearly linear. The tarsal spines in the Psaphidini and Nocloini are heavy and are not arranged in regular rows. Tarsal spines are arranged in three rows in the Triocnemidini, Azeniini, and Grotellini. 2) The vesica is elongate and covered with large, long cornuti, which generally do not fall into definite patterns. These cornuti occur on weakly sclerotized, rugose sections of the vesica in the Nocloini and Triocnemidini. The cornuti in the Grotellini, although reduced in number and strength, are also contained in rugose areas. Rugose areas are absent in the Psaphidini. An elongate vesica with massive groups of cornuti is also found in the Oncocnemidinae, but in this subfamily the vesica does not contain rugose areas. 3) The seventh tergum of the male abdomen is heavily sclerotized, enlarged, and its caudal margin is sometimes weakly bilobed. This condition is present, and is fairly characteristic, of the Psaphidini, Nocloini, and Triocnemidini. It is absent in the Grotellini, Azeniini, and Feraliini. 4) The external structure of the tympanic region is characterized by a large to medium-sized hood followed by a large bulla in the intersegmental membrane of the first two abdominal segments. This structure is found in the Psaphidini, Nocloini, and Triocnemidini but is absent in the Feraliini, Azeniini, and Grotellini. However, this structure of the tympanic region is also found in the Cuculliinae and may be plesiomorphic for the "larva liberae" branch of the "trifid" evolutionary tree. 5) The known larvae of the Psaphidini and Feraliini fall in the "larvae liberae" class. None of the larvae in the Nocloini, Triocnemidini, Azeniini, or Grotellini is known. 6) Prothoracic tibial claws are present in all of the tribes except the Feraliini and Azeniini. These claws are associated with a sclerotized plate (text figure 126). This plate may be the strongest character uniting the Psaphidini, Nocloini, and Triocnemidini as a monophyletic group. I cannot determine if a plate exists in the Grotellini.

The tribe Feraliini is treated in this fascicle as part of the Psaphidinae, but the irregular spining of the tarsal segments and the lashed eyes are the only significant characters uniting the Feraliini with the Psaphidini. The genus *Feralia* also shares an early spring flight period and general "hairiness" with the Psaphidini. The larvae were placed by Crumb (1956) in his broad definition of the Cuculliinae and occur in the same portion of his key as genera in the Psaphidini. The adults of *Feralia,* on the other hand, do not have a prothoracic tibial claw, the seventh tergum of the male abdomen is not enlarged, and the vesica is not elongate nor massively spiny. The Feraliini may not be properly placed in the Psaphidinae, but I have no better suggestion for their placement.

The Grotellini are placed in the Psaphidinae because I do not know where else to place them. The species in the tribe are small, often almost entirely white or off-white. The prothoracic and mesothoracic legs of *Grotella* have spines on the tibia. There is a strong frontal process with a large central tubercle. The male and female genitalia are simplified and uninformative. The tribe may eventually prove to be a sister group of the Heliothinae, but the spiny tibiae and intuition are my only reasons for uniting the two groups.

The Azeniini are a heterogeneous group of genera and probably do not represent a monophyletic group. The species of the tribe are all small, and the male and female genitalia are simplified and uninformative. A frontal process is present and usually complicated. No good synapomorphies defining the tribe exist, and the purpose of the tribe is to provide a place for a series of enigmatic genera of small species within the noctuid phylogeny.

The following statements summarize the phylogeny of the Psaphidinae as defined in this work: 1) The Psaphidini, Nocloini, and Triocnemidini form a monophyletic group and comprise the Psaphidinae in the strict sense. 2) The Feraliini share some characters with the Psaphidini and disagree in others. The placement of the Feraliini in the Psaphidinae is more strongly supported than anywhere else in the Noctuidae classification. 3) The Azeniini and Grotellini are probably not directly related to the Psaphidinae. Their position in the phylogeny of the Noctuidae is unknown. Both tribes are placed in the Psaphidinae as a matter of convenience.

TRIBE
Psaphidini Grote

The formal description of the tribe Psaphidini given below is based on the genus *Psaphida.* Several other genera are included in the tribe based on 1) the early flight period of the adults, 2) the irregular and heavy spining of the tarsal segments, and 3) the common occurrence of pectinate or strongly serrate male antennae. The male vesica is elongate and massively spiny, but the spines are not located on rugose patches. A large terminal cornutus usually is present at the apex of the vesica. The valve is rectangular and simple. The apex of the valve is usually heavily haired, and the corona is weak or absent. The uncus is often swollen over its entire length. The clasper is small, and no basal process of the sacculus is present. The caudal half of the ductus bursae is membranous, sometimes expanded, and its external margin is continuous with the intersegmental membrane of the seventh sternum. Consequently, it is easy to sever the bursa from the eighth abdominal segment and the ovipositor during dissection. More importantly this feature of the ductus bursae appears to be fairly characteristic of the tribe Psaphidini. The tibial claw of the prothoracic leg is strong and associated with a plate along the outer margin of the tibia. The larvae of the Psaphidini all feed on deciduous trees or shrubs.

The technical description follows. Head: antenna strongly pectinate in male, laminate or weakly pectinate in females; eyes naked, fully developed, or slightly reduced, weakly lashed at base of antenna, lashed with dark hairs posteriorly; scaling shaggy, consisting of long, often slightly toothed, scales; a moderately large mass of scales on vertex posterior to antennal bases, but head without distinct tufts; groups of long scales from bases of antennae almost meeting at midline of head; frons roughly scaled, but without a frontal prominence and not swollen; palpus slightly reduced, second segment fringed below with hairlike scales, third segment small, globular; proboscis reduced, apical ⅓ thickly covered with sensillae. Thorax: scaling either of long hair or long toothed scales; two weak scale tuft groups, one posterior to patagia and the second a double tuft at posterior margin of thorax, both tufts usually weak to indistinct. Prothoracic leg: femur and tibia unspined except for tibial claw; tibial claw (text figure 126) associated with a platelike structure; first four tarsal segments covered ventrally with long, strong, irregularly spaced spines; fifth segment with spines reduced, hairlike at apex of segment; two long hairlike setae at dorsodistal end of fifth segment overhanging claws; claw large with a strong tooth, pulvillus strongly setose. Mesothoracic and metathoracic legs: femur and tibia unspined, well developed, shaggy, but without specific tufts of hair. Internal tympanum: text figure 126 *a*; scutal phragma lipped; pocket I not lipped; pocket II short; pocket III short, triangular, not fused with tympanal air sack; pocket IV normal, with circular opening; posterior tendon plate large, well developed; nodular sclerite elongate and irregularly rounded. External tympanum: text figure 126 *d*; hood well developed, separated from large, triangular bulla in abdominal intersegmental membrane; anterolateral edge of first tergum without overhanging lip; tympanal groove strong; weak depression on ventral margin of second tergum. Wing venation: forewing as in the Stiriinae; M_2 of hindwing weak to moderately well developed. Male abdomen: no dorsal scale tufts on abdominal segments; basal hair pencils and Stobbe's glands absent; seventh tergum of male strongly sclerotized, enlarged; sclerotization of eighth sternum of male weak, U-shaped, no weak hair brush. Male genitalia: text figure 126 *b, c*; valve simple, corona present, but sometimes poorly developed; clasper present but weak; basal process of sacculus absent; uncus usually swollen, sometimes also flattened, usually with a hook at tip; free pleurite present, simple, broadly articulated with tegumen and not fused with vinculum; juxta broad; transtilla weak; anellus weakly spinose; vesica heavily spined, usually, but not always, with a large terminal cornutus; disposition of other cornuti variable. Female genitalia: plate U, figure 11; ovipositor neck short; ovipositor lobes square with spines, unmodified; apophyses anteriores and posteriores short; ostium and ductus bursae membranous, external edge of ostium con-

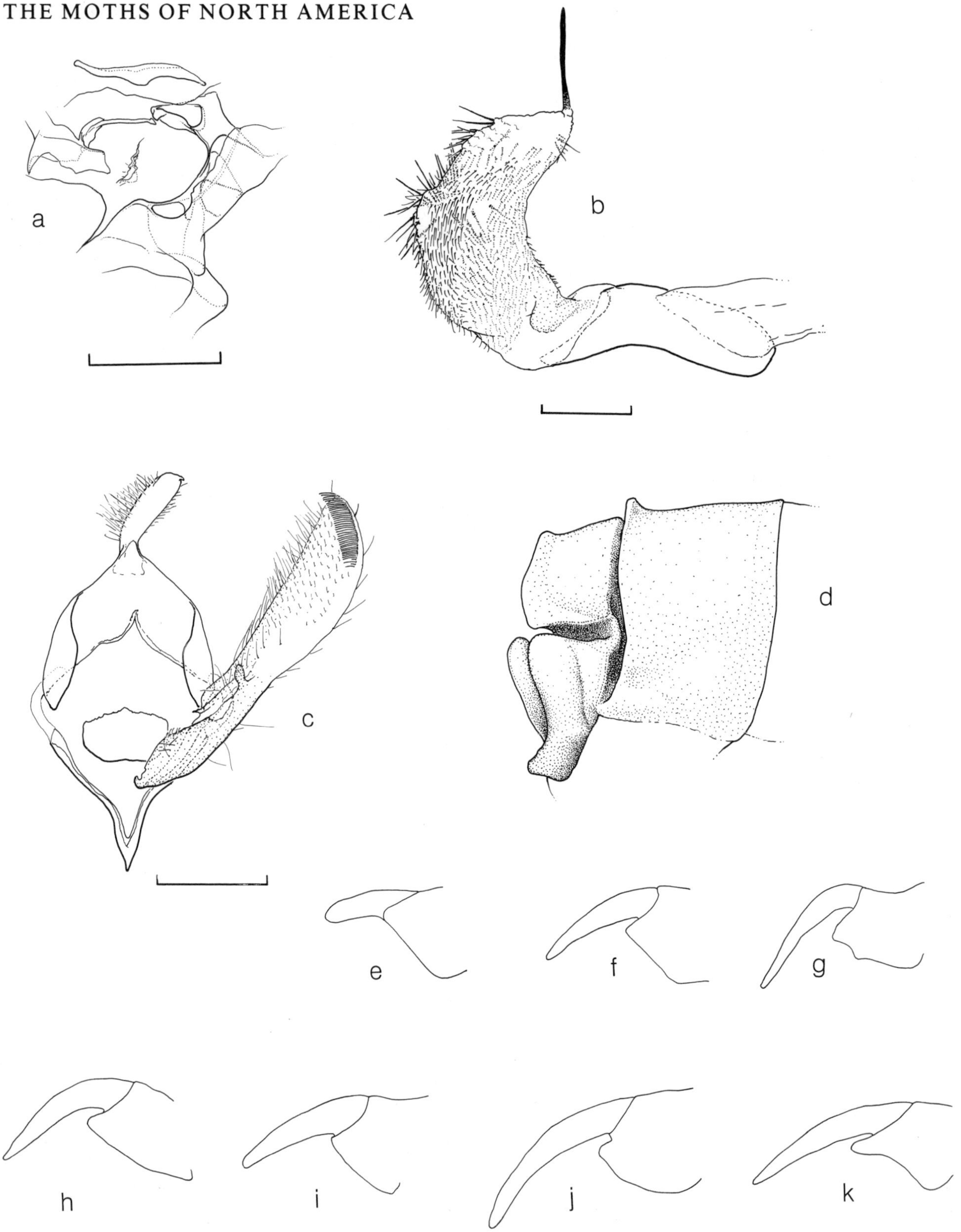

FIGURE 126: STRUCTURAL FEATURES OF *PSAPHIDA RESUMENS* AND THE PROTHORACIC TIBIAL CLAW OF SPECIES OF *PSAPHIDA*

a. Internal tympanic structure. *b.* Aedoeagus. *c.* Male genitalia. *d.* External tympanic structure. *e.* Prothoracic tibial claw of *Psaphida resumens*. *f. Psaphida thaxterianus*. *g. Psaphida styracis*. *h. Psaphida rolandi*. *i. Psaphida grandis*. *j. Psaphida electilis*. *k. Psaphida damalis*. Scale = 1 mm.

tinuous with intersegmental membrane of seventh sternum; ostium without structure; corpus bursae long, apical (caudal) tip usually somewhat sclerotized; corpus bursae usually with a round or slightly elongate dentate signum; ductus seminalis arising from cephalic end of corpus bursae.

KEY TO THE NORTH AMERICAN GENERA OF THE PSAPHIDINI

1. Prothoracic tibial claw with associated plate .... *Psaphida* p. 165
– Prothoracic tibial claw without associated plate 2
2. Forewing length greater than 20 mm *Brachionycha* p. 177
– Forewing length less than 20 mm 3
3. Dorsal surface of abdomen without tufts 4
– Dorsal surface of abdomen with tufts 5
4. Male antenna simple; forewing with silvery markings .... *Provia* p. 180
– Male antenna pectinate; forewing gray, without silvery markings .... *Pleromella* p. 178
5. Ductus seminalis arising from caudal end of bursa; eastern half of North America .. *Copivaleria* p. 176
– Ductus seminalis arising from cephalic end of bursa; west coast of North America .... *Pseudocopivaleria* p. 172

GENUS
Psaphida Walker

Psaphida Walker, 1865, *List of the Specimens of Lepidopterous Insects in the Collection of the British Museum,* **32**: 447.
Type species: *Psaphida resumens* Walker, 1865. Monotypy.

Dicopis Grote, 1874, *Ann. Rept. Trustees Peabody Acad. Sci.,* **6**: 23.
Type species: *Dicopis muralis* Grote, 1874. Monotypy.

Copipanolis Grote, 1874, *Ann. Rept. Trustees Peabody Acad. Sci.,* **6**: 25. NEW SYNONYMY.
Type species: *Copipanolis cubilis* Grote, 1874. Monotypy.

Eutolype Grote, 1874, *Proc. Acad. Nat. Sci. Philadelphia,* **1874**: 198. NEW SYNONYMY.
Type species: *Eutolype rolandi* Grote, 1874. Monotypy.

The diagnostic characters of the genus *Psaphida* are 1) the enlarged and irregularly spaced spines of the first four tarsal segments of all three pairs of legs, 2) the elongate forewing, 3) the early spring flight period of the adults, 4) the pectinate male antenna, 5) the shaggy appearance of the adults, 6) the reduced palpus, 7) the presence of a prothoracic tibial claw and associated plate (text figure 126), and 8) the other characters of the tribe.

The bursa copulatrix is difficult to interpret in *Psaphida* and its relative *Pseudocopivaleria.* The "typical" bursa usually consists of a corpus bursae and an appendix bursae. The corpus bursae is usually pleated and may have various forms of the signa. The appendix bursae is unpleated and arises from the distal (caudal) end of the corpus bursae near its juncture with the ductus bursae. The ductus seminalis arises from the distal end of the appendix bursae. A single pleated sack with a large signum is present in *Psaphida* and *Pseudocopivaleria,* however, and the ductus seminalis appears to arise from the cephalic (headward) end of the sack. I cannot explain this situation satisfactorily. Either what appears to be the corpus bursae is a strange appendix bursae with pleats, or the appendix bursae has migrated to the headward end of the corpus bursae from its usual caudal position and then disappeared without leaving a trace.

The description of the tribe is based on *Psaphida,* therefore, no formal description is needed.

Psaphida is most likely to be confused with *Copivaleria* and *Pseudocopivaleria. Psaphida* may be separated from these two genera by the absence of dorsal abdominal tufts and/or the presence of an associated plate with the fore tibial claw.

KEY TO THE SPECIES OF *PSAPHIDA*

1. Forewing light brown or red brown ...... *styracis* p. 170
– Forewing gray or gray black, not light brown 2
2. Hindwing white with accented veins and discal dot; forewing light gray; plate associated with tibial claw reduced, almost absent; California and Colorado .... *damalis* p. 171

— Hindwing not pure white, with at least some infuscation; forewing light gray or gray black; plate associated with tibial claw fully developed; east of the Rocky Mountains 3

3. Forewing orbicular and reniform marks obscure or not distinctly lighter than rest of forewing; forewing light gray 4
— Forewing orbicular and reniform marks not obscure, usually conspicuously lighter than ground color of forewing 5

4. A distinct light band following forewing antemedial line; both orbicular and reniform marks usually absent .... *grandis* p. 168
— No distinct light band following forewing antemedial line; both orbicular and reniform marks clearly present although never prominent .... *rolandi* p. 168

5. Scales of basal half of forewing without teeth or only minute teeth .... *electilis* p. 169
— Scales of basal half of forewing with large, distinct teeth 6

6. Forewing terminal area light green gray, sharply separated from blackish remainder of wing; terminal area without streaks in the fold; forewing claviform mark not distinct, never large and never partly fused with orbicular mark .... *thaxterianus* p. 167
— Forewing terminal area usually lighter than remainder of wing but not sharply set off from remainder of wing; terminal area with black streaks in fold; forewing claviform mark usually large, light, often partly fused with orbicular mark .... *resumens* p. 166

Psaphida resumens Walker

PL. 5, FIGS. 1–3; TEXT FIG. 126 *b, c* (♂ gen.); PL. S, FIGS. 5, 6 (♂ gen.); PL. U, FIG. 11 (♀ gen.); TEXT FIG. 126 *e* (tibial claw); TEXT FIG. 127 (map) (RWH 10019).

Psaphida resumens Walker, 1865, *List of the Specimens of Lepidopterous Insects in the Collection of the British Museum,* **32**: 448.
Type locality: [USA], Georgia. [BMNH]

Cymatophora viridescens Walker, 1865, *List of the Specimens of Lepidopterous Insects in the Collection of the British Museum,* **32**: 601.
Type locality: [USA], Florida. [BMNH]

Dicopis muralis Grote, 1874, *Sixth Ann. Rept. Trustees Peabody Acad. Sci. Year 1873,* p. 23.
Type locality: [USA], New York. [BMNH]

NOTE—*Dicopis muralis* was based on an indeterminate number of specimens from Pennsylvania and New York. Only males are mentioned in the original description. The male in the BMNH bearing Grote's type label is designated and has been labeled as the LECTOTYPE. The location of the Pennsylvania specimen or specimens has not been determined.

Psaphida resumens is a gray-black to green-black species. Its diagnostic characters are 1) the large, light claviform mark usually abutts and often partially fuses with the orbicular mark, 2) black streaks are present in the fold along the forewing outer margin, and 3) a finely spiculate basal diverticulum is present in the vescia. Most of the vesica is uniformly covered with long cornuti. *Psaphida resumens* superficially is most similar to *P. thaxterianus, P. electilis,* and *Copivaleria grotei. Psaphida thaxterianus* lacks black streaks in the fold along the forewing outer margin, and its claviform mark is not large, light, or abutting the orbicular mark. In addition the forewing terminal area of *thaxterianus* is green gray and contrasts with the basal part of the wing. The terminal area of *resumens* often is green gray as well, but it never contrasts with the rest of the wing. *Psaphida resumens* is easily separated from *electilis* by the forewing scales: the scale tips of *electilis* are rounded with only a slight hint of teeth, but the tips of the scales are distinctly toothed in *resumens. Copivaleria grotei* may be the closest superficial mimic of *resumens.* The forewing of *grotei,* however, is always black and is never suffused with the greenish cast of *resumens. Copivaleria grotei* has a black tuft on the third abdominal segment, a tuft is absent in all species of *Psaphida.* Wing length: mean = 17.38 mm, standard deviation = 0.65 mm, n = 10.

The larva has not been described in detail. Forbes (1954) lists maple (*Acer* sp.) (Aceraceae) and oak (*Quercus* sp.) (Fagaceae) as foodplants. *Psaphida resumens* is probably fairly polyphagous on various species of deciduous trees. I have not seen a larva, but Donald Lafontaine (personal communication) has provided the following brief description. The larva is whitish green. The pinacula below the lateral line are white, and a diagonal line is present on each abdominal segment between the subdorsal and lateral lines. The lateral line extends along the anal

FIGURE 127: DISTRIBUTION OF SPECIMENS EXAMINED OF *PSAPHIDA RESUMENS*

FIGURE 128: DISTRIBUTION OF SPECIMENS EXAMINED OF *PSAPHIDA THAXTERIANUS*

plate, not down the anal proleg. This last character is typical of Crumb's "larvae liberae" and occurs in all species of *Psaphida* with a visible lateral line.

Psaphida resumens is an early spring species of the eastern United States. It occurs from northern and central Florida to central Texas and Arkansas in the south. In the north *resumens* has been collected from southern New England and Quebec westward to southern Minnesota, Illinois, and eastern Kansas. Within population variability in coloration is significant, particularly in the amount of white or green suffusion over the gray-black forewing ground color. I have seen two "melanics" from South Carolina. They are completely suffused with dark brown green. Geographical variation does not seem to be significant. Populations from the southeastern United States appear to be slightly lighter and greener, on average, than normal, particularly specimens from the Padre Island region in southeastern Texas. Adults fly in April and May in the north and February and March in the south.

Psaphida thaxterianus (Grote)

PL. 5, FIGS. 4–6; PL. S, FIGS. 7, 8 (♂ gen.); PL. U, FIG. 12 (♀ gen.); TEXT FIG. 126 *f* (tibial claw); TEXT FIG. 128 (map) (RWH 10020).

Dicopis thaxterianus Grote, 1874, *Bull. Buffalo Soc. Nat. Sci.*, **2**: 196.
Type locality: [USA], Massachusetts, Newtonville. [MCZ]

Psaphida thaxterianus has a distinctive shaggy appearance. The forewing appears narrower and more sharply pointed than in any other species of *Psaphida* except *grandis*. *Psaphida thaxterianus* is superficially most similar to *resumens*; however, the forewing terminal area of *thaxterianus* is a contrasting green gray, and black dashes are not present in the fold along the outer margin. The claviform mark is not large and light, nor does it abutt the orbicular mark. The plate associated with the prothoracic tibial claw projects into a point below the claw (text figure 126 *f*) in *thaxterianus* but is flush with the claw in *resumens*. The vesica of *thaxterianus* lacks a basal diverticulum. The spining of the vesica is fairly uniform with moderate to long cornuti. A long apical cornutus is present. Wing length: mean = 16.87 mm, standard deviation = 0.73 mm, n = 10.

The larva has been described several times, e.g., Dyar (1901). Dyar lists white oak (*Quercus alba* L.) (Fagaceae). Dyar describes the larva as red brown with large subdorsal cream-white patches on "joints 2 to 14." The pinacula and setae are white. The head is brown dotted with white, and black patches are present on either side of the clypeus. Dyar's description portrays a rather striking caterpillar. I have not seen any larval specimens.

Psaphida thaxterianus has a spotty distribution and is not common anywhere. It occurs in a band running from Massachusetts through New York and Pennsylvania westward to Illinois and eastern Missouri. The northern extent of its range occurs in Ontario and southern Michigan. A single specimen in the USNM is labeled only "Tex." but the species is otherwise unknown from Texas. Forbes (1954) lists a specimen from North Carolina, but I have not seen it. Little variation exists in *thaxterianus* either among populations or individuals. The forewing terminal area is less distinct than normal in a few specimens. Adults fly in early spring from March to May.

Psaphida rolandi (Grote), NEW COMBINATION
PL. 5, FIGS. 16–20; PL. S, FIG. 13; PL. T, FIG. 1 (♂ gen.); PL. V, FIG. 2 (♀ gen.); TEXT FIG. 126 *h* (tibial claw); TEXT FIG. 129 (map) (RWH 10014).

Eutolype rolandi Grote, 1874, *Proc. Acad. Nat. Sci. Philadelphia,* **1874**: 198.
Type locality: [USA], Massachusetts. [BMNH]
NOTE—Grote described *Eutolype rolandi* from an indeterminate number of specimens from Massachusetts and Missouri. All of them presumably are in the BMNH. Hampson (1906) lists a male and female from Massachusetts with the notation "type." The male bearing Grote's type label from Massachusetts is designated and has been labeled as the LECTOTYPE.

Copipanolis vernalis Morrison, 1874 [1875], *Proc. Boston Soc. Nat. Hist.,* **17**: 133.
Type locality: [USA], Massachusetts. [MCZ]
NOTE—*Copipanolis vernalis* is based on a single specimen which I have been unable to locate with certainty. A specimen exists in the MCZ labeled "type 23727" and "vernalis Morrison type." The type label is not typical of a Morrison type. However, the type listed by Morrison came from Thaxter, and the Thaxter collection is in the MCZ. Therefore, this specimen is probably the type listed by Morrison.

Psaphida rolandi might be confused with *grandis* or *damalis*. *Psaphida damalis* lives in western North America and *rolandi* is restricted to the Great Plains and eastward. *Psaphida grandis* has a distinct light band or patch following the forewing antemedial line; this band is absent in *rolandi*. The forewing of *Psaphida rolandi* is slate gray. The maculation is obscure, but clearly visible, particularly the orbicular and reniform marks. A striking characteristic of *rolandi* is the presence of thoracic tufts of dull-gold metallic scales. One tuft is located near the base of the abdomen and the other toward the front of the thorax just behind the patagia. Specimens of *rolandi* become greasy easily, and the tufts are not always evident. The vesica of *rolandi* is elongate with a large basal diverticulum. The vesica is medially constricted with small cornuti in the middle and much larger cornuti at the base and apex. The apical cornutus is very long. Wing length: mean = 15.93 mm, standard deviation = 0.69 mm, n = 10.

The larva and foodplants of *rolandi* have not been described despite the species' abundance and wide distribution. John G. Franclemont (personal communciation) has reared *rolandi* on red and black oak (*Quercus* spp.) (Fagaceae).

Psaphida rolandi is a common eastern North American species ranging as far north as southern New England and southern Ontario, west to North Dakota and Manitoba, and south to Florida, Louisiana, and Texas. Individual variation in *rolandi* mostly involves differences in the darkness or lightness of the slate-gray forewing. This variation is moderately pronounced within populations. The orbicular and reniform marks are usually accented with light gray in darker specimens. Geographic variation also exists in *rolandi.* Specimens from South Carolina and Florida are consistently lighter gray and the maculation more pronounced than in material from other parts of its range. Specimens from eastern Texas and Louisiana tend to be more variable in size and forewing coloration than individuals from elsewhere in the species' range. One specimen from Manitoba is light gray and curiously like the southeastern population. Specimens from eastern North Dakota clearly are equivalent to northeast and north-central populations. I suspect the Manitoba specimen is mislabeled. Adults fly in the early spring; April and May in the north with earlier dates of emergence in the south.

Psaphida grandis (Smith), NEW COMBINATION
PL. 5, FIGS. 7–10; PL. S, FIGS. 9, 10 (♂ gen.); PL. U, FIG. 13 (♀ gen.); TEXT FIG. 126 *i* (tibial claw); TEXT FIG. 130 (map) (RWH 10013).

Eutolype grandis Smith, 1898, *Can. Ent.,* **30**: 321.
Type locality: [USA], Missouri, Kansas City. [USNM]

Psaphida grandis is a slate-gray to violet-gray species. It is most similar to *rolandi* but can be identified instantly by the presence of a patch or band of dull, light-blue scales following the forewing antemedial line. These light-blue scales may not be strong in a few specimens, but they are almost always present. The orbicular and reniform marks are usually obscure although the reniform mark may be

FIGURE 129: DISTRIBUTION OF SPECIMENS EXAMINED OF *PSAPHIDA ROLANDI*

FIGURE 130: DISTRIBUTION OF SPECIMENS EXAMINED OF *PSAPHIDA GRANDIS*

indicated by the presence of a dark patch. The median area is usually vaguely darker than the rest of the forewing. The forewing appears narrower than in any other species of *Psaphida* except *thaxterianus.* The vesica is of average length and evenly covered with short cornuti; patches of longer cornuti are absent. The apical cornutus is long. A basal diverticulum is absent. Wing length: mean = 18.38 mm, standard deviation = 0.58 mm, n = 10.

The life history and foodplants of *grandis* are unknown. Oak may be a reasonable guess for the host given the predilections of other members of the genus and the species' distribution.

Psaphida grandis ranges as far south as central Georgia and northern Florida and as far north as western Massachusetts and central New York. It appears to be commonest, although never abundant, in the central eastern United States, specifically Wisconsin, Michigan, Illinois, Indiana, Ohio, Pennsylvania, and south to Missouri and northeastern Texas. *Psaphida grandis* varies in size more than most. The range in wing length is 16–20 mm with a modal value of between 17 and 18 mm. Forewing coloration is also variable. A few specimens have matte-black basal and median areas. The amount of black in these specimens varies from small patches to extensive areas covering all of the basal and median areas. The amount of light-blue scaling between the forewing antemedial and medial lines is also variable ranging from prominent to almost absent. No discernible geographical variation is present in the limited series of specimens I have seen except, possibly, in three specimens from Marion County in northern Florida; they are strongly tinged with purple. Adults fly in April and May in the north, but as early as mid-February in the far south.

Psaphida electilis (Morrison), NEW COMBINATION

PL. 5, FIGS. 11–15; PL. S, FIGS. 11, 12 (♂ gen.); PL. V, FIG. 1 (♀ gen.); TEXT FIG. 126 *j* (tibial claw); TEXT FIG. 131 (map) (RWH 10012).

Dicopis electilis Morrison, 1875, *Proc. Boston Soc. Nat. Hist.,* **18**: 114.
Type locality: [USA], Pennsylvania, Easton. [MSU]

Dicopis depilis Grote, 1881, *Papilio,* **1**: 48.
Type locality: [USA], Ohio, Columbus. [USNM]

Eutolype bombyciformis Smith, 1892, *Proc. U. S. Natl. Mus.,* **15**: 59.
Type locality: [USA], Missouri. [USNM]; lectotype designated by Todd (1982).

Psaphida electilis is difficult to characterize because of its variability. However, the scales at the base of the forewing have round tips or have, at most, minute teeth. The only other species of *Psaphida* with this type of scale is *damalis,* a western U. S. species with light-gray to blue-gray forewings. *Psaphida electilis* occurs in three basic forms. The first, and

FIGURE 131: DISTRIBUTION OF SPECIMENS EXAMINED OF *PSAPHIDA ELECTILIS*

apparently most common, form is dark gray without prominent light patches or streaks. The second form is basically the same as the first, but with a prominent light-brown patch filling the claviform mark and stretching to the base of the wing. This second form typically has light-brown patches following the postmedial line. The third basic form has a prominent, continuous or interrupted black streak along the cubital vein from the base of the forewing to the postmedial line. The general forewing color in this streaked form may be either dark gray or some noticeably lighter color. Wing length: mean = 17.16 mm, standard deviation = 0.51 mm, n = 10.

Crumb (1956) has described the larva and records the foodplant as walnut (*Juglans* sp.) (Juglandaceae) while Forbes (1954) records hickory (*Carya* sp.) (Juglandaceae) as a foodplant. Crumb's description was based on an inflated specimen in the USNM. The body color is presumably green but has faded to straw yellow. A broad, but much broken, middorsal white line is present as well as subdorsal and subspiracular white lines. The setigerous tubercles are subconical, particularly the dorsal ones, and are large and white. No distinctive markings are present on the head.

Psaphida electilis occurs from southern New England and southern Ontario south to Maryland in the east, west to Wisconsin, Iowa, Kansas, and eastern and central Texas, and south to Alabama. *Psaphida electilis* has been collected in the southeastern United States but is known only from a few isolated localities. I have seen material from Louisiana and Mississippi not shown on the distribution map. Geographical variation is not important over most of this species' range, with some notable exceptions. Specimens from Mississippi and Louisiana collected by Vernon Brou and Bryant Mather are uniformly dark gray with obscure maculation, without even the light patches usually found along the forewing subterminal line. None of the various forms found in the northern populations are known from this part of the United States. Specimens from San Saba County in north-central Texas are also monomorphic, although slightly lighter in color and with some light shading along the subterminal line. In contrast at least two of the northern morphs occur in northeastern Texas. The male and female genitalia and other morphological characters of these southern populations identify them as *electilis,* despite the absence of the three forms. Adults fly in the early spring, slightly earlier in the south than in the north.

Psaphida styracis (Guenée), NEW COMBINATION

PL. 5, FIGS. 22–27; PL. T, FIGS. 4, 5 (♂ gen.); PL. V, FIG. 4 (♀ gen.); TEXT FIG. 126 *g* (tibial claw); TEXT FIG. 132 (map) (RWH 10016).

Taeniocampa styracis Guenée, 1852, *in* Boisduval and Guenée, *Histoire Naturelle des Insectes. Species General des Lépidoptères,* **5**: 357.
Type locality: Amerique Septentrionale [Georgia]. [type lost]

NOTE—Guenée based *Taeniocampa styracis* on a manuscript plate of Abbot.

Copipanolis cubilis Grote, 1874, *Sixth Ann. Rept. Trustees Peabody Acad. Sci. for 1873,* p. 26.
Type locality: [USA], Michigan, Lansing. [BMNH]

NOTE—Grote described *Copipanolis cubilis* from an indeterminate number of specimens from Lansing, Michigan and Massachusetts. A male specimen from Lansing, Michigan in the BMNH bearing Grote's type label is designated and has been labeled LECTOTYPE.

Copipanolis stigma Smith, 1890, *Ent. Americana,* **6**: 220.
Type locality: [USA], Florida. [AMNH]

Copipanolis borealis Smith, 1892, *Proc. U. S. Natl. Mus.,* **15**: 62.

Type locality: [USA], Minnesota, [St. Anthony's Park]. [lost]

Copipanolis fasciata Smith, 1892, *Proc. U. S. Natl. Mus.*, **15**: 62.
Type locality: [USA], Texas. [AMNH], lectotype designated by Todd (1982).

Psaphida styracis is the only brown species of the genus. The pointed wings, fuzzy appearance, pectinate male antennae, and fawn-brown coloration make it one of the more distinctive noctuids in North America. The vesica has the usual long apical cornutus, and there are two long, subapical cornuti. The cornuti tend to be arranged in distinct, but intersecting rows. No basal diverticulum is present. Wing length: mean = 17.13 mm, standard deviation = 0.71 mm, n = 10.

Crumb (1956) has given a description of the larva and records *Quercus alba* L. (Fagaceae) as a foodplant. This description was based on a single larva in alcohol. The larva is either brown or black. The body apparently lacks any markings except for black coloring on the lateral and posterior margins of the cervical shield. The setigerous tubercles are large and dark. Tietz (1972) lists *Fraxinus* sp. (Oleaceae) and *Quercus* sp. as foodplants.

Psaphida styracis occurs from southern Ontario and southern New England, west to Wisconsin, Minnesota, and Missouri, and south to eastern Texas and Florida. It has not been recorded from the central-southern states, e.g., West Virginia, western Virginia, Kentucky, Tennessee, or North Carolina. *Psaphida styracis* is apparently common at times. The most interesting aspect of *styracis* is its variability; it is particularly variable in the northern half of its range. Two chief factors typify its variability; the darkness of the brown color and the intensity of the maculation. Forewing color varies from dull straw yellow to dark fawn brown. Darker specimens seem to predominate. The orbicular mark, reniform mark, and the terminal third of the forewing of some specimens are contrastingly lighter and grayer, only the orbicular and reniform marks in others. The postmedial and antemedial lines are contrastingly darker than the forewing ground color in a few specimens. All possible intergrades between these basic types appear to exist. Sex linkage does not seem to be involved. Geographic variation is also present. Texas populations are uniform light brown and obscurely marked. Almost no variability is visible in these Texas populations in contrast to the marked variability of northern populations. The few Floridian specimens I have seen are, perhaps, even more

FIGURE 132: DISTRIBUTION OF SPECIMENS EXAMINED OF *PSAPHIDA STYRACIS*

variable than northern populations. *Psaphida styracis* overwinters as a pupa, the adult emerging and flying early in the spring.

Psaphida damalis (Grote), NEW COMBINATION

PL. 5, FIG. 21; PL. T, FIGS. 2, 3 (♂ gen.); PL. V, FIG. 3 (♀ gen.); TEXT FIG. 126 *k* (tibial claw); TEXT FIG. 133 (map) (RWH 10015).

Dicopis damalis Grote, 1879, *Bull. U. S. Geol. Geog. Surv. Terr.*, **5**: 208.
Type locality: [USA], California, Havilah. [AMNH]

Psaphida damalis is the only species of *Psaphida* in western North America. It is recognized by the light-gray forewings with obscure maculation and the white hindwings with accented veins. The plate associated with the tibial claw is poorly developed but clearly present. Wing length: mean = 16.44 mm, standard deviation = 0.61 mm, n = 10.

The larva has not been described. A single specimen in the USNM from Sacramento, California is labeled "reared from Oak."

Psaphida damalis has a peculiar distribution. Not only is it the only western North American species, but it also appears to have two disjunct populations. The majority of specimens in collections are from the vicinities of San Francisco and Sacramento, California. Two specimens in the USNM are from Silverton in southwestern Colorado. One specimen ex-

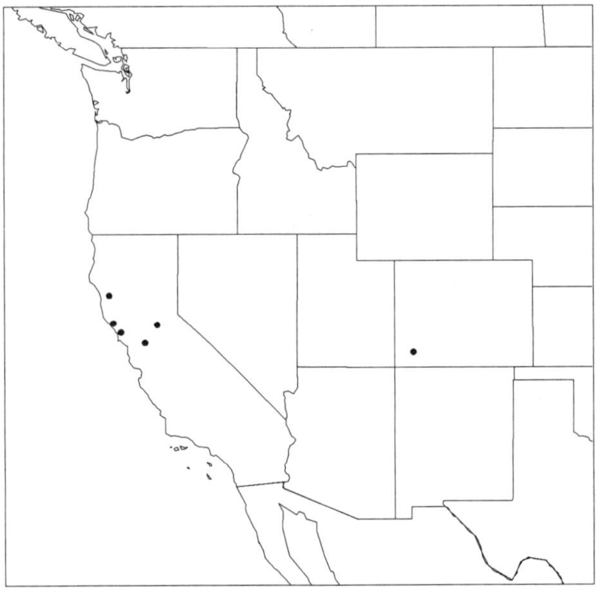

FIGURE 133: DISTRIBUTION OF SPECIMENS EXAMINED OF *PSAPHIDA DAMALIS*

ists in the USNM, undoubtedly mislabeled, from "New Haven, Ct." *Psaphida damalis* may have a wider distribution than indicated, but may not have been collected more commonly because of its early flight period. The darkness of the forewing varies slightly, but this species otherwise appears to be fairly constant individually and geographically. *Psaphida damalis* flies early in the spring.

GENUS
Pseudocopivaleria Buckett and Bauer

Pseudocopivaleria Buckett and Bauer, 1966, *Jour. Lep. Soc.*, **20**: 84.
Type species: *Psaphida sonoma* McDunnough, 1941. Original designation.

Pseudocopivaleria is an anomalous genus. It most resembles *Copivaleria* on external characters, but the male and female genitalia indicate a closer relationship to *Psaphida.* The origin of the ductus seminalis in the female genitalia is perhaps the most convincing character associating this genus with *Psaphida.* The ductus seminalis of *Psaphida* and *Pseudocopivaleria* seems to arise from the proximal (cephalic) end of the corpus bursae. The ductus seminalis of *Copivaleria* arises from a definite appendix bursae near the juncture of the corpus bursae and the ductus bursae. The anomolous origin of the ductus seminalis in both *Psaphida* and *Pseudocopivaleria* appears to be a strong, shared, derived character indicating a close affinity between the two genera.

Pseudocopivaleria can be separated from *Psaphida* by the presence of abdominal tufts on segments one, three, and four. Abdominal tufts are absent in *Psaphida.* The prothoracic tibial claw of *Pseudocopivaleria* is strong and recurved but lacks an associated plate. *Copivaleria* has the same type of tibial claw and tufting on the abdomen, but the tuft on segment four is not as evident as in *Pseudocopivaleria.* The forewing of *Pseudocopivaleria* is, superficially, not as elongate or pointed as in *Copivaleria.* The best character for separating the two genera is perhaps range. *Pseudocopivaleria* occurs only on the west coast and *Copivaleria* lives in the eastern half of North America.

This genus contains two species, *sonoma* and *anaverta.* They are superficially similar and cannot always be told apart solely by appearance. *Pseudocopivaleria sonoma* is the darker of the two species (compare the plate figures) and often has a vague, cream-colored streak in the tornus of the forewing. *Pseudocopivaleria anaverta,* on the other hand, is lighter gray, and the tornal marking is absent. The clasper of the male valve of *sonoma* is twice as long as wide, but the clasper is short and triangular in *anaverta.* Male abdominal basal hair pencils are present in *sonoma* but absent in *anaverta.* A constriction is present in the middle of the female bursa in *anaverta* roughly dividing it into upper and lower parts; this constriction is absent in *sonoma.*

The technical description follows. Head: male antenna pectinate with a row of cilia on each pectination; pectinations longest in middle, tapering in size toward apex and base; female antenna simple to slightly laminate; antenna base with tuft of hair directed toward front; eye strongly lashed from both base of antenna and rear margin of eye; head between antennae very shaggy in appearance, with a lower row of scales running between antennal bases and above that two vertical rows of scales; front roughly scaled; eye not reduced; ocellus present; pal-

→

FIGURE 134: MALE GENITALIA OF SOME GENERA OF THE PSAPHIDINAE TRIBE PSAPHIDINI
a. Pseudocopivaleria sonoma, genital capsule; *b.* Aedoeagus. *c. Copivaleria grotei,* genital capsule; *d.* Aedoeagus. *e. Provia argentata,* genital capsule; *f.* Aedoeagus. Scale = 1 mm.

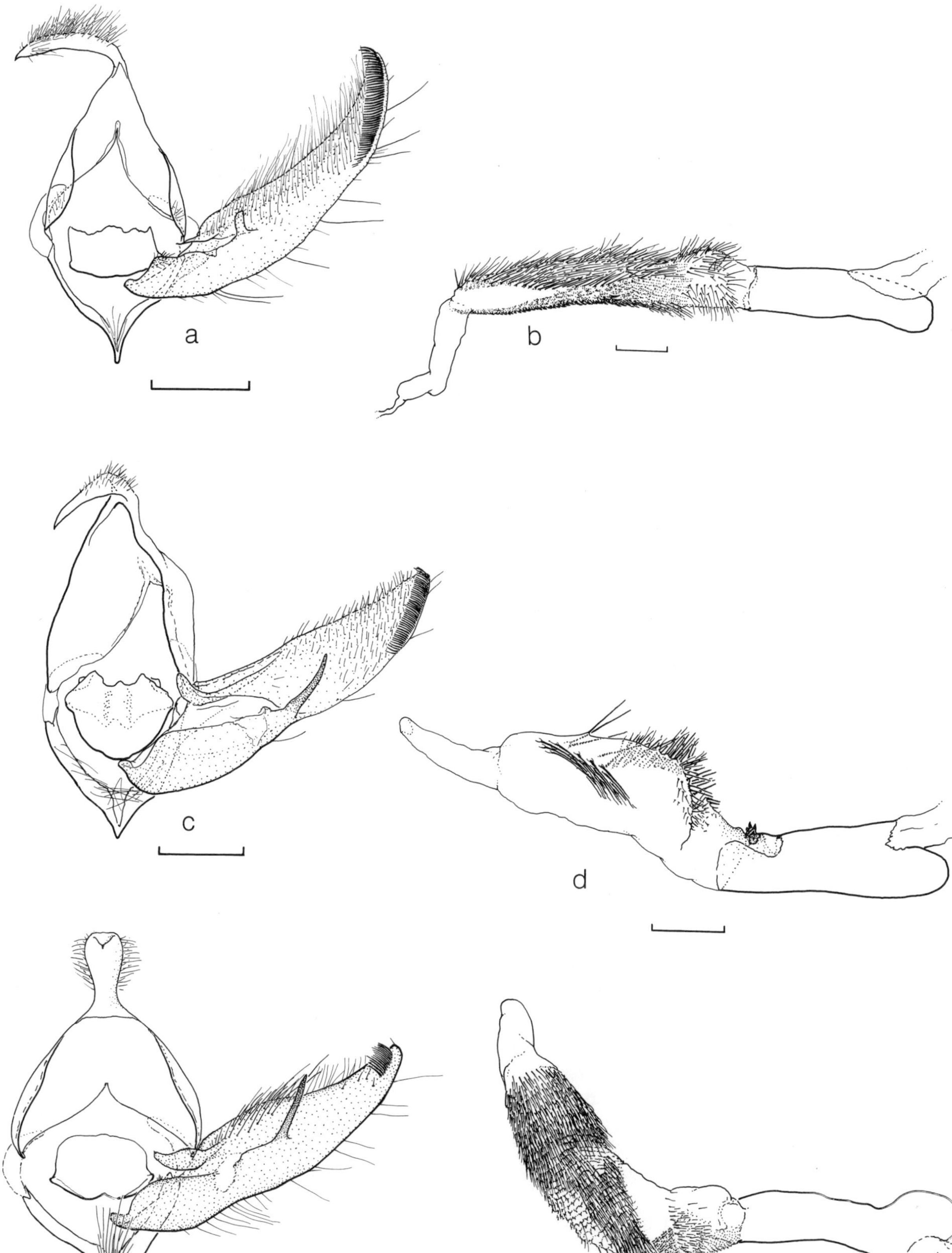
a
b
c
d
e
f

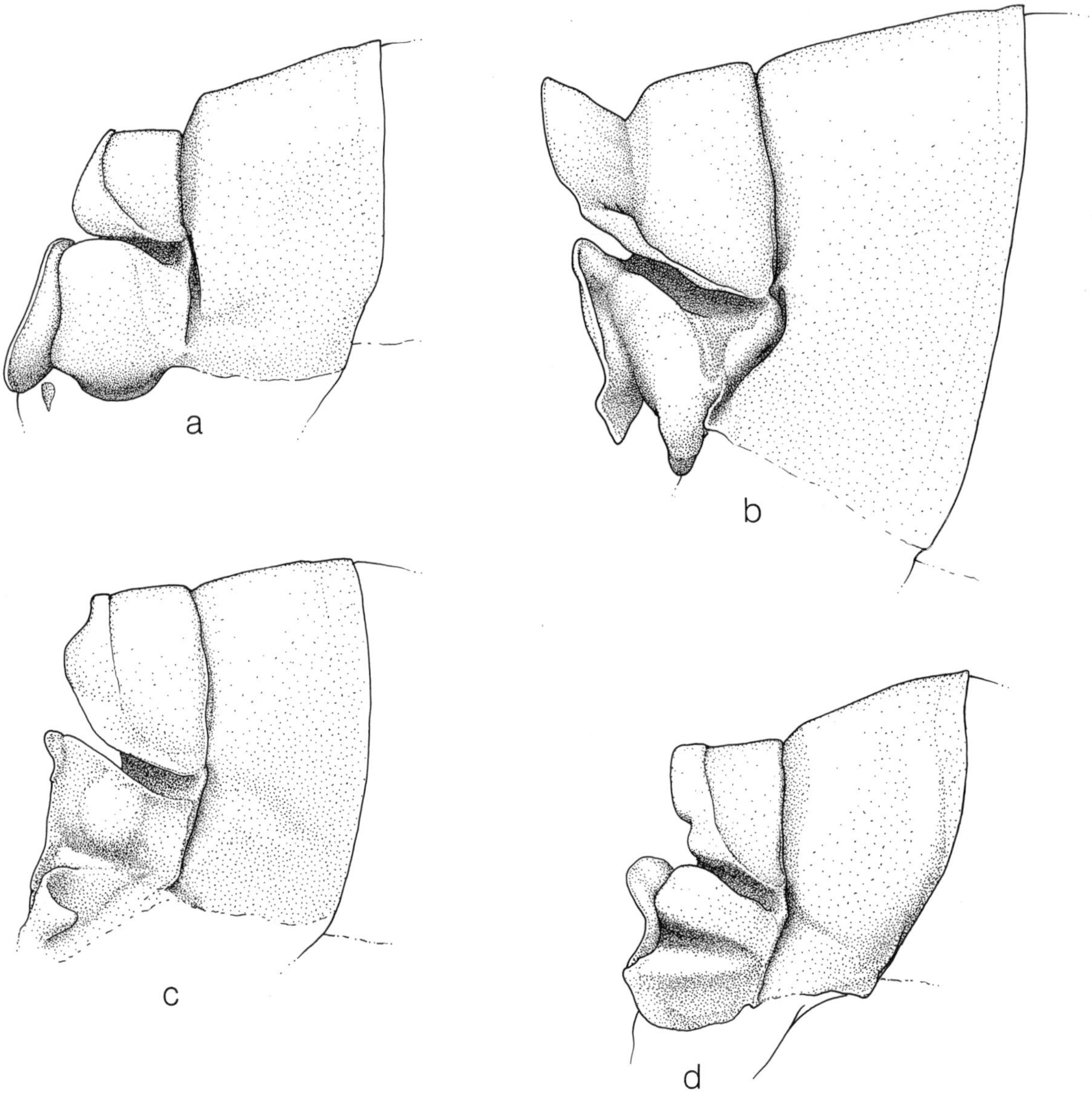

FIGURE 135: EXTERNAL TYMPANIC REGION OF FOUR GENERA OF THE PSAPHIDINAE TRIBE PSAPHIDINI

a. Copivaleria grotei. b. Brachionycha borealis. c. Pleromella opter. d. Provia argentata.

pus upturned, slightly reduced in size, with a ventral fringe of hair; proboscis normal. Thorax: dorsal vestiture of elongate, highly dentate scales; patagia not raised as a hood; two small tufts arising from distal margins of the patagia; metascutum with central tuft of curved scales and two lateral, elongate tufts; tuft of scales present at base of forewing; forewing scales unmodified; costa of ventral surface of forewing with long hairs projecting downward; scales of lower half of forewing longer than those of upper half; thorax covered with masses of hairlike scales ventrally; wing venation as in *Psaphida.* Prothoracic leg: femur with massive, ventral fringe of hair; large, recurved tibial claw present, associated plate absent; tibia about 3 × as long as wide; first tarsal segment as long as remaining four; tarsal spining as in tribe. Mesothoracic and metathoracic legs: femur with long, heavy fringe of hairs; upper half of tibia with strong tuft of scales; spines of first two tarsal segments particularly heavy. External tympanic region: as in *Copivaleria* (text figure 135 *c*), but hood covered with layers of flat scales giving it a fat appearance; a strong tuft of hairs arising from alula region projecting over the first abdominal segment. Abdomen: male with

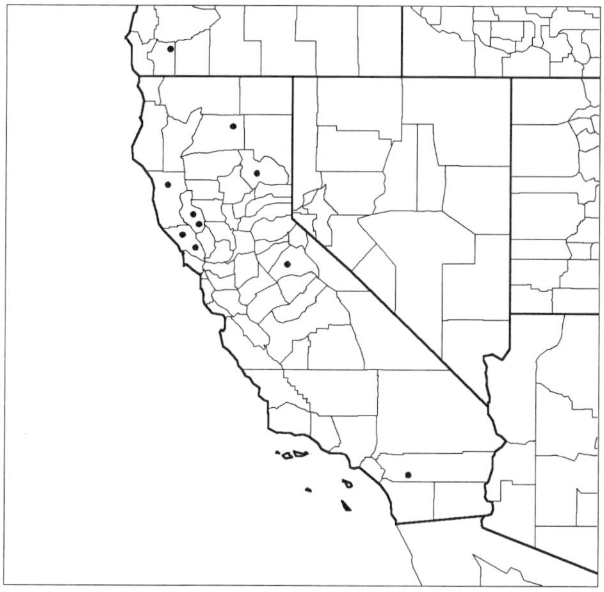

FIGURE 136: DISTRIBUTION OF SPECIMENS EXAMINED OF *PSEUDOCOPIVALERIA SONOMA*

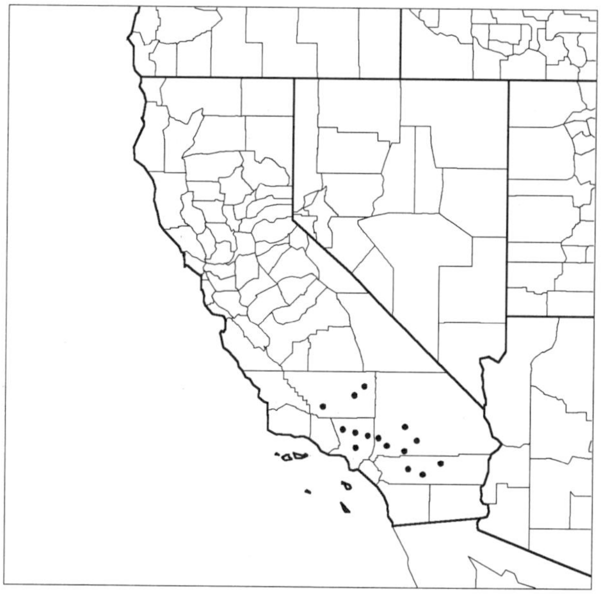

FIGURE 137: DISTRIBUTION OF SPECIMENS EXAMINED OF *PSEUDOCOPIVALERIA ANAVERTA*

or without tufts on segments one, three, and four, that on three largest; eighth tergum of male and seventh of female greatly swollen at apex and slightly bilobed. Male genitalia: text figure 134 *a*; valve elongate, with apices pointed; corona strong; clasper weak; juxta a single, quadrate plate; uncus swollen with an apical tooth; vesica elongate with two vague, parallel rows of long cornuti, one row located in a vaguely rugose area; apical cornutus absent. Female genitalia: ovipositor lobes unmodified, although proximal half more heavily sclerotized than distal half; both pairs of apophyses very short; eighth sclerite narrow, distally spiny; ostium and ductus bursae unmodified; ductus bursae very short; bursa elongate, weakly plicate; signum absent; ductus seminalis arising from proximal end of apparent corpus bursae.

The larvae and its foodplants are unknown for both species.

Pseudocopivaleria sonoma (McDunnough)
PL. 5, FIGS. 28, 29; TEXT FIG. 134 *a, b* (♂ gen.); TEXT FIG. 136 (map) (RWH 10017).

Psaphida sonoma McDunnough, 1941, *Can. Ent.*, **73**: 67.
Type locality: [USA], California, Sonoma Co., The Geysers. [CNC]

Pseudocopivaleria sonoma is, in general, a darker species than *anaverta*. A vague cream-colored streak usually is present in the forewing tornal region. The clasper is longer than in *anaverta* and basal abdominal hair pencils are present. The middle of the bursa is not constricted. The range of *sonoma* is more northern than that of *anaverta*, ranging from central California to southern Oregon, although an apparently disjunct, and very distinctive, southern population occurs in western Riverside County. Wing length: mean = 16.51 mm, standard deviation = 0.64 mm, n = 10.

This species has been recorded from a number of localities in central and northern California, and southern Oregon. The male hindwing is white with brown-tinged veins, but the female hindwing is evenly brown. Variation exists in the strength of the cream-colored tornal marking of the forewing. The isolated population from Riverside County has the forewing much darker and not as gray as the typical population from middle to northern California. The forewing also appears smoother and slightly velvety. There is a patch of tan scales in the lower half of the basal area. Tan scales may be present in the typical population, but are never as distinct. The male hindwing of the Riverside population has less dark scaling on the veins and along the outer margin. Despite the superficial differences, the male genitalia of the two populations are identical. Adults fly in the spring.

Pseudocopivaleria anaverta Buckett and Bauer

PL. 5, FIGS. 30, 31; PL. T, FIGS. 6, 7 (♂ gen.); TEXT FIG. 137 (map) (RWH 10018).

Pseudocopivaleria anaverta Buckett and Bauer, 1966, *Jour. Lep. Soc.*, **20**: 84.
Type locality: [USA], California, Los Angeles County, SSW of Valyermo. [USNM]

Pseudocopivaleria anaverta is a lighter and grayer species than *sonoma.* Other distinguishing characters are given under *sonoma.* Wing length: mean = 16.03 mm, standard deviation = 0.63 mm, n = 6.

This species appears to be restricted to southern California and northern Baja California, mostly at higher elevations. Too few specimens are known to discuss variation except for the sexual dimorphism in hindwing color. Adults fly in the spring.

GENUS
Copivaleria Grote

Copivaleria Grote, 1883, *Proc. Amer. Philosophical Soc.* **21**: 146, 168.
Type species: *Valeria grotei* Morrison, 1875. Original designation.

Copivaleria superficially resembles the genus *Psaphida.* However several significant structural differences exist. The clasper of the valve is long, recurved, and strongly developed. The vesica lacks the terminal cornutus found in most species of *Psaphida.* The ductus bursae is heavily sclerotized and pleated near the ostium, and both pairs of apophyses are very small. A definite corpus bursae and appendix bursae are present in *Copivaleria,* and the ductus seminalis arises from a caudal appendix bursae. The ductus seminalis arises from the cephalic end of the apparent corpus bursae in *Psaphida* and *Pseudocopivaleria.* A prothoracic tibial claw is present, but no associated plate exists. This plate is present in *Psaphida* and missing in *Pseudocopivaleria.* A tuft is present on the third abdominal tergum and a smaller one on the fourth tergum in *Copivaleria* and *Pseudocopivaleria,* but not in *Psaphida.* The bulla following the hood of the tympanal region is square, not triangular as in *Psaphida.* The genus *Pseudocopivaleria* occurs along the Pacific Coast; *Copivaleria* is known only from eastern and central North America.

The technical description follows. Head: vertex and front covered with erect, long brown scales, tufts separated by a cleft between antennal bases; male antenna pectinate, female antenna simple to slightly prismatic; tufts of scales from anterior and posterior margins of antennal scape; lashes from near base of antenna weak, black; lashes along posterior margin of eye strong, long, black; palpus slightly reduced, otherwise as in rest of tribe; proboscis normal. Thorax: dorsum of thorax with long, forked scales; posterior margin of thorax with diffuse tuft of curved scales; patagia not capable of being raised into a hood; tuft of hair from lower basal margin of forewing present; tufts of long black hairs from alulas of tympanal region. Prothoracic leg: tibia short, with a strong, curved apical claw, claw without associated plate; first tarsal segment about 2.5 × as long as second segment; tarsal spining irregular and weak; tarsal claw toothed. Mesothoracic and metathoracic legs: tarsal segments essentially as in prothoracic leg, but spining slightly stronger; femur and tibia of metathoracic leg covered with long hair. External tympanic region: text figure 135 *a*; alula moderate; hood moderate, followed by a large, square bulla; anterior margin of first tergum membranous but without an overhanging lip; margin of second tergum with strong depression running almost all the way across tergum. Abdomen: dorsum of third segment with small black tuft; male basal abdominal hair pencils and Stobbe's gland present; seventh tergum of male enlarged and sclerotized; eighth sternum of male U-shaped with weak hair brush. Male genitalia: text figure 134 *b*; corona present; clasper long, recurved pointed; juxta a single plate; aedoeagus with toothed region at apex of shaft; vesica with two fields of cornuti, no apical cornutus. Female genitalia: plate V, figure 5; ductus bursae heavily sclerotized and ridged near ostium; apophyses anteriores and posteriores very short, almost absent; ovipositor lobes unmodified.

Copivaleria grotei (Morrison)

PL. 5, FIG. 32; TEXT FIG. 134 *c, d* (♂ gen.); PL. V, FIG. 5 (♀ gen.); TEXT FIG. 138 (map) (RWH 10021).

Valeria grotei Morrison, 1875, *Bull. Buffalo Soc. Nat. Sci.*, **1**: 274.
Type locality: [USA], Massachusetts, Cambridge. [MSU]

Copivaleria grotei is black with a conspicuous white forewing reniform mark. It may be separated from all species of *Psaphida* by the presence of a tuft on the third, and sometimes the fourth, abdominal tergum. The prothoracic tibial claw lacks the associated plate always found in *Psaphida.* The only species of *Copivaleria, grotei,* lives in eastern and central North

America. Both species of *Pseudocopivaleria,* in contrast, are known only from the Pacific Coast region of the United States. Wing length: mean = 18.35 mm, standard deviation = 0.62 mm, n = 10.

The larva has been described by Crumb (1956). The larva is recorded from ash (*Fraxinus* sp.) (Oleaceae). His description is terse because all coloration and markings of the larvae he examined were obliterated.

Copivaleria grotei is common in eastern and central North America. It has been collected in southern New England, western Quebec, Ontario, and ranges as far south as South Carolina and Mississippi. It occurs throughout the Midwest from eastern and central Texas and Missouri in the south, and northward through the Great Plains as far north as North Dakota, Manitoba, and Saskatchewan. The western limit of its range is in western Nebraska and central Saskatchewan. Geographical variation does not appear to be significant, although western specimens tend to be somewhat lighter in color than eastern ones. Individual variation is not great although variation exists in the darkness of the forewing and the amount of olive green in the anal angle of the forewing. The adults are early spring fliers.

FIGURE 138: DISTRIBUTION OF SPECIMENS EXAMINED OF *COPIVALERIA GROTEI*

GENUS
Brachionycha Hübner

Brachionycha Hübner, [1819] 1816, *Verzeichniss Bekannter Schmettlinge* [sic], p. 144. Type species: *Bombyx nubeculosa* Esper, [1785]. Subsequent designation by Hampson, 1906, *Catalogue of the Lepidoptera Phalaenae in the British Museum,* **6**: 202.

Asteroscopus Boisduval, 1829, *Europaeorum Lepidopterorum Index Methodicus,* p. 59. Type species: *Bombyx cassinia* [Denis and Schiffermüller], 1775. Subsequent designation by Duponchel, 1829, *Histoire Naturelle Lépidoptères ou Papillons de France,* **7**(2): 71.

Petasia Stephens, 1829, *Illustrations of British Entomology, Haustellata,* **2**: 31. Type species: *Bombyx cassinia* [Denis and Schiffermüller], 1775. Monotypy.

Selenoscopus Heinemann, 1859, *Die Schmetterlinge Deutschlands und der Schweiz,* **1**: 445. Type species: *Bombyx nubeculosa* Esper, [1785]. Monotypy.

Brachionycha is a small, but distinctive, genus of primarily Palearctic species. One species, *Brachionycha borealis,* occurs in the boreal forests of North America. The phylogenetic relationships of *Brachionycha* are not entirely clear although several characters point to a relationship somewhere on the fringes of the Psaphidini. In particular the adults have a superficial resemblance to the species of *Psaphida. Brachionycha* shares the following characters with *Psaphida*: 1) a spring flight period, 2) a claw on the apex of the prothoracic tibia, 3) a massively hairy body, 4) irregular spination on the tarsal segments of all three pairs of legs, 5) a relatively simple valve, and 6) a massively spiny vesica. The possibility of shared primitive characters or convergent evolution in response to an early spring flight period exists, however. The male antennae are pectinate. The female antennae are serrate in the type species *nubeculosa* and simple to prismatic in the North American species *borealis.*

The technical description follows. Head: front smooth, covered with long, shaggy hair; male antenna pectinate, the pectinations covered with fine hairs; female antenna serrate or simple; palpus slightly reduced in size, third segment small; lashes from base of eye heavy, black; lashes from base of antenna weak to almost absent. Thorax: dorsum covered with long, hairlike scales and with no noticeable tufts; forewing covered with flat black or white scales intermixed with long, hairlike scales, most heavily near base and costa; ventral surface of thorax massively hairy, hair merging with hair of mesothoracic and metathoracic legs. Prothoracic leg:

FIGURE 139: DISTRIBUTION OF SPECIMENS EXAMINED OF *BRACHIONYCHA BOREALIS*

tibia with long claw but without associated plate; spining of tarsal segments irregular, not in rows; tarsal claws without a tooth; first tarsal segment as long as remaining four. Mesothoracic and metathoracic legs: metathoracic tibia slightly swollen and tibial spurs greatly reduced in size; first metathoracic tarsal segment slightly swollen at base. Abdomen: dorsal and ventral surfaces hairy; male basal abdominal hair pencils and Stobbe's gland absent; seventh male tergum not swollen or sclerotized; eighth male sternum and tergum lightly sclerotized. External tympanic area: text figure 135 *b*; first abdominal tergum with an extremely long lip in front; abdominal bulla oblong, not round, slanted downward distally; hood strong and large; alula well developed. Internal tympanic structure: not examined. Male genitalia: plate T, figures 8, 9; genitalia greatly simplified, valve simple in *borealis*; clasper present in some Palearctic species although greatly reduced in type species *nubeculosa*; uncus swollen; vesica with large masses of cornuti as in rest of tribe. Female genitalia: plate V, figure 6; in *nubeculosa* ovipositor lobes weak, unmodified; ostium a V-shaped depression with minute spines and granules; caudal half of ductus bursae unsclerotized, expanded caudally into a membranous sack continuous with intersegmental membrane between seventh and eighth sterna; lower half of ductus bursae ribbed, slightly sclerotized; corpus bursae unilobed, triangular; seventh tergum longer than wide; seventh sternum with caudal margin neither more strongly sclerotized than rest of sclerite or invaginated.

Brachionycha borealis (Smith)

PL. 5, FIG. 33; PL. T, FIGS. 8, 9 (♂ gen.); PL. V, FIG. 6 (♀ gen.); TEXT FIG. 139 (map) (RWH 10011).

Asteroscopus borealis Smith, 1899, *Can. Ent.*, **31**: 25.

Type locality: [Canada], Manitoba, Cartwright. [USNM]

Brachionycha borealis is a distinctive species with a number of peculiarities. First it is the only species of *Brachionycha* in North America, flying in the early spring in the boreal forest of the United States and Canada. Females are very rare in collections. The large gray, rounded wings and the pectinate antennae in the male should readily separate it from any other North American species of Noctuidae. The dorsum of the abdomen is covered with long, very coarse hairs. Wing length: ♂ mean = 21.62 mm, standard deviation = 0.41 mm, n = 10; ♀ mean = 26.02 mm, standard deviation = 4.24 mm, n = 2.

Specimens in the CNC were reared from *Vaccinium angustifolium* Ait. (Ericaceae). J. D. Lafontaine has kindly supplied me with color slides of the larva. The larva is large, translucent green. The abdominal segments are distinctly separated and segment eight has a noticeable hump. The segments are covered with white spots. The spiracles are white with red rims. The most distinctive marking is a white line running from the eighth abdominal spiracle back along the anal shield and forward to the spiracle on the other side. The thoracic legs are red. The head is translucent green. The closely related Old World species *nubeculosa* feeds on a variety of deciduous trees, apparently preferring birch (Betulaceae).

Specimens of *borealis* are rare in collections. I have seen it from Michigan, Wisconsin, Manitoba, Alberta, Saskatchewan, Quebec, Ontario, Pennsylvania, and northern West Virginia. *Brachionycha borealis* is possibly more common than it seems, but its early spring flight habits may keep it from being noticed more. The species has a tendency for melanism. "Normal" specimens are a medium gray while melanics are suffused with black.

GENUS
Pleromella Dyar

Pleromella Dyar, 1921, *Insecutor Inscitiae Menstruus,* **9**: 40.

Type species: *Pleromella opter* Dyar, 1921. Monotypy.

Pleromella is a monotypic genus from California. It is characterized by 1) the early spring flight period of the adults, 2) the heavy, irregular spining of the tarsal segments, and 3) the morphology of the external tympanic area, particularly the abdominal bulla and the deep depression following the tympanal groove between the first and second abdominal terga. The female genitalia are distinctive. The caudal half of the ductus bursae is membranous and expands into a membranous pocket continuous with the intersegmental membrane between the seventh and eighth abdominal segments. The corpus bursae is globular with large, oval, heavily sclerotized plates near the appendix bursae and the junction of the corpus and ductus bursae.

The phylogenetic relationships of *Pleromella* are unclear. *Pleromella* has no clear relatives. The early spring adult flight period suggests a placement in the Psaphidini. The external tympanic structure is reminiscent of *Provia* but is intermediate between that of *Provia* and the rest of the Psaphidini. The heavy, irregular spining of the tarsal segments is characteristic of the Psaphidini; however, the spining may be a convergence caused by the early adult flight period.

Head: male antenna bipectinate, female antenna simple to slightly prismatic; vertex and frons covered with bushy, hairlike scales, scales of vertex not separated from scales of frons; antennal scape with only a weak tuft of scales; no black lashes from base of antenna; posterior margin of eye fringed with black hairs; palpus normal for subfamily, third segment short, globular; proboscis normal. Thorax: dorsum covered with long, hairlike scales; patagia not capable of being raised into a hood; no evident thoracic tufts; wing venation as in subfamily. Prothoracic leg: tibia massively hairy with a long, thin claw; tibial claw without an associated lateral plate; prothoracic, mesothoracic, and metathoracic tarsal segments all with irregular masses of spines, not arranged in three rows; tarsal claw toothed; dorsal apex of last tarsal segment with two, medial, long setae. Internal tympanic structure: not examined. External tympanic structure: text figure 135 *c*; tympanal groove strong; anterior margin of first tergum membranous, without anterior overhang; strong depression present at end of tympanal groove between first and second terga, depression continuing across second tergum; hood very weak; abdominal bulla as in text figure 135 *c*; allula strong. Abdomen: untufted; males without basal hair pencils or Stobbe's glands on sternum 1+2; male seventh tergum not enlarged. Male genitalia: plate T, figures 12, 13; valve as in the figure; bottom of sacculus projecting forward as a spine; clasper present, but weakly developed; uncus swollen, with an apical spine; vesica as in figure, heavily spined, toothed at base. Female genitalia: plate V, figure 8; ovipositor lobes unmodified, slightly elongate; setae from cephalic margin of ovipositor lobes long, fine; eighth sternum and tergum spined; caudal half of ductus bursae membranous, expanded into a large, membranous sac continuous with intersegmental membrane between seventh and eighth sterna; cephalic half of ductus bursae sclerotized with obscure ridges and continuous with large, oval, heavily scelerotized plate in caudal end of corpus bursae; a second, oval, sclerotized plate in region of appendix bursae; corpus bursae ovate; seventh sternum with caudal margin heavily sclerotized and U-shaped; intersegmental membrane between seventh and eighth sterna extensive and continuous with cephalic margin of ductus bursae.

Pleromella opter Dyar

PL. 5, FIG. 34; PL. T, FIGS. 12, 13 (♂ gen.); PL. V, FIG. 8 (♀ gen.); TEXT FIG. 140 (map) (RWH 10026).

Pleromella opter Dyar, 1921, *Insecutor Inscitiae Menstruus*, **9**: 40.

Type locality: [USA], California, Eldridge. [USNM]

NOTE—Dyar described *Pleromella opter* from "one male and two females." The " male type" with the male genitalia on slide JGFC 1715 is designated and has been labeled the LECTOTYPE.

Pleromella opter is an early, spring flying species from California. It is characterized by the light-gray forewing, black-accented veins, and a dull diagonal line running from the wing apex to the middle of the inner margin. All other maculation is absent. The hindwing of both males and females is pure white. No apparent variation exists within or among populations. Wing length: mean = 16.75 mm, standard deviation = 0.64 mm, n = 10.

McFarland (1975) recorded that captive larvae avidly accepted young leaves of *Arctostaphylos glauca* Lindl. (Ericaceae).

Pleromella opter is known only from southern Oregon, California, and Baja California. Adults have been collected from February to June, but are apparently commonest in April.

FIGURE 140: DISTRIBUTION OF SPECIMENS EXAMINED OF *PLEROMELLA OPTER*

GENUS
Provia Barnes and McDunnough

Provia Barnes and McDunnough, 1910, *Can. Ent.*, **42**: 248.
Type species: *Provia argentata* Barnes and McDunnough, 1910. Original designation.

Provia is a distinctive, monotypic genus of the western United States. The genus may be recognized by the following combination of characters: 1) heavily lashed eyes, 2) a strong tibial claw without an accompanying plate, 3) external tympanic structure (text figure 135 *d*) of a long, lateral abdominal bulla separated by a lateral depression from a second ventral bullalike area, 4) simple male antennae, and 5) adult spring flight period.

Provia is another distinctive genus of uncertain affinities with which the subfamily abounds. The closest relative of *Provia* is probably *Pleromella* Dyar. The external and internal tympanic features of both genera appear similar. The tarsal segments of *Provia* tend to be irregularly spined, although not as distinctively so as in *Pleromella.* The adults of both genera are early spring fliers. Both genera share the irregular tarsal spining, the spring flight pattern, and the general external tympanic structure of the Psaphidini. The structure of the abdominal bulla of the external tympanic region is different; other similarities may be the result of convergence due to the spring flight habits of the adults. Lacking a better place to put them, I have placed these two genera at the end of the Psaphidini. These two genera are possibly relics linking the Psaphidini with the Triocnemidini.

Head: male and female antenna simple, scaled dorsally, ciliated ventrally; antennal base with a tuft of white and brown scales overhanging eye; large, black tuft of lashes present at base of antenna; rear margin of eye with long, black lashes; front without tubercles or other processes, closely covered with white scales; vertex with bushy tuft of long, white scales; third palpal segment short, globular; second and first palpal segments covered with a mixture of short white and brown scales, and long black hairs; proboscis normal. Thorax: dorsum covered with long, hairlike white and brown scales; no noticeable thoracic tufts except for weak, metascutal tuft of slightly curled scales; wing venation as in subfamily. Prothoracic leg: tibia short with a long tibial claw, claw without associated plate; first tarsal segment long, about 3 × as long as second segment, only apical third spined; segments 2–4 with three rows of spines, but middle row irregular with many small narrow spines; fifth tarsal segment almost bare, only a few scattered, small spines; tarsal claw with tooth; dorsal apex of fifth segment with two lateral, long hairs. Mesothoracic and metathoracic legs: tarsal segments as in prothoracic leg, but middle row of spines more irregular, not in a discernible row; fifth segment more heavily spined than in prothoracic leg, but spines not in discernible pattern. Internal tympanum: scutal phragma present; pocket 1 not lipped; pocket 2 short; pocket 3 not fused to tympanal air sack; pocket 4 unmodified; tendon plate large; nodular sclerite poorly developed. External tympanum: text figure 135 *d*; alula moderate; abdominal bulla long, lateral, separated from ventral bulla by long lateral depression; cephalic margin of first tergum not lipped; tympanal groove strong, clearly continued on second tergum; hood moderate, not strong. Abdomen: dorsal abdominal tufts absent; male sternum 1+2 without basal hair pencils or Stobbe's gland; male seventh tergum enlarged, sclerotized; eighth male sternum with U-shaped sclerotized area, with weak, medial hair brush. Male genitalia: text figure 134 *c*; valve simple, narrow, with long, well-developed clasper; uncus short, flattened, with pointed tip; vesica oblong, heavily covered with spines in a band widening from base to apex; no terminal, larger spine present. Female genitalia: plate W, figure 1; ovipositor lobes unmodified; bursa large, globular, faintly bilobed with sclerotized crenulations in corpus bursae; corpus bursae slightly plicate.

Provia argentata Barnes and McDunnough
PL. 5, FIG. 35; TEXT FIG. 134 *e, f* (♂ gen.); PL. W, FIG. 1 (♀ gen.); TEXT FIG. 141 (map) (RWH 10025).

Provia argentata Barnes and McDunnough, 1910, *Can. Ent.,* **42**: 248.
Type locality: [USA], Utah, Eureka. [USNM]
NOTE—*Provia argentata* was described from "two males and one female." The specimen bearing the notation "type ♂" is designated and has been labeled as the LECTOTYPE.

Provia argentata is a pretty species from the western United States. The moth's forewing has a silvery-white wash over a brown background. A white, V-shaped mark radiates from the bottoms of the reniform and claviform marks forming a conspicuous white triangle. The male hindwing is dull white with brown accented veins. The female has a uniformly brown hindwing. Wing length: mean = 15.95 mm, standard deviation = 0.71 mm, n = 10.

McFarland (1975) records the foodplant of the larvae as *Purshia glandulosa* Curran, a woody shrub of the family Rosaceae. He states that the larva is colorful, "marked with vivid red and pure white over a translucent green ground color."

The distribution of *argentata* is spotty. The type series was from central Utah, but I have also seen specimens from southern California, and Yavapai County, Arizona. No geographical variation is present. The adults of this species fly in late spring.

FIGURE 141: DISTRIBUTION OF SPECIMENS EXAMINED OF *PROVIA ARGENTATA*

TRIBE

Feraliini Poole, NEW TRIBE

Feraliini Poole.
Type genus: *Feralia* Grote, 1874.
NOTE—The tribe Feraliini was first proposed by Franclemont and Todd *in* Hodges et al. (1983) but was not described. No earlier description has been found. Therefore, the tribal name Feraliini is treated as new.

The tribe contains, for certain, only the genus *Feralia.* The Feraliini bear several similarities to the Psaphidini, but sufficient differences and uncertainties exist to treat them as a separate entity. The similarities are 1) an almost identical external tympanic structure, 2) the massive and irregular spining of the tarsal segments, 3) the spring flight habits of the adults, and 4) the hairiness of the adults. All of these characters except the first may be convergences because of the spring flight habits of the adults in both tribes. Major differences between the Psaphidini and the Feraliini are the absence of a prothoracic tibial claw in *Feralia* and the highly modified male and female genitalia of *Feralia.*

The phylogenetic affinities of *Feralia* are, as often seems to be the case, uncertain. *Feralia* bears a strong external and superficial resemblance to the type species of *Daseochaeta* Warren, the Japanese species *Daseochaeta viridis* (Leech). The external structure of the tympanum, in particular, is nearly identical in both genera; however, the two types of male genitalia are widely divergent. Another candidate for sister group is the primarily Asian genus *Diphtherocome* Warren. Superficial and external similarities are again persuasive, but widely different types of male genitalia are not.

The species of *Feralia* are early spring fliers feeding, as larvae, on either coniferous or deciduous trees. All of the known species, except one, live in northern or montane parts of North America. A single species occurs in Japan and the eastern regions of Russia, *Feralia sauberi* (Graeser).

The formal description of the tribe and the genus *Feralia* follow. Head: male antenna bipectinate with pectinations ciliate; female antenna ciliate; antenna of both sexes with bushy hair tufts from scape of antenna; front of head bushy with long hairlike scales; ocellus present, but sometimes well hidden by hairs, ocellus located well back from eye margin; eye, in some species, slightly reduced in size; front not protuberant and without a frontal processes; proboscis present, well developed; palpus unmodified. Thorax: dorsum covered with hairlike scales; two ill-defined tufts arising from mesothorax and metathorax; wing venation as in "trifid" Noctuidae. Prothoracic leg: tibia with extraordinary large epiphysis arising from base and projecting past distal end, almost as large as tibia itself; tibial claw absent; tarsal segments heavily and irregularly spined; tarsal claw toothed. Mesothoracic leg: unremarkable; first tarsal segment approximately as long as remaining four segments; spining of tarsal segments heavy and irregular. Metathoracic leg: similar to mesothoracic leg; all three pairs of legs with tibia and femur heavily covered with hair. Male abdomen: abdomen covered with hairlike scales and, in *februalis* and *meadowsi,* hairs mixed with quill-like spines; last tergum and sternum of male poorly sclerotized, without apophyses; basal hair pencils and Stobbe's glands absent in all species. External tympanic region: hood large; alula well developed; bulla in intersegmental membrane rectangular to ovoid, laterally stretching from hood to second abdominal segment. Internal tympanic structure: identical with *Psaphida.* Male genitalia: plate T, figures 14, 15; aberrant, clasper absent; possible digitus present as projection on inner margin of valve; uncus swollen, with a hook and a protuberance of undetermined origin from its inner face; juxta simple, V-shaped; vesica ovoid with elongated tail, cornuti occurring in patches and often with patches of small hairy spines. Female genitalia: plate W, figure 2; ovipositor lobes with long spines interspersed with minute spines; lamella postvaginalis and antevaginalis spiny; ostium and upper part of ductus bursae rectangular, internally granulose to spiny; lower half of ductus bursae clearly delimited from upper half, heavily granulose internally; internal granules of ductus bursae continued into intersection of appendix and corpus bursae; appendix bursae small and triangular; corpus bursae ovate-elongate, without signum. Larva: larva of "larva liberae" type; crochets uniordinal; white lateral stripe continued onto anal shield; larva typically green with stripes; an external feeder on coniferous and deciduous trees and shrubs.

GENUS

Feralia Grote

Feralia Grote, 1874, *Bull. Buffalo Soc. Nat. Sci.,* **2**: 6.
Type species: *Diphtera jocosa* Guenée, 1852. Original designation.

Momophana Grote, 1875, *Stettiner Ent. Zeit.,* **36**: 195.
Type species: *Feralia comstocki* Grote, 1874. Monotypy.

Arthrochlora Grote, 1875, *Stettiner Ent. Zeit.,* **36**: 197.
Type species: *Feralia februalis* Grote, 1874. Original designation.

The green coloration and hirsute appearance of *Feralia* and the stucture of the male and female genitalia are very distinctive, especially the digituslike projection of the male valve and the highly modified uncus.

Feralia is divided into three, well-defined, species groups. The first group contains *jocosa* and *deceptiva* as sister species with *major* coming off the next lower node. The second group consists of *comstocki,* and possibly the Old World species *sauberi,* and is defined by the long, pendulant process of the uncus and the divergent shape and spining of the vesica. The third group consists of *februalis* and its sister species *meadowsi.* Both have quill-like spines on the terga of the adult abdomen. This group has changed foodplants from coniferous to deciduous plants. Also these two species represent a shift from the mesic habits of the other species to a more xeric environment. All of the species are early spring fliers, in April and May in the north, to as early as January in southern California.

The systematic position of the only Old World species, *Feralia sauberi* Graeser, has not been determined. Its superficial appearance is nearly identical to that of *comstocki.*

A key to species based on superficial appearance follows. The prevalence of melanism in this genus should be borne in mind when using this key because melanics can cause problems in identification. A combination of the male genitalia and locality is usually sufficient to identify a species in cases of aberrant specimens.

KEY TO THE SPECIES OF *FERALIA* IN NORTH AMERICA

1. Hindwing white with at most a discernible median line, not suffused with brown or brownish shades; terga of abdomen with quill-like spines 2
— Hindwing with at least some brown or other dark shades; abdomen without quill-like spines on terga 3

2. Forewing lighter with a bluer, more hoary appearance; dash in tornos of forewing generally stronger; dot in anal margin of hindwing gen-

erally stronger; valve of male genitalia broader, ratio of greatest width of sacculus to length from apex of valve to base of sacculus 0.21; known only from Channel Islands off California .... *meadowsi* p. 187

— Forewing usually less blue, clearer, without a hoary appearance; dash in tornos of forewing generally weaker; dot in anal margin of hindwing generally weaker; valve of male genitalia narrower, ratio of greatest width of sacculus to length from apex of valve to base of sacculus approximately 0.15; mainland California, not known from Channel Islands .... *februalis* p. 186

3. Hindwing entirely suffused with dull black brown with a dull-white outer margin; lower outer margin of hindwing not suffused with green 4

— Hindwing various, but with lower outer margin of hindwing suffused with green 5

4. Larger, average wing length 16–18 mm; outer side of forewing postmedial line defined by black; claviform mark usually degenerate; West Coast from central California to British Columbia and Alberta .... *deceptiva* p. 184

— Smaller, average wing length from base to apex 14–16 mm; outer side of forewing postmedial line usually not defined by black; claviform mark usually distinctly present; mostly eastern North American although occurring in eastern and central British Columbia .... *jocosa* p. 183

5. Forewing color green, black, and white, without a hoary appearance; black patches usually present between orbicular and reniform marks, and between lower end of reniform mark and postmedial line; contrasting black and green-white patches usually present between outer margin of reniform mark and postmedial line .... *comstocki* p. 185

— Forewing color hoary green with at most only diffuse black or green-white patches; patch between orbicular and reniform marks not strong or distinct; no distinct black patch from lower end of reniform mark to postmedial line; no contrasting green-white and black patches between outer margin of reniform mark and postmedial line .... *major* p. 185

Feralia jocosa (Guenée) (Féralie Joyeuse (f.), Fr.)

PL. 5, FIGS. 36–39; PL. T, FIGS. 14, 15 (♂ gen.); PL. W, FIG. 2 (♀ gen.); TEXT FIG. 142 (map) (RWH 10005).

Diphtera jocosa Guenée, 1852, *in* Boisduval and Guenée, *Histoire Naturelle des Insectes, Lépidoptères,* **5**: 37.
Type locality: "Amerique Septentrionale, female." [BMNH]

NOTE—Guenée specifies only a single female in the original description. Two females marked type exist in the BMNH, neither with Guenée's original type label. The second female is a form with a black median area. This black median area is not mentioned in the original description, and I consider this "type" to be spurious. The first specimen is considered to be the holotype.

Feralia furtiva Smith, 1909, *Jour. New York Ent. Soc.,* **17**: 57.
Type locality: [Canada], Ontario, Sudbury. [AMNH]

Feralia jocosa is a variable species of the forests of northern North America. The commonest form is uniformly apple green with thick antemedial and postmedial lines, white bordered with black on both inner and outer sides. The black markings on the outer side of the postmedial line are never as strong or as distinctive as in *deceptiva.* Variable amounts of black obscure the markings in other specimens, particularly in the basal and median areas. The subterminal area seems immune to the black blotches. The subterminal area is more often yellow green rather than apple green in specimens with extensive black areas. A distinctive third form exists with a rusty-green forewing and strongly contrasting whitish markings. The green colors of the forewing are strongly susceptible to fading through age or moisture. The eyes of both *jocosa* and *deceptiva* are reduced in size, and a wide space exists between the upper margin of the eye and the ocellus. Wing length: mean = 15.07 mm, standard deviation = 0.52 mm, n = 10.

The closest relative of *jocosa* is *deceptiva. Feralia deceptiva* is a larger species than *jocosa* and has a nearly disjunct range. *Feralia jocosa* occurs through most of northern North America from Nova Scotia in the east, westward to eastern and central British Columbia. *Feralia deceptiva* is apparently restricted to the coastal areas of the west coast. *Feralia deceptiva* has a wing length of 17–18 mm; the corresponding measurement in *jocosa* is 14–16 mm. *Fe-*

FIGURE 142: DISTRIBUTION OF SPECIMENS EXAMINED OF *FERALIA JOCOSA*

ralia deceptiva appears to lack the variability in forewing coloration of *jocosa* and is uniformly apple green without black forms. The forewing postmedial line of *deceptiva* is strongly accented with black. Black scaling also occurs along the outer margin of the postmedial line of *jocosa,* but never as distinctly so. The median area of *jocosa* always seems to have at least some black scaling. The median area of *deceptiva* lacks any black scaling. I have found no consistent differences between the male or female genitalia of the two species.

The larva has been described by Crumb (1956) who lists black spruce (*Picea mariana* (Mill.) B.S.P.), balsam fir (*Abies balsamea* (L.) Mill.), and hemlock (*Tsuga canadensis* (L.) Carrière) as hosts. Prentice (1962) lists many host plants of the genera *Abies, Picea,* and *Tsuga.* All hosts are Pinaceae. The commonest foodplant was *Abies balsamea.* McCabe (1991) published a photograph of the larva. It is green with conspicuous longitudinal white lines. The spiracles are white and margined dorsally with a strong red line. Yellow spots are present on the prolegs. The head is green with a few black flecks. The larva, according to McCabe, is indistinguishable from that of *comstocki,* but may be separated from *major* by the evenness of the subdorsal line in *major* and its periodic constriction in *jocosa* and *comstocki.*

Feralia jocosa is most common in the forests of the northeastern United States and Canada. *Feralia jocosa* is common from Nova Scotia south as far as western Maryland and southern Ohio. The distribution map shows specimens I have seen, but Prentice (1962) records a distribution stretching throughout almost all of the Canadian Provinces south of the taiga zone from Newfoundland in the east to central British Columbia in the west. These records are based on larvae collected by the Canadian Forest Survey. Prentice (1962) does not record *Feralia deceptiva,* and thus I suspect most records of *jocosa* from the Pacific Coast region of British Columbia refer to *deceptiva.* A single specimen of *jocosa* labeled Victoria, British Columbia is present in the USNM; it may be mislabeled. The line of demarcation between *jocosa* and *deceptiva* needs to be studied further. The percentage of melanic forms may be a function of location, but I cannot determine any details of geographic differences from the specimens available. I have not detected any other geographical variation. The adults fly in April and May and probably overwinter as pupae.

Feralia deceptiva McDunnough

PL. 5, FIG. 40; PL. U, FIGS. 1, 2 (♂ gen.); PL. W, FIG. 3 (♀ gen.); TEXT FIG. 143 (map) (RWH 10006).

Feralia deceptiva McDunnough, 1920, *Can. Ent.,* **52**: 162, fig. 2 (male genitalia).
Type locality: [Canada], British Columbia, Vancouver. [CNC]

Feralia deceptiva is the sister species of *jocosa.* No consistent differences have been found in the male or female genitalia of the two species. *Feralia deceptiva* is significantly larger than *jocosa*; forewing length is 17–18 mm in *deceptiva* but only 14–16 mm in *jocosa.* Secondly *deceptiva* appears to lack the individual variability of *jocosa* and is always uniformly apple green. Other differences in superficial appearance were discussed under *jocosa.* Wing length: mean = 17.98 mm, standard deviation = 0.47 mm, n = 10.

Crumb (1956) describes the larvae and records the foodplant as Douglas-fir (*Pseudotsuga menziesii* (Mirbel) Franco) (Pinaceae). The larva is essentially the same as in *jocosa,* but the subdorsal and subventral lines are continuous and not broken up into segments as in *jocosa.*

Feralia deceptiva is found primarily along the Pacific Coast from Vancouver southward through Washington and Oregon to central California. A specimen has been seen from western Alberta. Adults fly in the spring.

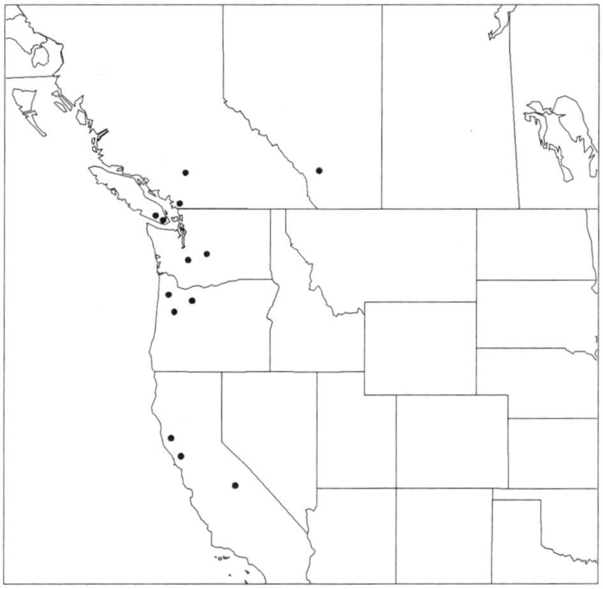

FIGURE 143: DISTRIBUTION OF SPECIMENS EXAMINED OF *FERALIA DECEPTIVA*

FIGURE 144: DISTRIBUTION OF SPECIMENS EXAMINED OF *FERALIA MAJOR*

Feralia major Smith
PL. 5, FIGS. 41–43; PL. U, FIGS. 3, 4 (♂ gen.); PL. W, FIG. 4 (♀ gen.); TEXT FIG. 144 (map) (RWH 10007).

Feralia major Smith, 1890, *Ent. Americana,* **6**: 26.
Type locality: [USA], District of Columbia. [USNM]

Feralia major has a wide distribution in eastern North America but is apparently not as common as most of the other species. *Feralia major* is superficially similar to *jocosa* and is sympatric with it in many localities. The presence of green along the lower outer margin of the hindwing in *major* should easily separate it from *jocosa. Feralia major* is slightly larger than *jocosa,* a fact usually obvious in mixed series of the two species. This species has a more hoary appearance than *jocosa,* and the green form of *major* appears as if it has a slight dusting of powder. Melanic and partially melanic specimens of *major* exist, but the black is more diffuse than in *jocosa.* The reniform and orbicular marks are not contrastingly lighter in melanic specimens of *major. Feralia comstocki* is a brighter appearing species, and there are distinctive black patches in the forewing between the orbicular and reniform marks and between the reniform mark and the postmedial line. The forewing of *major* is duller in appearance, and there are no distinctive black patches as described above. Wing length: mean = 16.65 mm, standard deviation = 1.12 mm, n = 10.

Rockburne and Lafontaine (1976) record the larva from spruce (*Picea* spp.) (Pinaceae). McCabe (1991) records the species from *Pinus strobus* L. (white pine). *Feralia major* may be separated from *jocosa* and *comstocki* by the evenness of the subdorsal lines.

This species has a wide distribution. It occurs from Nova Scotia, south to northern Florida, and west to eastern Texas. In the north it ranges west to Wisconsin. Melanic specimens are common, ranging from a light dusting of black to totally black specimens. There is no apparent geographical variation. The adults fly in April and May in the north to as early as December in Florida and Georgia.

Feralia comstocki Grote (Féralie de Comstock (f.), Fr.)
PL. 5, FIGS. 44, 45; PL. U, FIGS. 5, 6 (♂ gen.); PL W, FIG. 5 (♀ gen.); TEXT FIG. 145 (map) (RWH 10008).

Feralia comstocki Grote, 1874, *Bull. Buffalo Soc. Nat. Sci.,* **2**: 59.
Type locality: [USA], New York, Ithaca. [BMNH]

Feralia columbiana Smith, 1903, *Can. Ent.,* **35**: 9.

FIGURE 145: DISTRIBUTION OF SPECIMENS EXAMINED OF *FERALIA COMSTOCKI*

Type locality: [Canada], British Columbia, Westminster. [AMNH]

Momaphana comstocci Hampson, 1906, *Catalogue of the Lepidoptera Phalaenae in the British Museum*, **6**: 199.

NOTE—*Momaphana comstocci* is an unjustified emendation of *Feralia comstocki.*

Feralia comstocki is distinctive. The forewing pattern consists of a patchwork of green, black, and green-white patches. Black patches are prominent between the orbicular and reniform marks and between the reniform mark and the postmedial line. A distinctive green-white patch usually runs from the bottom of the reniform mark to the postmedial line. The hindwing is commonly lighter than in either *jocosa* or *major* and may have dull-white areas. The green areas along the lower outer margin of the hindwing will separate this species from *jocosa* and the strong black and green-white patches will distinguish it from either *jocosa* or *major.* Wing length: mean = 16.82 mm, standard deviation = 0.67 mm, n = 10.

The larva has been figured by McCabe (1991), but he considers it indistinguishable from *jocosa.* Ross (1955) described the larva and pupa, although his comparisons of the pupa of this species with *jocosa* may be based on a misidentification of *deceptiva.* Prentice (1962) records this species from many localities in British Columbia and western Alberta, and a few from eastern Canada from *Abies, Pseudotsuga,* and *Picea* (Pinaceae)

Feralia comstocki occurs from Nova Scotia to British Columbia, Oregon, and Washington. In the east it occurs as far south as the mountains of North Carolina. *Feralia comstocki* is not as variable as *jocosa* or *major,* and no true melanic specimens have been seen. The strength of the black patches varies slightly. The green of the forewing ranges from light green to green to an olive green. A few specimens have a strong yellow tint in the hindwing. The black forewing patches are slightly more diffuse in British Columbia and Washington material than in eastern specimens. The name *columbiana* is available for these western populations if a subspecies concept is used. Adults have been collected in May and June.

Feralia februalis Grote

PL. 5, FIGS. 46–48; PL. U, FIGS. 7, 8 (♂ gen.); PL. W, FIG. 6 (♀ gen.); TEXT FIG. 146 (map) (RWH 10009).

Feralia februalis Grote, 1874, *Bull. Buffalo Soc. Nat. Sci.,* **2**: 60.
Type locality: [USA], California, Sanzalito [Sausalito]. [BMNH]

Feralia februalis and *meadowsi* are easily recognized by the presence of quill-like spines intermixed with the hairs on the abdominal terga. The hindwing is white without the dark suffusions or patches of the other species in *Feralia* and with only a hint of a postmedial line and discal dot. Both species usually have a distinctive blue tint to the forewing. *Feralia februalis* and *meadowsi* are superficially very similar. The forewing of *meadowsi* is usually distinctly lighter and often bluer than *februalis.* The forewing of *meadowsi* also has a distinctive hoary appearance not found in *februalis.* Good characters separating the two species can be found in the male genitalia. In particular the valve of *februalis* is much narrower than that of *meadowsi.* The various ratios of valve length to sacculus width are given in the key to species. Wing length: mean = 15.93 mm, standard deviation = 0.92 mm, n = 10.

The larva has been described by Crumb (1956) who lists *Cercocarpus betuloides* Nutt. (Rosaceae) and *Quercus douglasi* Hook. & Arn. (Fagaceae) as hosts. Buckett [1968] lists other deciduous trees and shrubs. The larva has a very conspicuous conical hump on abdominal segment 8 and the pedicel of the antenna is yellow, two characters separating this

FIGURE 146: DISTRIBUTION OF SPECIMENS EXAMINED OF *FERALIA FEBRUALIS*

FIGURE 147: DISTRIBUTION OF SPECIMENS EXAMINED OF *FERALIA MEADOWSI*

species from other species in *Feralia.* This species, and presumably *meadowsi,* have distinctly different habits than the remaining species of the genus, feeding on deciduous shrubs and trees rather than conifers and in distinctly more xeric habitats.

Feralia februalis ranges from Oregon to southern California. A great deal of variation in forewing color exists within populations. Forewing color ranges from green to yellow green to blue green. A few specimens have black patches in the basal area and as a thick black line following the postmedial line. I have not noted any correlation between individual variability and geography. This species flies early in the spring.

Feralia meadowsi Buckett

PL. 5, FIG. 49; PL. U, FIGS. 9, 10 (♂ gen.); TEXT FIG. 147 (map) (RWH 10010).

Feralia meadowsi Buckett, [1968] 1967, *Jour. Research Lep.,* **6**: 44, figs. 1, 2 (adults), figs. 7, 8 (male genitalia).
Type locality: [USA], California, Santa Catalina Island, Avalon Club House. [USNM]

Feralia meadowsi is the sister species of *februalis* and is apparently endemic to Santa Catalina Island off the southern coast of California. The characters separating the two species are discussed under *februalis.* Only a single, unquestionable female of *meadowsi* is known, the specimen made allotype by Buckett in his original description. When the type material was returned to the USNM, the associated genitalia slides apparently were not. I have been unable to find these slides, and they may be lost. Consequently I have been unable to illustrate the female genitalia of *meadowsi* nor did Buckett figure them. Buckett described the female genitalia as being smaller than those of *februalis* with virtually no large setae on the ovipositor lobes in contrast to a number of large setae in *februalis.* The ductus bursae is said to be less strongly sclerotized than that of *februalis.* Wing length: mean = 15.93 mm, standard deviation = 0.78 mm, n = 10.

The larva and its foodplants are unknown, although presumably they are not much different than those of *februalis.*

Feralia meadowsi is apparently limited to Santa Catalina Island. A single female exists in the USNM from Santa Cruz Island that I have attributed to this species. However, without a male, it cannot be definitely attributed to *meadowsi.* There is little or no individual variability in the species.

LITERATURE

BARNES, W., and J. H. MCDUNNOUGH. 1917. *Checklist of the Lepidoptera of Boreal America.* 392 pp. Decatur: Herald Press.

BROWN, F. M. 1964. The dates of publication of *Lepidoptera, Rhopaloceres-Heteoceres* by Herman Strecker. *Jour. Lep. Soc.,* **18**: 43–44.

BUCKETT, J. S., and W. R. BAUER. 1966. Key to the genera of the Psaphidini, with descriptions of a new genus and species from western North America. *Jour. Lep. Soc.,* **20**: 83–91.

COMSTOCK, J. A., and C. DAMMERS. 1935. Notes on the early stages of three butterflies and six moths from California. *Bull. So. California Acad. Sci.,* **34**: 120–142.

COMSTOCK, J. A., and C. HENNE. 1940. Notes on the early stages of *Xanthothrix ranunculi. Bull. So. California Acad. Sci.,* **9**: 198–200.

COMSTOCK, J. A., and C. HENNE. 1943. The larva of *Copicucullia basipuncta* B. & McD. (Lepidoptera). *Bull. So. California Acad. Sci.,* **42**: 45–46.

COOK, W. C. 1930. An ecologically annotated list of the Phalaenidae of Montana (Lepid.). *Can. Ent.,* **62**: 265–277.

CRUMB, S. E. 1956. The Larvae of the Phalaenidae. *U. S. Dept. Agric. Tech. Bull.,* **1135**: 1–356, pl. 1–11.

DETHIER, V. G. 1944. Observations on the life history of *Cucullia asteroides* Gn. *Can. Ent.,* **76**: 161–162, 1 plate.

DYAR, H. G. 1901. Note on the larva of *Psaphidia thaxterianus. Jour. N.Y. Ent. Soc.,* **9**: 84–85.

FARRIS, J. S. 1988. *Hennig 86.* [A computer program for phylogenetic estimation].

FORBES, W. T. M. 1954. Lepidoptera of New York and neighboring States. Part III. Noctuidae. *Cornell Univ. Agric. Experiment Station, Mem.,* **68**: 1–433.

FORSTER, W., and T. A. WOHLFAHRT. 1971. *Die Schmetterlinge Mitteleuropas. Eulen (Noctuidae).* 329 pp., 32 plates. Stuttgart: Franckhsche Verlagshandlung.

FRANCLEMONT, J. G., and E. L. TODD. 1983. Noctuidae. *In* HODGES, R. W., et al., *Check List of the Lepidoptera of America North of Mexico,* pp. 120–159. London: E. W. Classey Ltd. and The Wedge Entomological Research Foundation.

GODFREY, G. L., and F. H. STEHR. 1985. Note on Crumb's "liberae et confluentae" couplet (Noctuidae). *Jour. Lep. Soc.,* **39**: 57–59.

GROTE, A. R. 1874. List of the Noctuidae of North America. *Bull. Buffalo Soc. Nat. Sci.,* **2**: 1–77, 1 plate.

GROTE, A. R. 1882. *New Checklist of North American Moths.* 73 pp. [no publisher given].

GROTE, A. R. 1883. Introduction to a study of the North American Noctuidae. *Proc. Amer. Philosophical Soc.,* **21**: 134–176.

GROTE, A. R. 1890. *North American Lepidoptera. Revised Checklist of the North American Noctuidae. Part I. Thyatirinae–Noctuinae.* 52 pp. Bremen: Homeyer and Meyer.

HAMPSON, G.F. 1903. *Catalogue of the Lepidoptera Phalaenae in the British Museum,* **4**: 1–689.

HAMPSON, G. F. 1906. *Catalogue of the Lepidoptera Phalaenae in the British Museum,* **6**: 1–532.

HAMPSON, G. F. 1910. *Catalogue of the Lepidoptera Phalaenae in the British Museum,* **9**: 1–552.

HARDWICK, D. F. 1970. A Generic Revision of the North American Heliothidinae (Lepidoptera: Noctuidae). *Ent. Soc. Canada, Mem.* **73**: 1–59.

HODGES, R. W., et al., eds. 1983. *Check List of the Lepidoptera of America North of Mexico.* xxiv + 284 pp. London: E. W. Classey Ltd. and The Wedge Entomological Research Foundation.

HOGUE, C. L. 1963. A definition and classification of the tribe Stiriini (Lepidoptera: Noctuidae). *Contributions in Science,* **64**: 1–129, pl. 1–32.

HOLLAND, W. J. 1903. *The Moth Book.* xxiv + 479 pp., 48 plates. New York: Doubleday, Page and Company.

JONES, F. M. 1937. *Cirrhophanus triangulifer* Grt. (Lepidoptera, Noctuidae) in Delaware. *Ent. News,* **48**: 121–124, 1 plate.

LAFONTAINE, J. D., and R. W. POOLE. 1991. Noctuidae, Plusiinae. *In* DOMINICK, R. B., et al., *The Moths of America North of Mexico.* Fasc. **25.1**: 1–182.

LINTNER, J. A., 1876. Notes on some Lepidoptera. *Entomological Contributions,* **4**: 121–128.

MATTHEWS, M. 1991. Classification of the Heliothinae. *Natural Resources Institute, Bull.*, **44**: 1–198.

MCCABE, T. L. 1991. Atlas of Adirondack Caterpillars with a Host List, Rearing Notes and a selected Bibliography of Works Depicting Caterpillars. *New York State Museum Bull.*, **470**: 1–113.

MCCABE, T. L., and L. M. JOHNSON. 1980. Catalogue of the Types in the New York State Museum Insect Collection. *New York State Museum Bull.*, **434**: 1–38.

MCDUNNOUGH, J. H. 1938. Check List of the Lepidoptera of Canada and the United States of America. Part I. Macrolepidoptera. *Mem. So. California Acad. Sci.*, **1**: 1–275.

MCFARLAND, N. 1975. Larval foodplant records for 106 species of North American moths. *Jour. Lep. Soc.*, **29**: 112–125.

PRENTICE, R. M. (compiler) 1962. Nycteolidae, Noctuidae, Notodontidae, Liparidae. *Forest Lepidoptera of Canada recorded by the Forest Insect Survey.* **2**: 1–281. (Dept. of Forestry of Canada, Bull. 128.)

RICHARDS, A. G. 1933. Comparative skeletal morphology of the noctuid tympanum. *Ent. Americana*, **13**: 1–84.

ROSS, D. A. 1955. Differences in the pupae of *Feralia comstocki* Grt. and *F. jocosa* (Gn.) (Lepidoptera: Noctuidae). *Can. Ent.*, **87**: 275–276.

SMITH, J. B. 1882. A synopsis of the North American genera of the Noctuidae). *Bull. Brooklyn Ent. Soc.*, **4**: 47–52.

TIETZ, H. M. 1972. *An Index to the described life Stages, early Stages and Hosts of the Macrolepidoptera of the continental United States and Canada.* Volume 1, pp. 1–535 + iv; Volume 2, pp. 536–1041. Los Angeles: Entomological Reprint Specialists.

TODD, E. L. 1982. The Noctuid Type Material of John B. Smith (Lepidoptera). *U. S. Dept. Agric. Tech. Bull.*, **1645**: 1–228.

MONOCHROME PLATES

PLATE A: STRUCTURAL FEATURES OF TRIFID NOCTUIDS

1. Hair pencils on eighth abdominal sternum. *Acontia tetragona* Walker. ELT 1208.
2. Basal hair pencils and Stobbe's gland. *Cucullia lilacina* Schaus. USNM 44619. (p. 40).
3. Pockets for basal hair pencils. *Cucullia lilacina* Schaus. USNM 44619. (p. 40).
4. Digitus present. *Mniotype* sp. USNM 44579.
5. Clasper oriented along axis of valve. *Andropolia* sp. USNM 44595.
6. Clasper oriented perpendicular to axis of valve. *Cucullia omissa* Dod. USNM 44629. (p. 36).
7. Tegumen with shoulders. *Epidemas* sp. USNM 44617.
8. Tegumen without shoulders. *Oncocnemis deceptiva* Barnes and Lindsey. USNM 41224.
9. Free pleurite present. *Cucullia intermedia* Speyer. USNM 44642. (p. 43).
10. Free pleurite modified. *Apamea cinefacta* Grote. JGFC 71.
11. Free pleurite absent. *Condica charada* Schaus. USNM 44837.
12. Corona present. *Cucullia lethe* Poole. USNM 44652. (p. 50).

PLATE B: MALE GENITALIA OF CUCULLIINAE SPECIES

1. *Cucullia lilacina* Schaus; Guadaloupe Mts., Texas. USNM 44619;
2. Aedoeagus. (p. 40)
3. *Cucullia eucaena* Dyar; Sierra Vista, Arizona. USNM 44621;
4. Aedoeagus. (p. 40).
5. *Cucullia oribac* Barnes; Zacualpan, Mexico. USNM 44623;
6. Aedoeagus. (p. 41).
7. *Cucullia convexipennis* Grote and Robinson; Woodhaven, New York. USNM 44625;
8. Aedoeagus. (p. 42).
9. *Cucullia asteroides* Guenée; Newtown, New York. USNM 44627;
10. Aedoeagus. (p. 34).
11. *Cucullia montanae* Grote; Denver, Colorado. USNM 44637;
12. Aedoeagus. (p. 35).
13. *Cucullia similaris* Smith; Colorado Springs, Colorado. USNM 44639;
14. Aedoeagus. (p. 36).
15. *Cucullia omissa* Dod; West Dover, Nova Scotia. USNM 44629. (p. 36).

PLATE C: MALE GENITALIA OF CUCULLIINAE SPECIES

1. *Cucullia omissa* Dod; aedoeagus, West Dover, Nova Scotia. USNM 44629. (p. 36).
2. *Cucullia florea* Guenée; Debert, Nova Scotia. USNM 44635;
3. Aedoeagus. (p. 37).
4. *Cucullia postera* Guenée; Petite Riviere, Nova Scotia. USNM 44633;
5. Aedoeagus. (p. 38).
6. *Cucullia obscurior* Smith; Routt Co., Colorado. USNM 44641; (p. 43).
7. Aedoeagus. (p. 39).
8. *Cucullia intermedia* Speyer; Flushing, New York. USNM 44642;
9. Aedoeagus. (p. 43).
10. *Cucullia speyeri* Lintner; Pittsburgh, Pennsylvania. USNM 44644;
11. Aedoeagus. (p. 46).
12. *Cucullia styx* Poole; Lomas, Distrito Federal, Mexico. USNM 44646;
13. Aedoeagus. (p. 48).
14. *Cucullia dorsalis* Smith; Glenwood Springs, Colorado. USNM 44648;
15. Aedoeagus. (p. 48).

PLATE D: MALE GENITALIA OF CUCULLIINAE SPECIES

1. *Cucullia laetifica* Lintner; Culberson Co., Texas. USNM 44650;
2. Aedoeagus. (p. 49).
3. *Cucullia lethe* Poole; Lake Brownwood State Park, Texas. USNM 44652;
4. Aedoeagus. (p. 50).
5. *Cucullia charon* Poole; Chisos Mts., Brewster Co., Texas. USNM 44654;
6. Aedoeagus. (p. 51).
7. *Cucullia alfarata* Strecker; Tryon, North Carolina. USNM 38494;
8. Aedoeagus. (p. 53).
9. *Cucullia eccissica* Dyar; Santa Catalina Island, California. USNM 44656;
10. Aedoeagus. (p. 53).
11. *Cucullia antipoda* Strecker; Eureka, Utah. USNM 44658;
12. Aedoeagus. (p. 55).
13. *Cucullia eurekae* Poole; Eureka, Utah. USNM 44660;
14. Aedoeagus. (p. 56).
15. *Cucullia incresa* (Smith); San Bernardino Co., California. USNM 44661. (p. 58).

PLATE E: MALE GENITALIA OF CUCULLIINAE SPECIES

1. *Cucullia incresa* (Smith); aedoeagus, San Bernardino Co., California. USNM 44661. (p. 58).
2. *Cucullia heinrichi* Barnes and Benjamin; Mohave Co., Arizona. JGFC 2608;
3. Aedoeagus. (p. 59).
4. *Cucullia astigma* (Smith); Tulare County, California. USNM 44664;
5. Aedoeagus. (p. 60).
6. *Cucullia eulepis* (Grote); Alamosa County, Colorado. USNM 44669;
7. Aedoeagus. (p. 61).
8. *Cucullia cuculliodes* (Barnes and Benjamin); San Bernardino County, California. USNM 44671;
9. Aedoeagus. (p. 61).
10. *Cucullia basipuncta* (Barnes and Benjamin); Clark County, Nevada. USNM 44686;
11. Aedoeagus. (p. 62).
12. *Cucullia mcdunnoughi* (Henne); Angel Creek, Elko County, Nevada. USNM 44688;
13. Aedoeagus. (p. 63).
14. *Cucullia luna* Morrison; Scottsbluff, Nebraska. USNM 44690;
15. Aedoeagus. (p. 64).

PLATE F: MALE GENITALIA OF CUCULLIINAE SPECIES

1. *Cucullia serraticornis* Lintner; Visalia, California. USNM 44692;
2. Aedoeagus. (p. 65).
3. *Cucullia comstocki* (McDunnough); Pinyon Hills, San Bernardino County, California. USNM 44694;
4. Aedoeagus. (p. 67).
5. *Cucullia strigata* (Smith); Osoyoos, British Columbia. USNM 44696;
6. Aedoeagus. (p. 67).
7. *Cucullia albida* Smith; Rounthwaite, Saskatchewan. USNM 44698;
8. Aedoeagus. (p. 68).
9. *Cucullia pulla* (Grote); Pinyon Crest, Riverside County, California. USNM 44700;
10. Aedoeagus. (p. 69).
11. *Cucullia dammersi* (McDunnough); near Perris, California. USNM 44702;
12. Aedoeagus. (p. 70).
13. *Dolocucullia dentilinea* (Smith); Cherry County, Nebraska. USNM 44706;
14. Aedoeagus. (p. 72).
15. *Dolocucullia minor* (Barnes and McDunnough); Fort Davis, Texas. USNM 44704. (p. 72).

PLATE G: GENITALIA OF CUCULLIINAE SPECIES

1. *Dolocucullia minor* (Barnes and McDunnough); aedoeagus; Fort Davis, Texas. USNM 44704. (p. 72).
2. *Cucullia lilacina* Schaus; female genitalia; Big Bend National Park, Texas. USNM 44620. (p. 40).
3. *Cucullia eucaena* Dyar; female genitalia; Big Bend National Park, Texas. USNM 44622. (p. 40).
4. *Cucullia oribac* Barnes; female genitalia; Jalapa, Mexico. USNM 44624. (p. 41).
5. *Cucullia convexipennis* Grote and Robinson; female genitalia; East New York. USNM 44626. (p. 42).
6. *Cucullia asteroides* Guenée; female genitalia; Wayne Co., New York. USNM 44628. (p. 34).
7. *Cucullia montanae* Grote; female genitalia; Glenwood Springs, Colorado. USNM 44638. (p. 35).
8. *Cucullia similaris* Smith; female genitalia; Denver, Colorado. USNM 44640. (p. 36).
9. *Cucullia omissa* Dod; female genitalia; Hammond's Plains, Nova Scotia. USNM 44630. (p. 36).
10. *Cucullia florea* Guenée; Tatamagouche, Nova Scotia. USNM 44636. (p. 37).
11. *Cucullia postera* Guenée; West Dover, Nova Scotia. USNM 44634. (p. 38).
12. *Cucullia obscurior* Smith; Glenwood Springs, Colorado. USNM 44631. (p. 39).

PLATE H: FEMALE GENITALIA OF CUCULLIINAE SPECIES

1. *Cucullia intermedia* Speyer; Catskills, New York. USNM 44643. (p. 43).
2. *Cucullia speyeri* Lintner; Pittsburg, Pennsylvania. USNM 44645. (p. 46).
3. *Cucullia styx* Poole; Paradise, Arizona. USNM 44647. (p. 48).
4. *Cucullia dorsalis* Smith; Glenwood Springs, Colorado. USNM 44649. (p. 48).
5. *Cucullia laetifica* Lintner; Alpine, Brewster Co., Texas. USNM 44651. (p. 49).
6. *Cucullia lethe* Poole; Brownsville, Texas. USNM 44653. (p. 50).
7. *Cucullia charon* Poole; Chisos Mts., Brewster Co., Texas. USNM 44655. (p. 51).
8. *Cucullia alfarata* Strecker; Washington Co., Missouri. USNM 38495. (p. 53).
9. *Cucullia eccissica* Dyar; Loma Linda, California. USNM 44657. (p. 53).
10. *Cucullia antipoda* Strecker; Colorado. USNM 44659. (p. 55).
11. *Cucullia eurekae* Poole; Eureka, Utah. USNM 38590. (p. 56).
12. *Cucullia incresa* (Smith); San Diego, California. USNM 44662. (p. 58).

PLATE I: FEMALE GENITALIA OF CUCULLIINAE SPECIES

1. *Cucullia heinrichi* Barnes and Benjamin; Newberry Springs, California. USNM 44663. (p. 59).
2. *Cucullia astigma* (Smith); Tulare County, California. USNM 44665. (p. 60).
3. *Cucullia eulepis* (Grote); Alamosa County, Colorado. USNM 44670. (p. 61).
4. *Cucullia cuculliodes* (Barnes and Benjamin); San Bernardino County, California. USNM 44685. (p. 62).

5. *Cucullia basipuncta* (Barnes and McDunnough); Clark County, Nevada. USNM 44687. (p. 62).
6. *Cucullia mcdunnoughi* (Henne); Redington, Arizona. USNM 44689. (p. 63).
7. *Cucullia luna* Morrison; Colorado. USNM 44691. (p. 64).
8. *Cucullia serraticornis* Lintner; Loma Linda, California. USNM 44693. (p. 65).
9. *Cucullia comstocki* (McDunnough); Palmdale, California. USNM 44695. (p. 67).
10. *Cucullia strigata* (Smith); Eureka, Utah. USNM 44697. (p. 67).
11. *Cucullia albida* Smith; Saskatchewan. USNM 44699. (p. 68).
12. *Cucullia pulla* (Grote); Palmdale, California. USNM 44701. (p. 69).

PLATE J: FEMALE GENITALIA OF CUCULLIINAE SPECIES AND MALE GENITALIA OF STIRIINAE SPECIES

1. *Cucullia dammersi* (McDunnough); female genitalia; Riverside, California. USNM 44703. (p. 70).
2. *Dolocucullia dentilinea* (Smith); female genitalia; Cherry Co., Nebraska. USNM 44707. (p. 72).
3. *Dolocucullia minor* (Barnes and McDunnough); female genitalia; Alpine, Texas. USNM 44705. (p. 72).
4. *Fala ptychophora* Grote; male genitalia; Mohave Desert, California. USNM 44734;
5. Aedoeagus. (p. 86).
6. *Plagiomimicus olvello* (Barnes); male genitalia; Douglas, Arizona. USNM 44736;
7. Aedoeagus. (p. 89).
8. *Plagiomimicus dimidiata* (Grote); male genitalia; Green Gulch, Big Bend National Park, Brewster County, Texas. USNM 44738;
9. Aedoeagus. (p. 90).
10. *Plagiomimicus aureolum* (Hy. Edwards); male genitalia; Santa Catalina Mts., Pinal County, Arizona. USNM 44740;
11. Aedoeagus. (p. 90).
12. *Plagiomimicus heitzmani* Poole; male genitalia; Southern Illinois University Campus, Madison County, Illinois. USNM 44744. (p. 91).

PLATE K: MALE GENITALIA OF STIRIINAE SPECIES

1. *Plagiomimicus heitzmani* Poole; aedoeagus; Southern Illinois University Campus, Madison County, Illinois. USNM 44744. (p. 91).
2. *Plagiomimicus caesium* (Blanchard and Knudson); Welder Wildlife Refuge, San Patricio County, Texas. USNM 44742;
3. Aedoeagus. (p. 94).
4. *Plagiomimicus manti* (Barnes); Garner State Park, Uvalde County, Texas. USNM 44746;
5. Aedoeagus. (p. 95).
6. *Plagiomimicus curiosum* (Neumoegen); Arizona. USNM 44749;
7. Aedoeagus. (p. 95).
8. *Plagiomimicus astigmatosum* (Dyar); Paradise, Cochise County, Arizona. USNM 44750;
9. Aedoeagus. (p. 96).
10. *Plagiomimicus hilli* (Barnes and Benjamin); Riverside County, California. USNM 44751;
11. Aedoeagus. (p. 96).
12. *Plagiomimicus mimica* Poole; Alpine, Brewster County, Texas. USNM 44681;
13. Aedoeagus. (p. 98).
14. *Plagiomimicus triplagiatus* Smith; Wemple, New Mexico. USNM 44755;
15. Aedoeagus. (p. 101).

PLATE L: MALE GENITALIA OF STIRIINAE SPECIES

1. *Plagiomimicus expallidus* Grote; Denver, Colorado. USNM 44753;
2. Aedoeagus. (p. 103).
3. *Lineostriastiria sexseriata* (Grote); Paradise, Cochise Co., Arizona. USNM 44791;
4. Aedoeagus. (p. 108).
5. *Lineostriastiria biundulalis* (Zeller); Dallas, Texas. CLH 610407-3;
6. Aedoeagus. (p. 108).
7. *Lineostriastiria hutsoni* (Smith); San Diego, California. USNM 43770;
8. Aedoeagus. (p. 111).
9. *Xanthothrix neumoegeni* Hy. Edwards; Argus Mts., California. USNM 42327;
10. Aedoeagus. (p. 112).
11. *Xanthothrix ranunculi* Hy. Edwards; Havilah, California. USNM 42328;
12. Aedoeagus. (p. 113).
13. *Chrysoecia gladiola* (Barnes); Redington, Arizona. CLH 610427-3;
14. Aedoeagus. (p. 115).
15. *Basilodes pepita* Guenée; El Rancho Cima, Comal County, Texas. USNM 44805. (p. 120).

PLATE M: MALE GENITALIA OF STIRIINAE SPECIES

1. *Basilodes pepita* Guenée; aedoeagus; El Rancho Cima, Comal County, Texas. USNM 44805. (p. 120).
2. *Hoplolythrodes arivaca* (Barnes); Deming, New Mexico. ELT 1270;
3. Aedoeagus. (p. 125).
4. *Cirrhophanus dyari* Cockerell; Mount Locke, Davis Mts., Texas. AB 4251;
5. Aedoeagus. (p. 126).
6. *Cirrhophanus pretiosa* (Morrison); Welder Refuge, Sinton, Texas. AB 4252;
7. Aedoeagus. (p. 129).

8. *Stiria rugifrons* Grote; Bottineau Co., North Dakota. USNM 44792;
9. Aedoeagus. (p. 138).
10. *Stiria intermixta* Dyar; Huachuca Mts., Arizona. USNM 44794;
11. Aedoeagus. (p. 140).
12. *Stiria dyari* Hill; Kingman, Arizona. USNM 41970;
13. Aedoeagus. (p. 141).
14. *Stiria blanchardi* (Hogue); Sierra Diablo Wildlife Management Area, Culberson Co., Texas. USNM 44796;
15. Aedoeagus. (p. 141).

PLATE N: MALE GENITALIA OF STIRIINAE SPECIES

1. *Stiria sulphurea* Neumoegen; SW Arizona. USNM 44798;
2. Aedoeagus. (p. 142).
3. *Stiria satana* Poole; Arizona. USNM 41968;
4. Aedoeagus. (p. 143).
5. *Stiria consuela* Strecker; Palm Springs, Riverside Co., California. USNM 44806;
6. Aedoeagus. (p. 143).
7. *Chalcopasta howardi* (Hy. Edwards); Brewster Co., Texas. USNM 44799;
8. Aedoeagus. (p. 147).
9. *Chalcopasta fulgens* Barnes and McDunnough; Sierra Diablo Wildlife Management Area, Culberson Co., Texas. USNM 44801;
10. Aedoeagus. (p. 147).
11. *Chalcopasta acema* (Druce); Hereford, Arizona. USNM 44803;
12. Aedoeagus. (p. 148).
13. *Neumoegenia poetica* Grote; Arizona. USNM 41841;
14. Aedoeagus; Madera Canyon, Santa Cruz Co., Arizona. USNM 41840. (p. 150).
15. *Argentostiria koebelei* (Riley); Indio, California. USNM 44723. (p. 152).

PLATE O: MALE GENITALIA OF STIRIINAE SPECIES

1. *Argentostiria koebelei* (Riley); aedoeagus; Indio, California. USNM 44723. (p. 152).
2. *Chamaeclea pernana* (Grote); Paradise, Cochise Co., Arizona. USNM 44811;
3. Aedoeagus. (p. 154).
4. *Chamaeclea basiochrea* Barnes and McDunnough; Hidalgo Co., Texas. USNM 44729;
5. Aedoeagus. (p. 155).
6. *Bistica noela* (Druce); Madera Canyon, Santa Cruz Co., Arizona. USNM 44716;
7. Aedoeagus. (p. 156).
8. *Hemioslaria pima* Barnes and Benjamin; Baboquivari Mts., Pima Co., Arizona. USNM 44812;
9. Aedoeagus. (p. 158).
10. *Heminocloa mirabilis* (Neumoegen); Baboquivari Mts., Pima Co., Arizona. USNM 44813;
11. Aedoeagus. (p. 159).
12. *Thurberiphaga diffusa* (Barnes); Santa Catalina Mts., Arizona. USNM 44814;
13. Aedocagus. (p. 161).

PLATE P: FEMALE GENITALIA OF STIRIINAE SPECIES

1. *Fala ptychophora* Grote; Southern California. USNM 44735. (p. 86).
2. *Plagiomimicus olvello* (Barnes); Douglas, Arizona. USNM 44737. (p. 89).
3. *Plagiomimicus dimidiata* (Grote); Green Gulch, Big Bend National Park, Brewster County, Texas. USNM 44739. (p. 90).
4. *Plagiomimicus aureolum* (Hy. Edwards); S.W. Arizona. USNM 44741. (p. 90).
5. *Plagiomimicus heitzmani* Poole; Urbana, Illinois. USNM 44745. (p. 91).
6. *Plagiomimicus caesium* (Blanchard and Knudson); Welder Wildlife Refuge, San Patricio County, Texas. USNM 44743. (p. 94).
7. *Plagiomimicus manti* (Barnes); Fort Davis, Jeff Davis County, Texas. USNM 44747. (p. 95).
8. *Plagiomimicus curiosum* (Neumoegen); Huachuca Mts., Arizona. USNM 44748. (p. 95).
9. *Plagiomimicus hilli* (Barnes and Benjamin); Indio, California. USNM 44752. (p. 96).
10. *Plagiomimicus mimica* Poole; San Bernardino County, California. USNM 44679. (p. 98).
11. *Plagiomimicus triplagiatus* Smith; Deming, New Mexico. USNM 44756. (p. 101).
12. *Plagiomimicus expallidus* Grote; Canadian, Hemphill County, Texas. USNM 44754. (p. 103).

PLATE Q: FEMALE GENITALIA OF STIRIINAE SPECIES

1. *Cuahtemoca unicum* (Barnes and Benjamin); holotype; Paradise, Cochise Co., Arizona. ELT 1271. (p. 105).
2. *Lineostriastiria sexseriata* (Grote); Paradise, Cochise Co., Arizona. CLH 610517-1. (p. 108).
3. *Lineostriastiria biundulalis* (Zeller); Laguna Park, Texas. USNM 44772. (p. 108).
4. *Lineostriastiria hutsoni* (Smith); Quartzite, Yuma Co., Arizona. USNM 44773. (p. 111).
5. *Xanthothrix neumoegeni* Hy. Edwards; Los Angeles, California. USNM 44575. (p. 112).
6. *Xanthothrix ranunculi* Hy. Edwards; Havilah, California. USNM 44576. (p. 113).
7. *Chrysoecia gladiola* (Barnes); Santo Domingo Valley, Baja California. USNM 44777. (p. 115).
8. *Basilodes pepita* Guenée; Holly Grove, West Virginia. CLH 610517-5. (p. 120).
9. *Basilodes straminea* Poole; Cochise Co., Arizona. USNM 43773. (p. 122).
10. *Cirrhophanus dyari* Cockerell; Paradise, Cochise Co., Arizona. USNM 44808. (p. 126).
11. *Cirrhophanus pretiosa* (Morrison); Welder Wildlife Refuge, San Patricio Co., Texas. USNM 44809. (p. 129).
12. *Stiria rugifrons* Grote; Torrington, Wyoming. USNM 44793. (p. 138).

PLATE R: FEMALE GENITALIA OF STIRIINAE SPECIES

1. *Stiria intermixta* Dyar; Ash Canyon Road, Huachuca Mts., Arizona. USNM 44795. (p. 140).
2. *Stiria dyari* Hill; Kingman, Arizona. USNM 44810. (p. 141).
3. *Stiria blanchardi* (Hogue); Sierra Diablo Wildlife Management Area, Culberson Co., Texas. USNM 44797. (p. 141).
4. *Stiria sulphurea* Neumoegen; Palmerlee, Arizona. CLH 610519-6. (p. 142).
5. *Stiria consuela* Strecker; Paradise, Cochise Co., Arizona. USNM 44807. (p. 143).
6. *Chalcopasta howardi* (Hy. Edwards); Deming, New Mexico. CLH 610517-7. (p. 147).
7. *Chalcopasta fulgens* Barnes and McDunnough; Paradise, Cochise Co., Arizona. CLH 610518-1. (p. 147).
8. *Chalcopasta acema* (Druce); Iguala, Mexico. CLH 610518-3. (p. 148).
9. *Neumoegenia poetica* Grote; lectotype; Arizona. USNM 44726. (p. 150).
10. *Argentostiria koebelei* (Riley); Indio, California. USNM 44724. (p. 152).
11. *Chamaeclea pernana* (Grote); Paradise, Cochise Co., Arizona. USNM 44732. (p. 154).
12. *Chamaeclea basiochrea* Barnes and McDunnough; female genitalia; Hidalgo Co., Texas. USNM 44730. (p. 155).

PLATE S: FEMALE GENITALIA OF STIRIINAE SPECIES, MALE GENITALIA OF PSAPHIDINAE SPECIES

1. *Bistica noela* (Druce); female genitalia; Madera Canyon, Santa Cruz Co., Arizona. USNM 44717. (p. 156).
2. *Hemioslaria pima* Barnes and Benjamin; female genitalia; Baboquivari Mts., Pima Co., Arizona. CLH 610420-1. (p. 158).
3. *Heminocloa mirabilis* (Neumoegen); female genitalia; Baboquivari Mts., Pima Co., Arizona. USNM 44594. (p. 159).
4. *Thurberiphaga diffusa* (Barnes); female genitalia; Baboquivari Mts., Pima Co., Arizona. JGF 751. (p. 161).
5. *Psaphida resumens* Walker; male genitalia; Deutschburg, Jackson Co., Texas. USNM 44766;
6. Aedoeagus. (p. 166).
7. *Psaphida thaxterianus* (Grote); male genitalia; New Brighton, Pennsylvania. USNM 44815;
8. Aedoeagus. (p. 167).
9. *Psaphida grandis* (Smith); Dayton, Ohio. USNM 44712;
10. Aedoeagus. (p. 168).
11. *Psaphida electilis* (Morrison); Ithaca, New York. USNM 44780;
12. Aedoeagus. (p. 169).
13. *Psaphida rolandi* (Grote); Ithaca, New York. USNM 38545. (p. 168).

PLATE T: MALE GENITALIA OF PSAPHIDINAE SPECIES

1. *Psaphida rolandi* (Grote); aedoeagus; Houston, Texas. USNM 44710. (p. 168).
2. *Psaphida damalis* (Grote); Cloverdale, Sonoma Co., California. USNM 44816;
3. Aedoeagus. (p. 171).
4. *Psaphida styracis* (Guenée); Bear Mt., New York. USNM 44714;
5. Aedoeagus. (p. 170).
6. *Pseudocopivaleria anaverta* Buckett and Bauer; Wrightwood, California. USNM 44817;
7. Aedoeagus. (p. 176).
8. *Brachionycha borealis* (Smith); Roscommon, Michigan. USNM 41548;
9. Aedoeagus. (p. 178).
10. *Viridemas galena* Smith; Ash Canyon Road, Huachuca Mts., Cochise Co., Arizona. USNM 44818;
11. Aedoeagus.
12. *Pleromella opter* Dyar; San Diego, California. USNM 41404;
13. Aedoeagus. (p. 179).
14. *Feralia jocosa* (Guenée); Mt. Uniacke, Nova Scotia. USNM 38548;
15. Aedoeagus. (p. 183).

PLATE U: GENITALIA OF PSAPHIDINAE SPECIES

1. *Feralia deceptiva* McDunnough; male genitalia; Marion Co., Oregon. USNM 41542;
2. Aedoeagus. (p. 184).
3. *Feralia major* Smith; male genitalia; Atlanta, Georgia. USNM 41544;
4. Aedoeagus. (p. 185).
5. *Feralia comstocki* Grote; male genitalia; Claremont, New Hampshire. USNM 41537;
6. Aedoeagus. (p. 185).
7. *Feralia februalis* Grote; male genitalia; Inverness, Marin Co., California. USNM 43215;
8. Aedoeagus. (p. 186).
9. *Feralia meadowsi* Buckett; male genitalia; Santa Catalina Island, California. USNM 43217;
10. Aedoeagus. (p. 187).
11. *Psaphida resumens* Walker; female genitalia; Bethany, New Haven Co., Connecticut. USNM 44820. (p. 166).
12. *Psaphida thaxterianus* (Grote); female genitalia; Quincy, Illinois. USNM 44819. (p. 167).
13. *Psaphida grandis* (Smith); female genitalia; Marion Co., Florida. USNM 44713. (p. 168).

PLATE V: FEMALE GENITALIA OF PSAPHIDINAE SPECIES

1. *Psaphida electilis* (Morrison); Ithaca, New York. USNM 44781. (p. 169).
2. *Psaphida rolandi* (Grote); Houston, Texas. USNM 44711. (p. 168).
3. *Psaphida damalis* (Grote); Eldridge, California. USNM 41385. (p. 171).

4. *Psaphida styracis* (Guenée); New York City, New York. USNM 44715. (p. 170).
5. *Copivaleria grotei* (Morrison); Ithaca, New York. USNM 41547. (p. 176).
6. *Brachionycha borealis* Smith; Roscommon Co., Michigan. USNM 44790. (p. 178).
7. *Viridemas galena* Smith; Palmerlee, Cochise Co., Arizona. JFGC 2555.
8. *Pleromella opter* Dyar; Ward Co., California. USNM 44789. (p. 179).

PLATE W: FEMALE GENITALIA OF PSAPHIDINAE SPECIES

1. *Provia argentata* Barnes and McDunnough; Stockton, Utah. JGF 1808. (p. 181).
2. *Feralia jocosa* (Guenée); New York. USNM 38415. (p. 183).
3. *Feralia deceptiva* McDunnough; Parkland, Washington. USNM 41543. (p. 184).
4. *Feralia major* Smith; Massapequa, Long Island, New York. USNM 44787. (p. 185).
5. *Feralia comstocki* Grote; Wellington, British Columbia. USNM 41540. (p. 185).
6. *Feralia februalis* Grote; Zuma Beach, Los Angeles Co., California. USNM 44788. (p. 186).

PLATE A: STRUCTURAL FEATURES OF TRIFID NOCTUIDAE

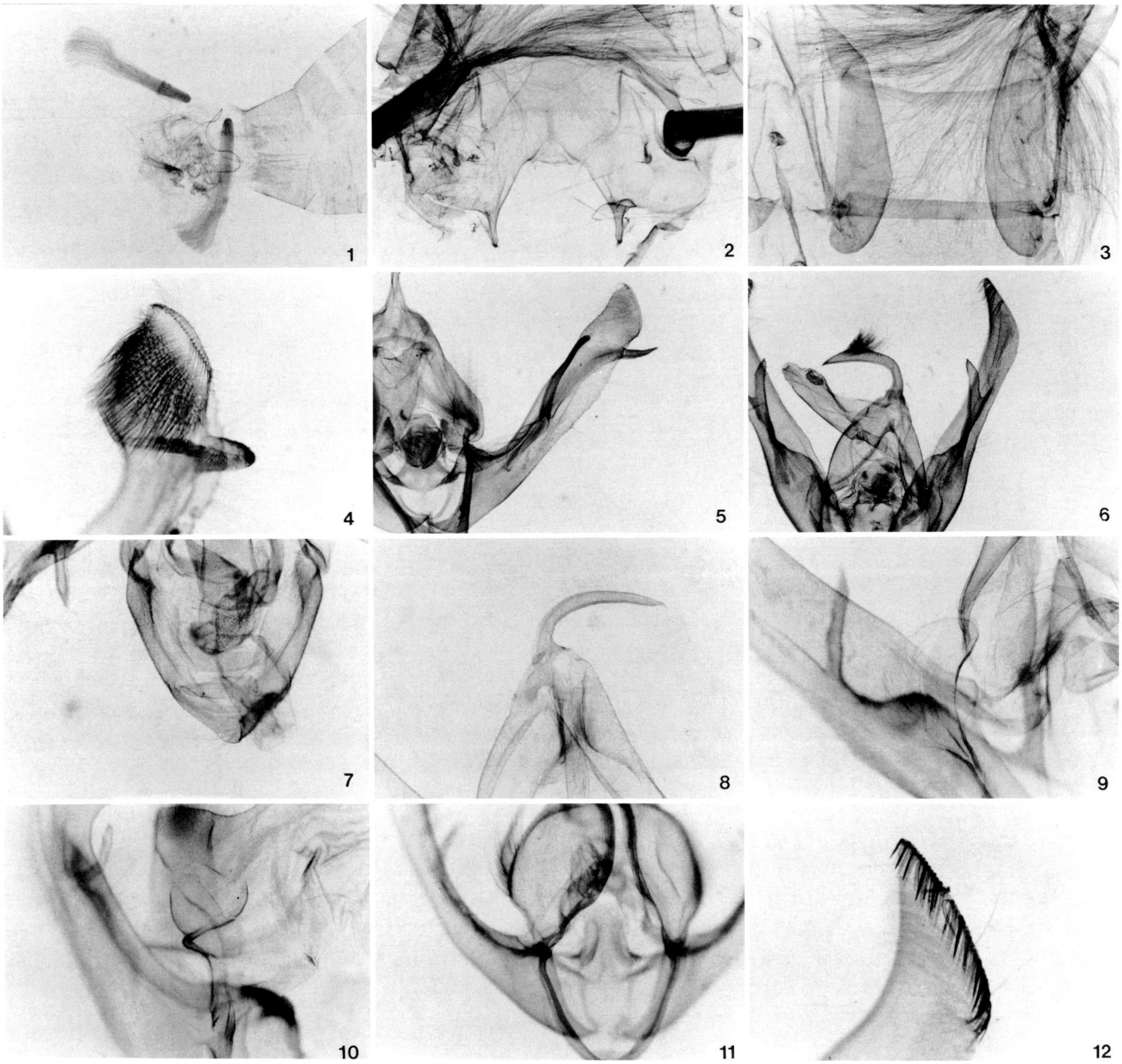

PLATE B: MALE GENITALIA OF CUCULLIINAE SPECIES

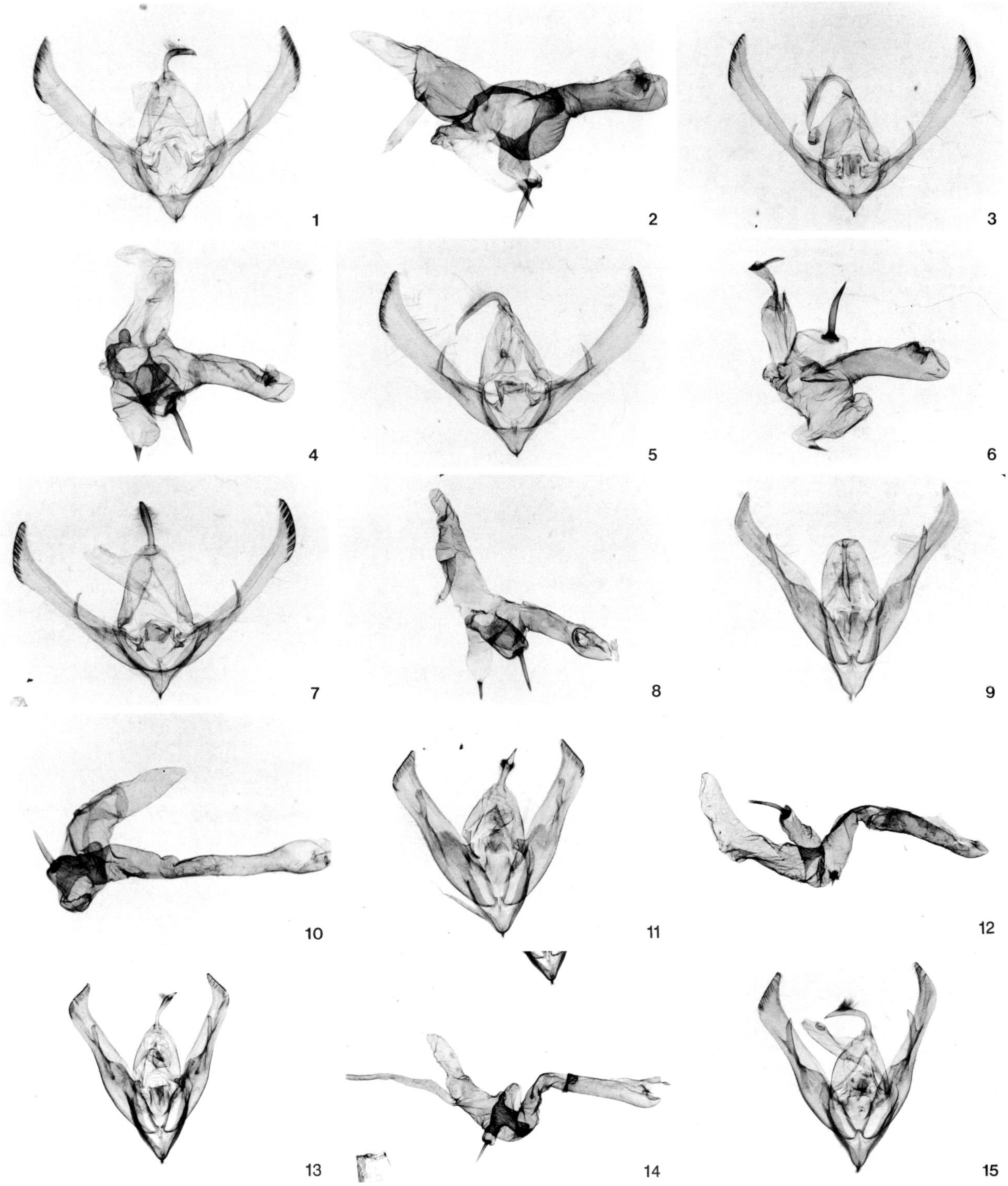

PLATE C: MALE GENITALIA OF CUCULLIINAE SPECIES

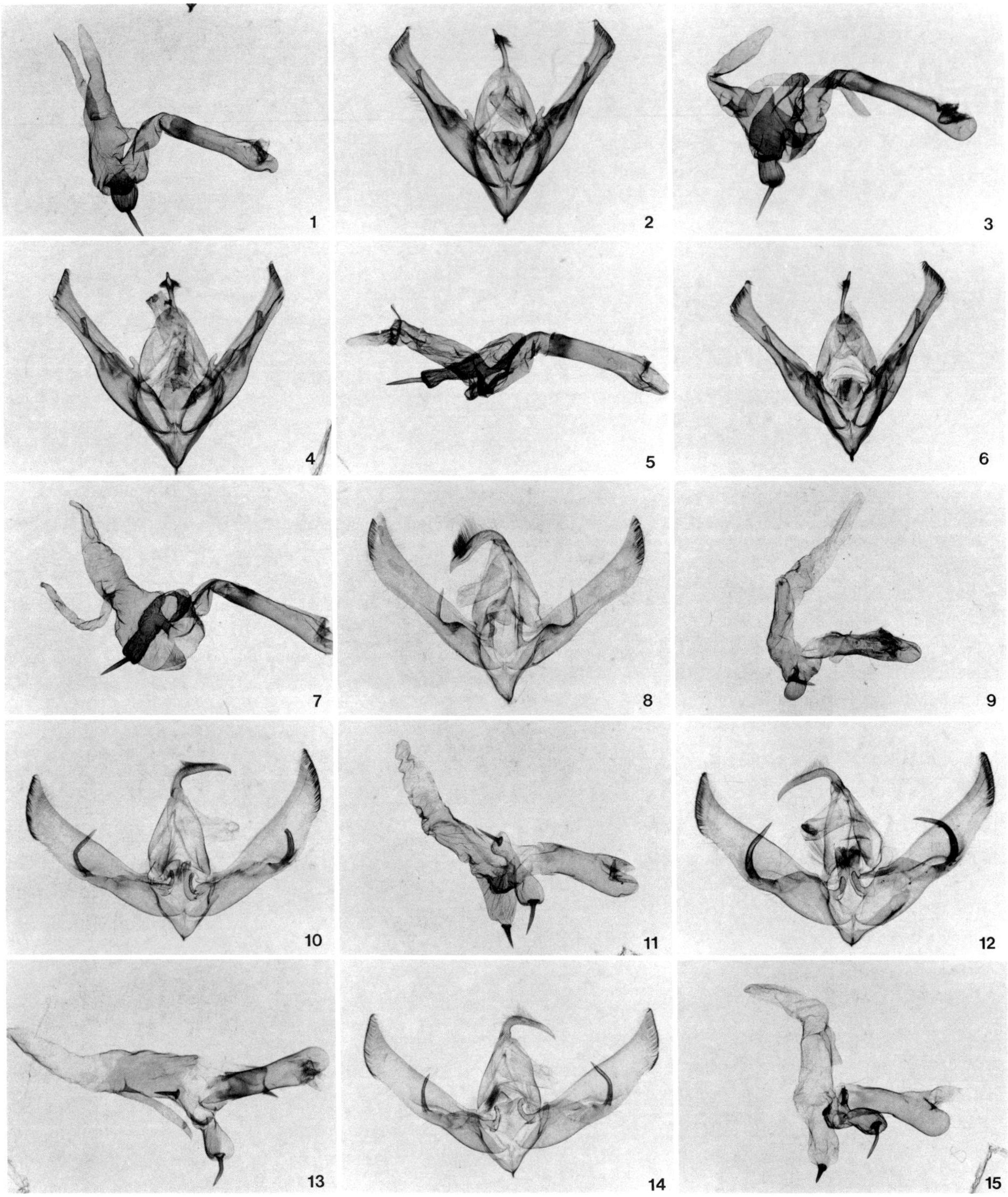

PLATE D: MALE GENITALIA OF CUCULLIINAE SPECIES

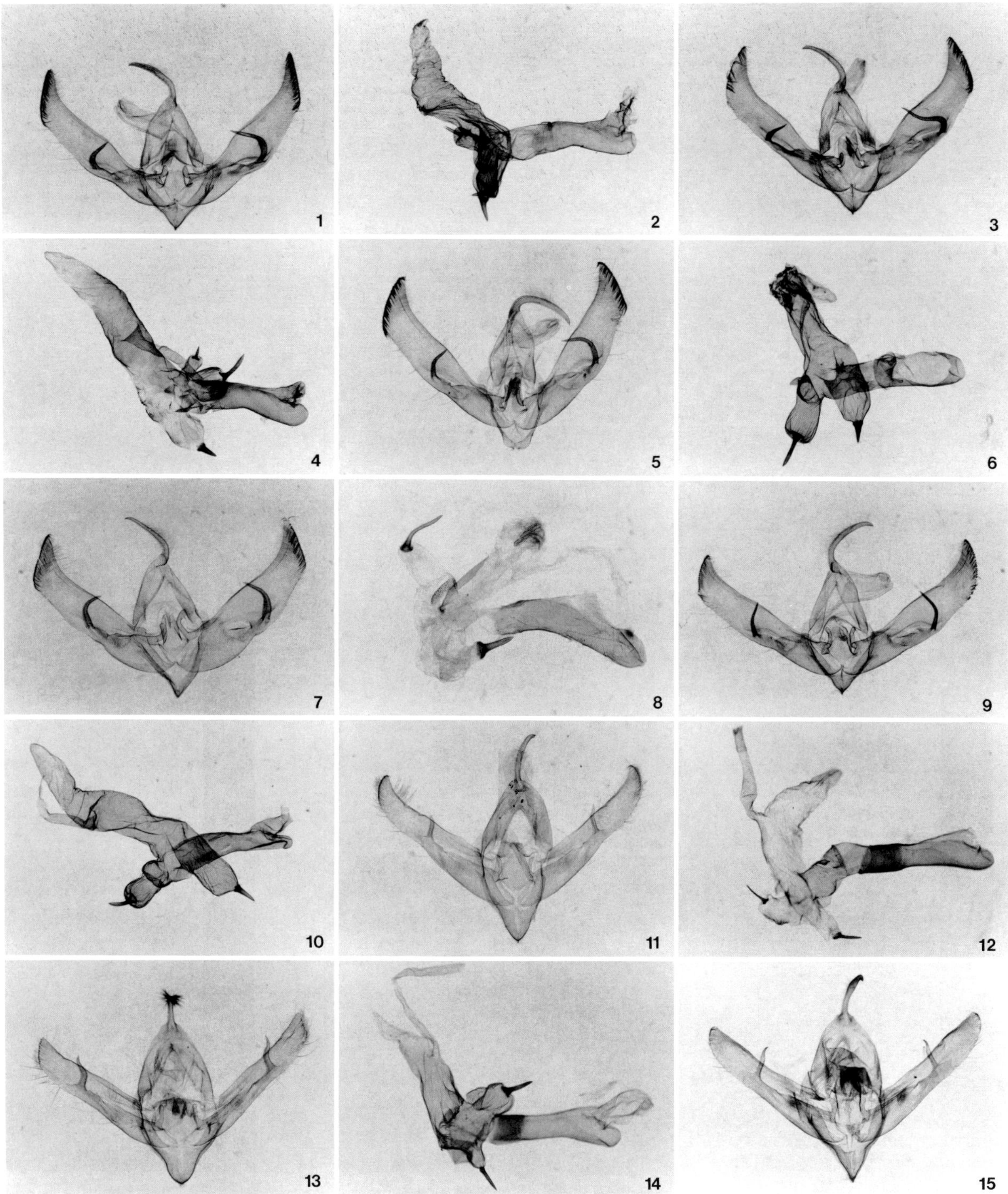

PLATE E: MALE GENITALIA OF CUCULLIINAE SPECIES

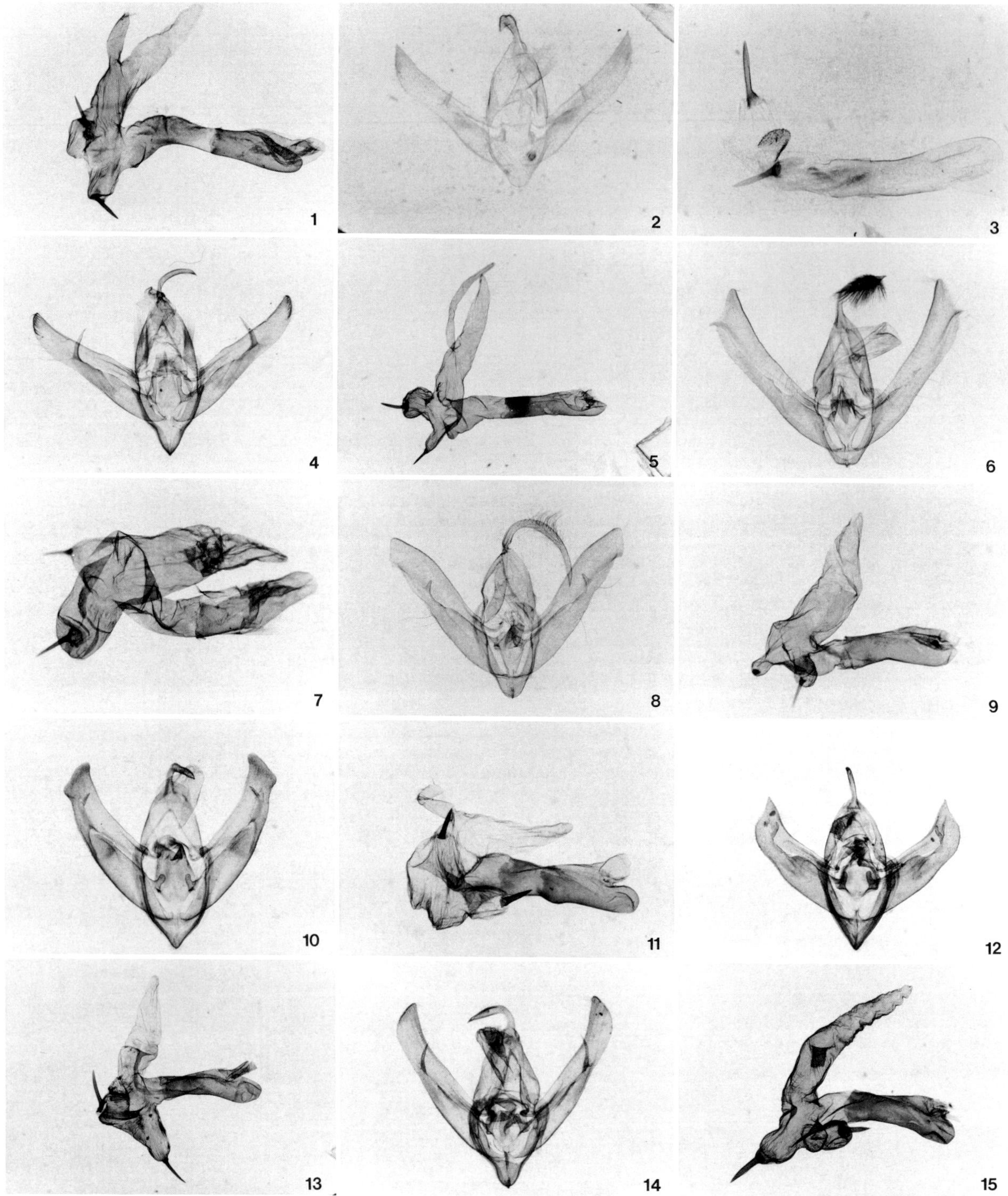

PLATE F: MALE GENITALIA OF CUCULLIINAE SPECIES

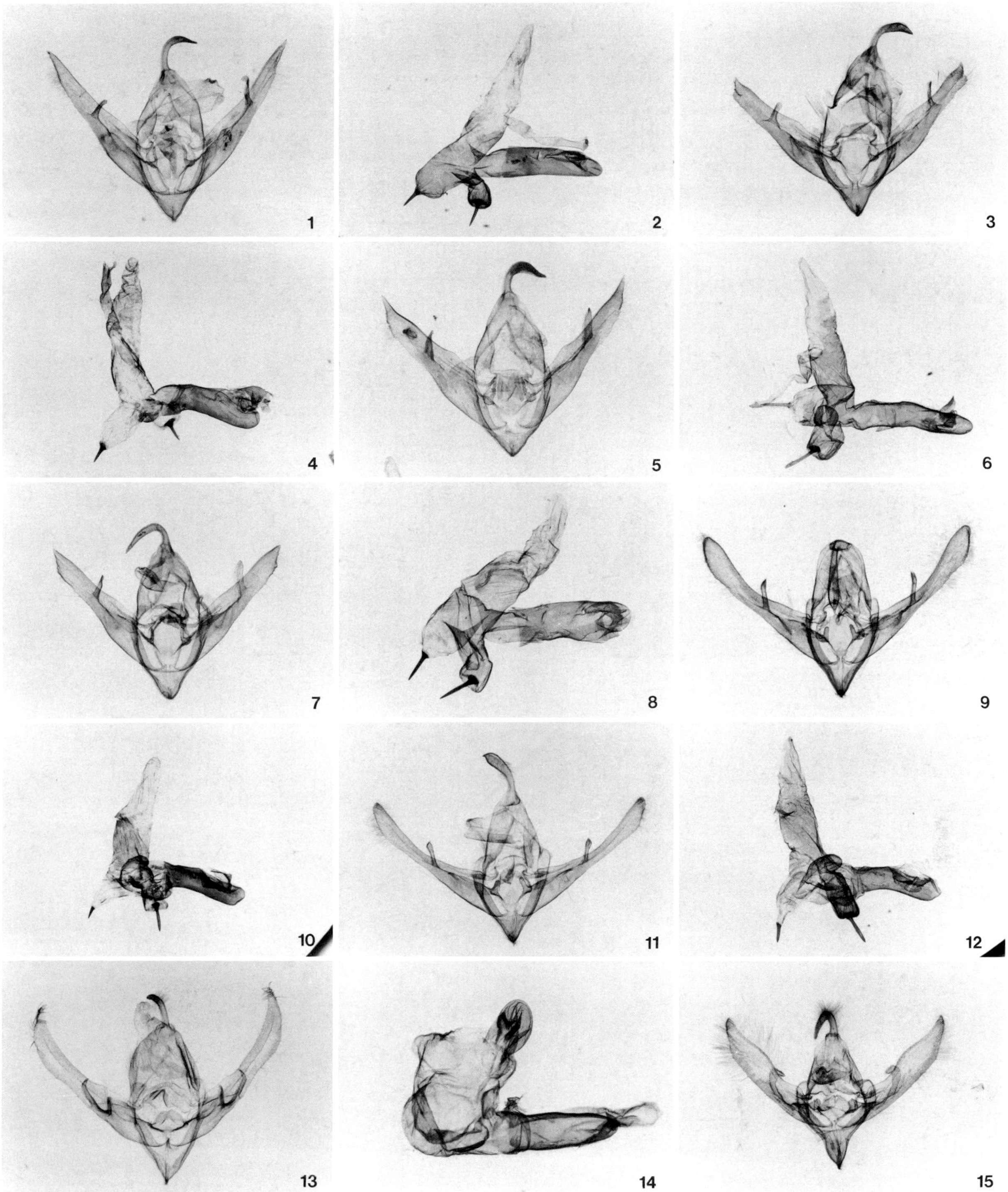

PLATE G: MALE AND FEMALE GENITALIA OF CUCULLIINAE SPECIES

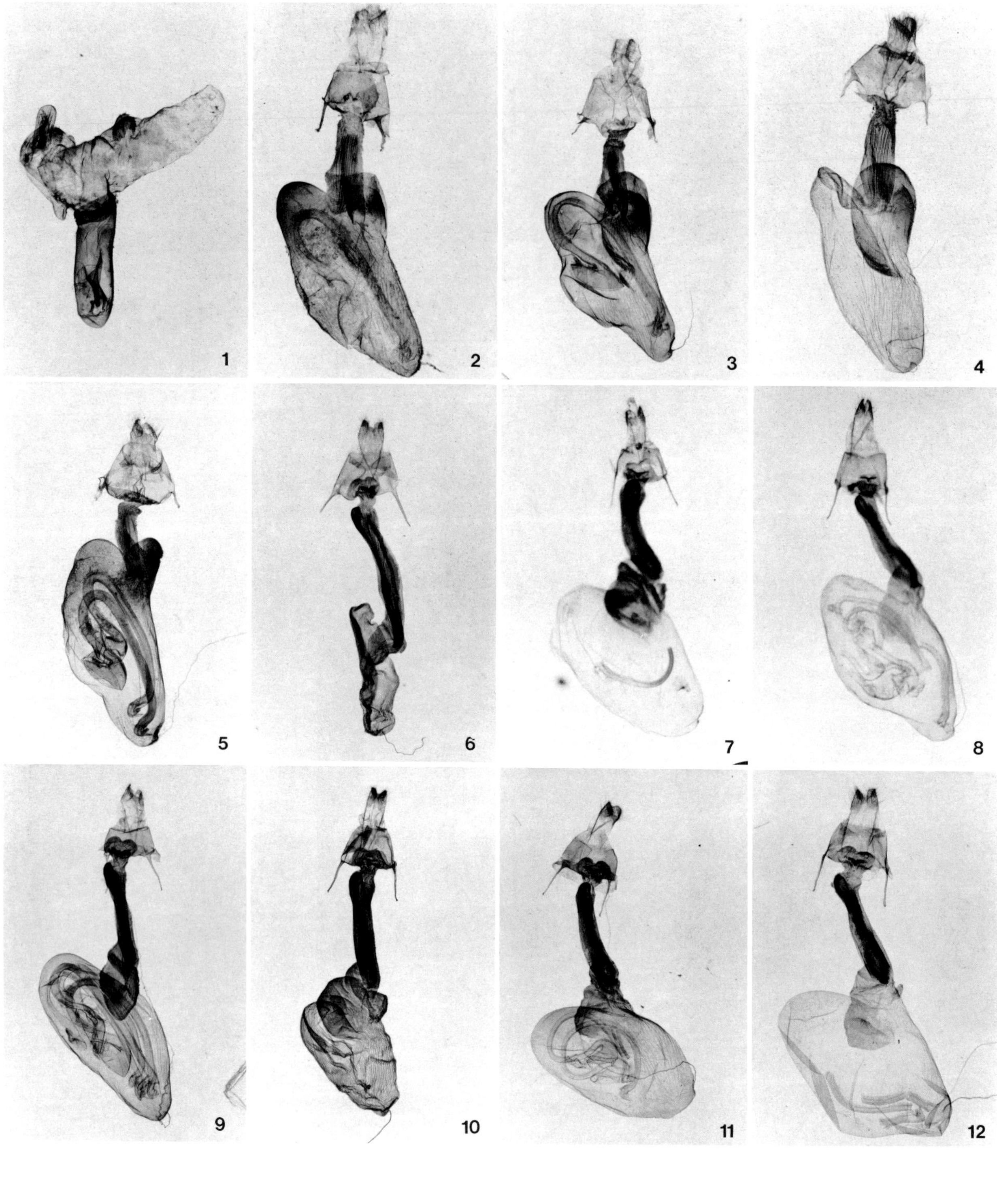

PLATE H: FEMALE GENITALIA OF CUCULLIINAE SPECIES

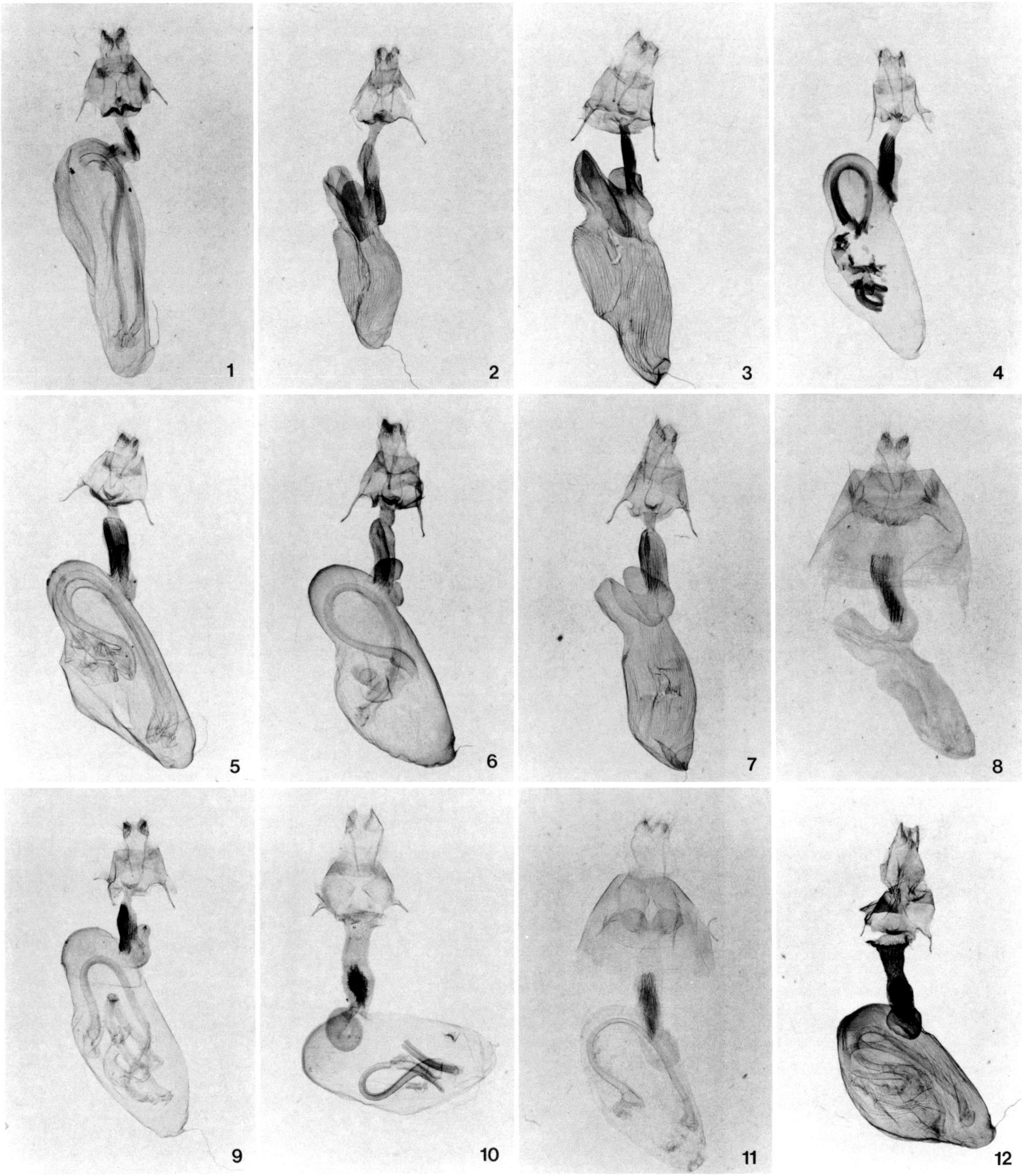

PLATE I: FEMALE GENITALIA OF CUCULLIINAE SPECIES

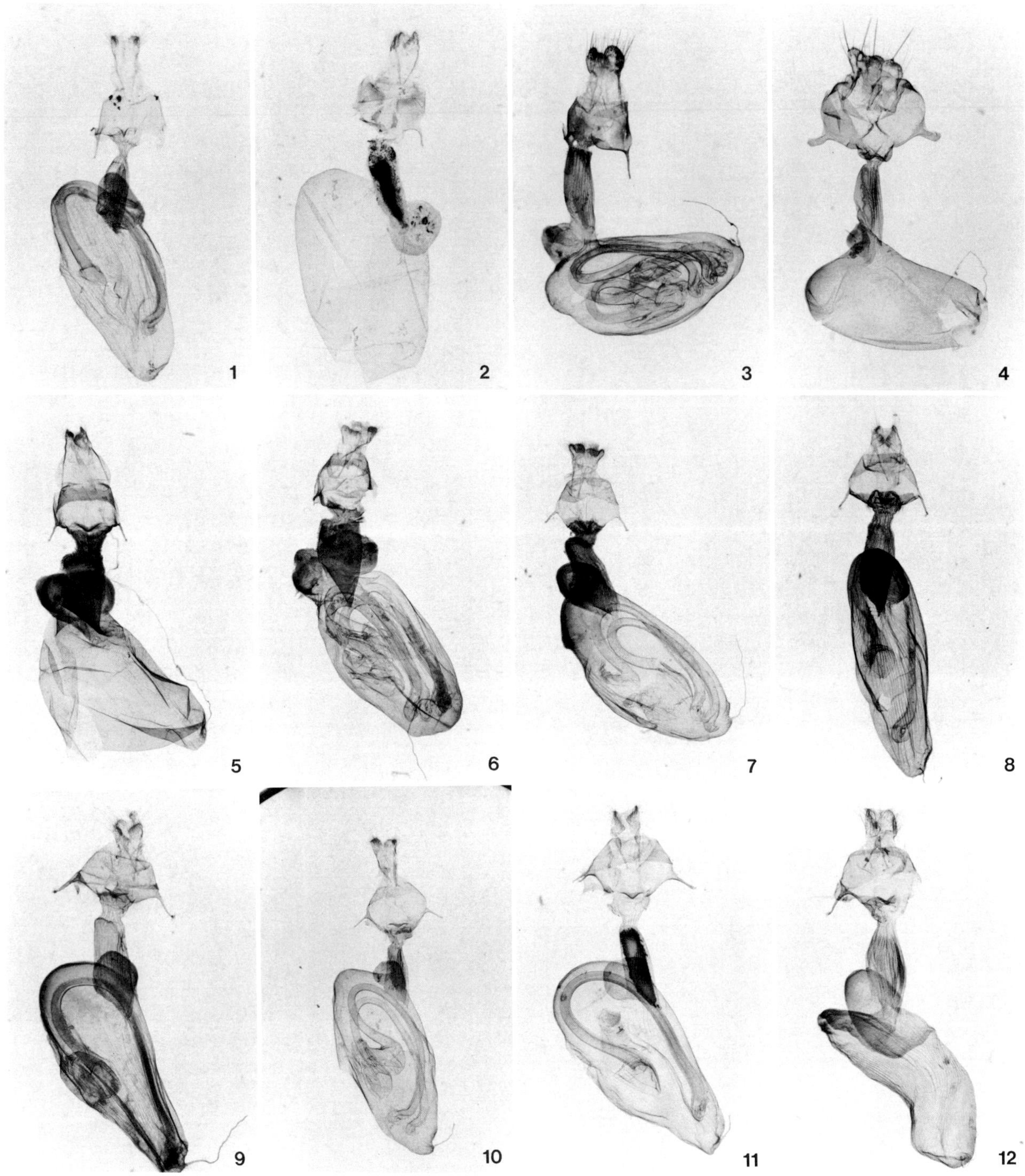

PLATE J: FEMALE GENITALIA OF CUCULLIINAE SPECIES
MALE GENITALIA OF STIRIINAE SPECIES

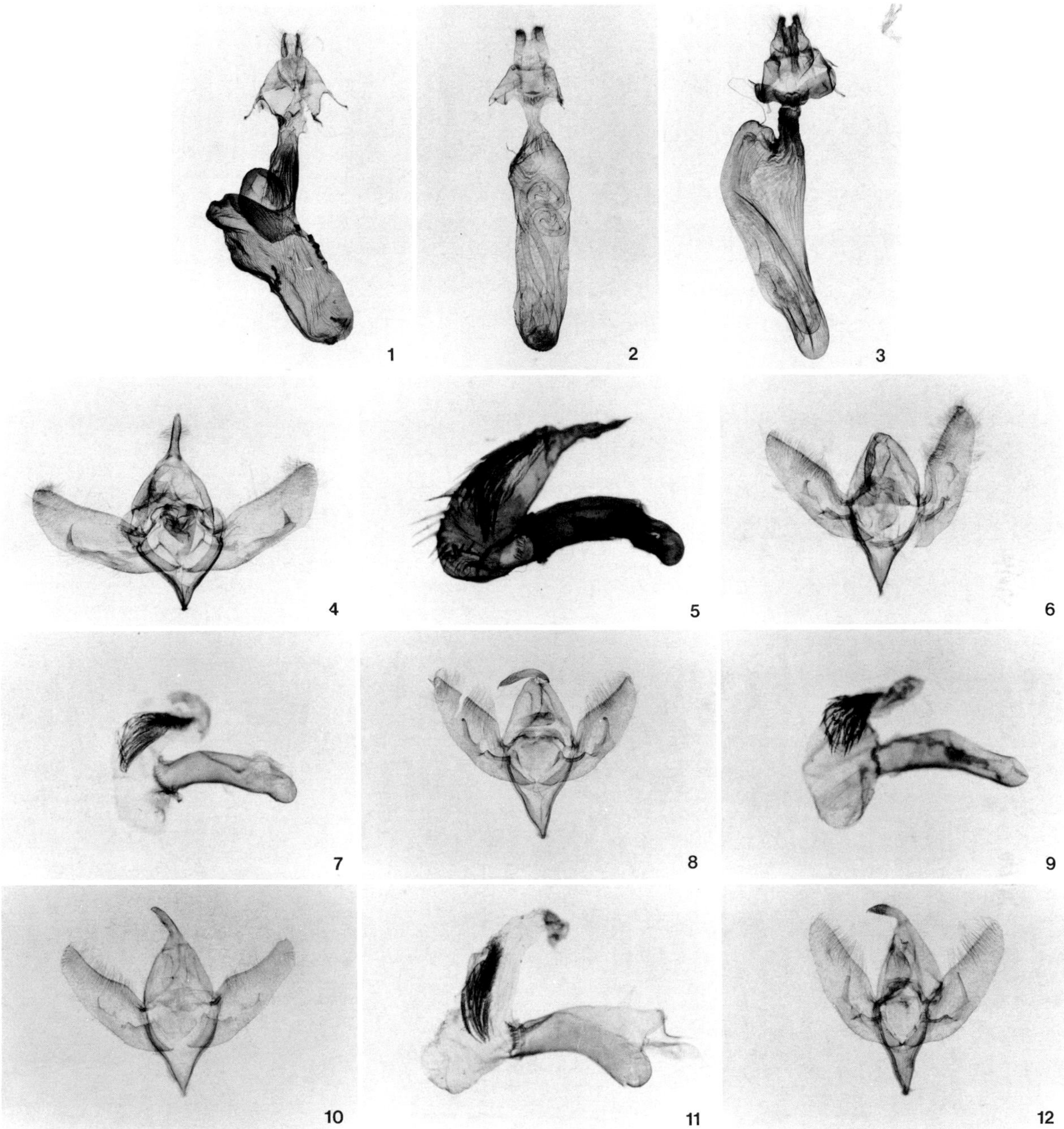

PLATE K: MALE GENITALIA OF STIRIINAE SPECIES

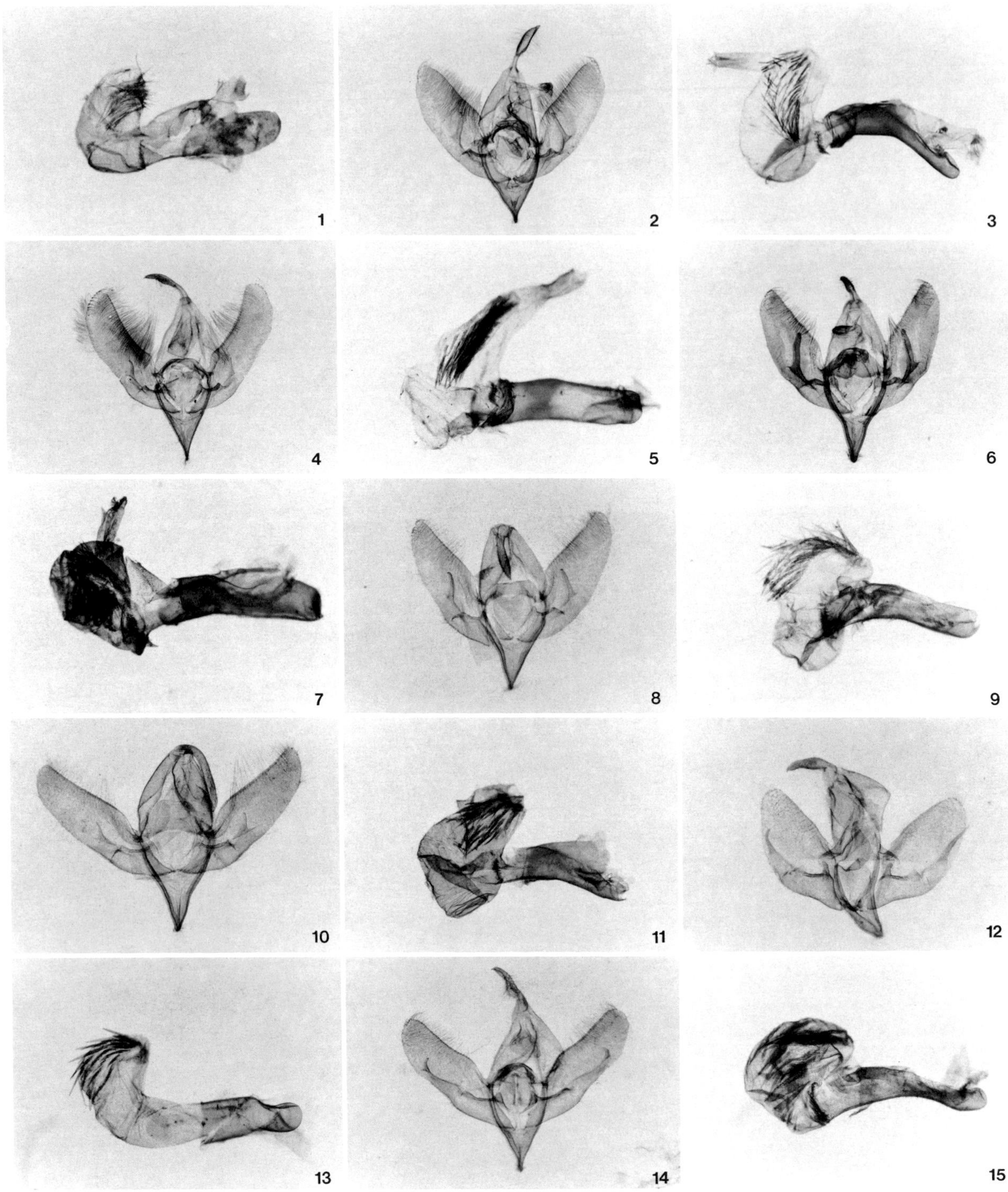

PLATE L: MALE GENITALIA OF STIRIINAE SPECIES

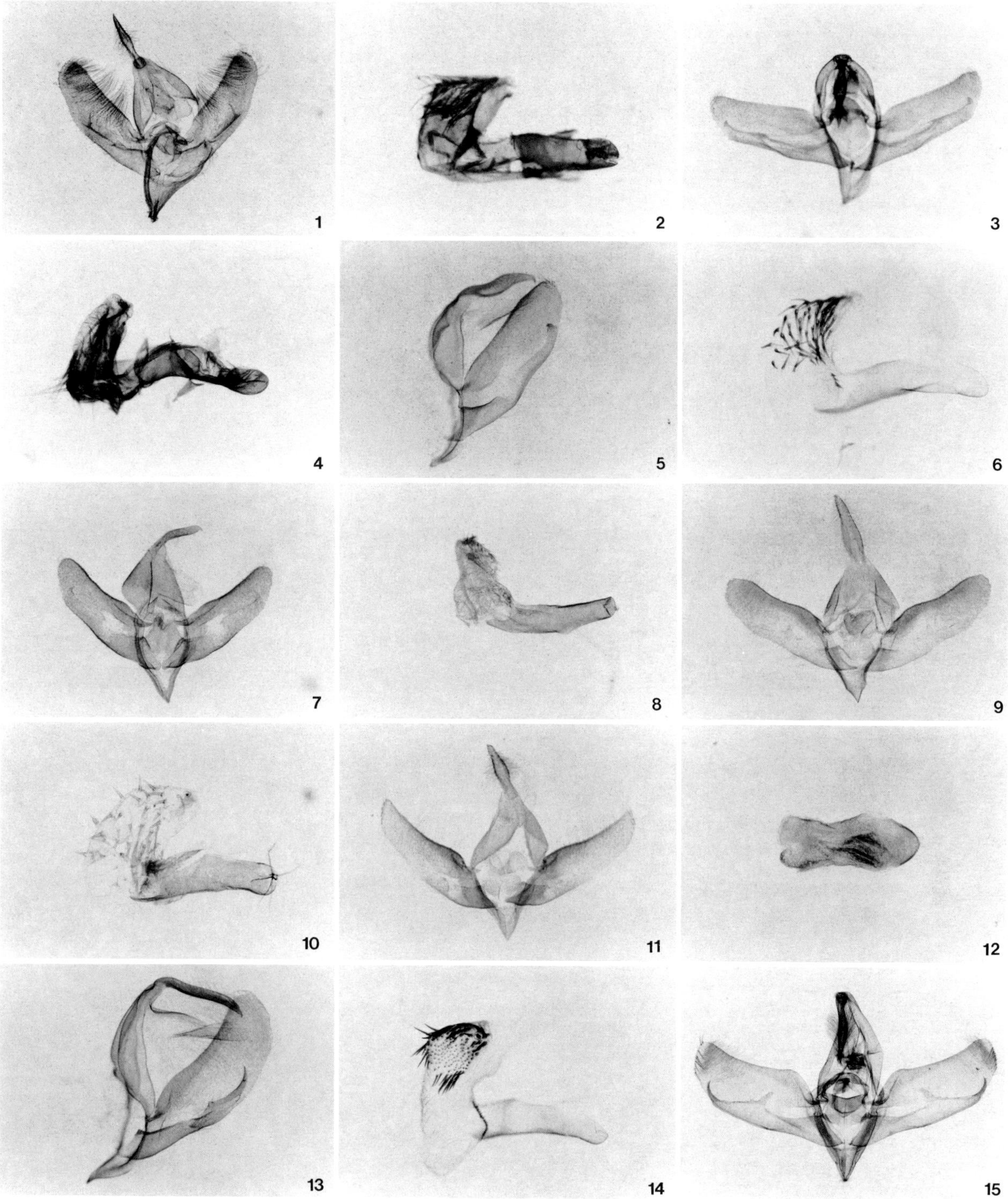

PLATE M: MALE GENITALIA OF STIRIINAE SPECIES

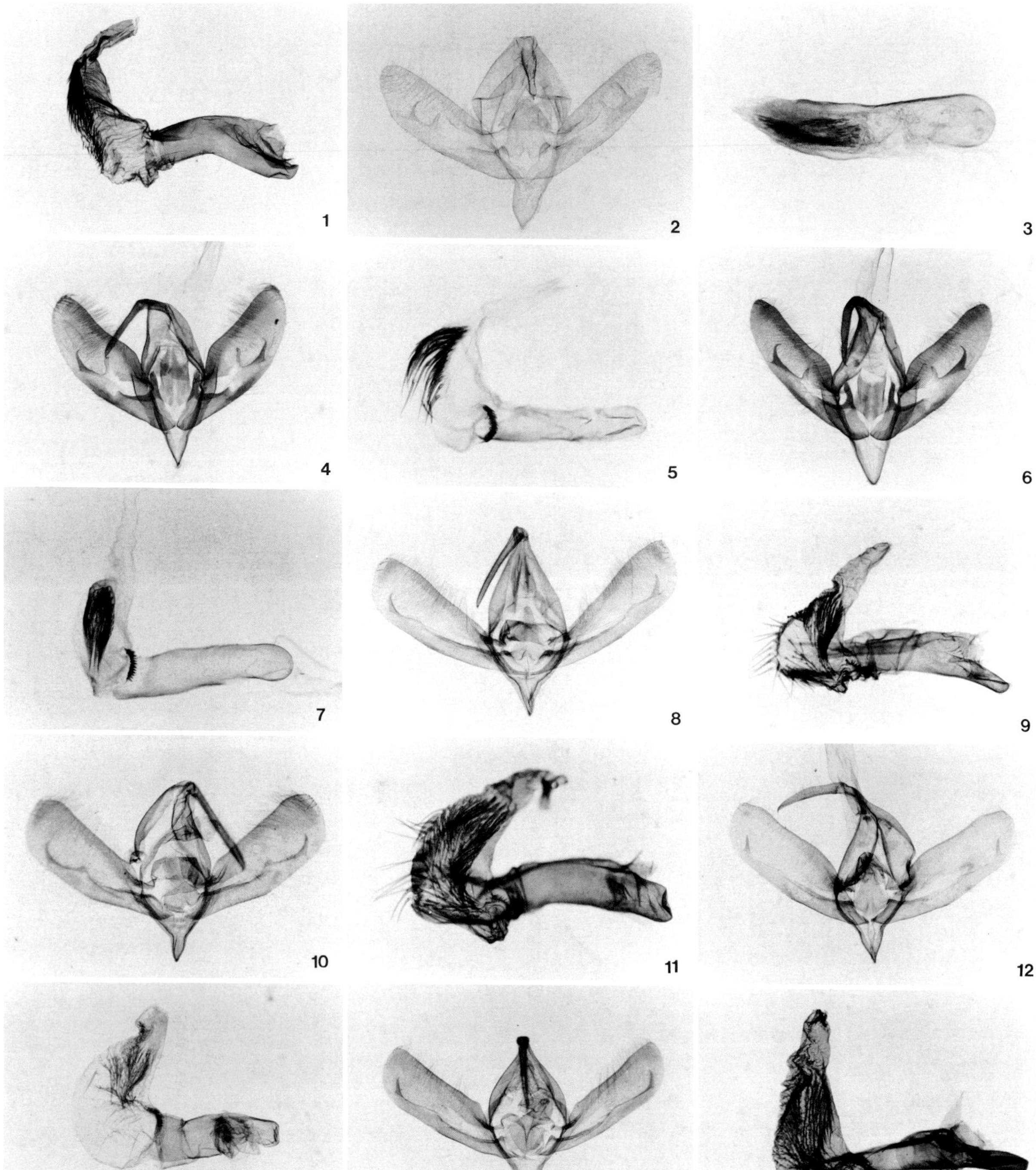

PLATE N: MALE GENITALIA OF STIRIINAE SPECIES

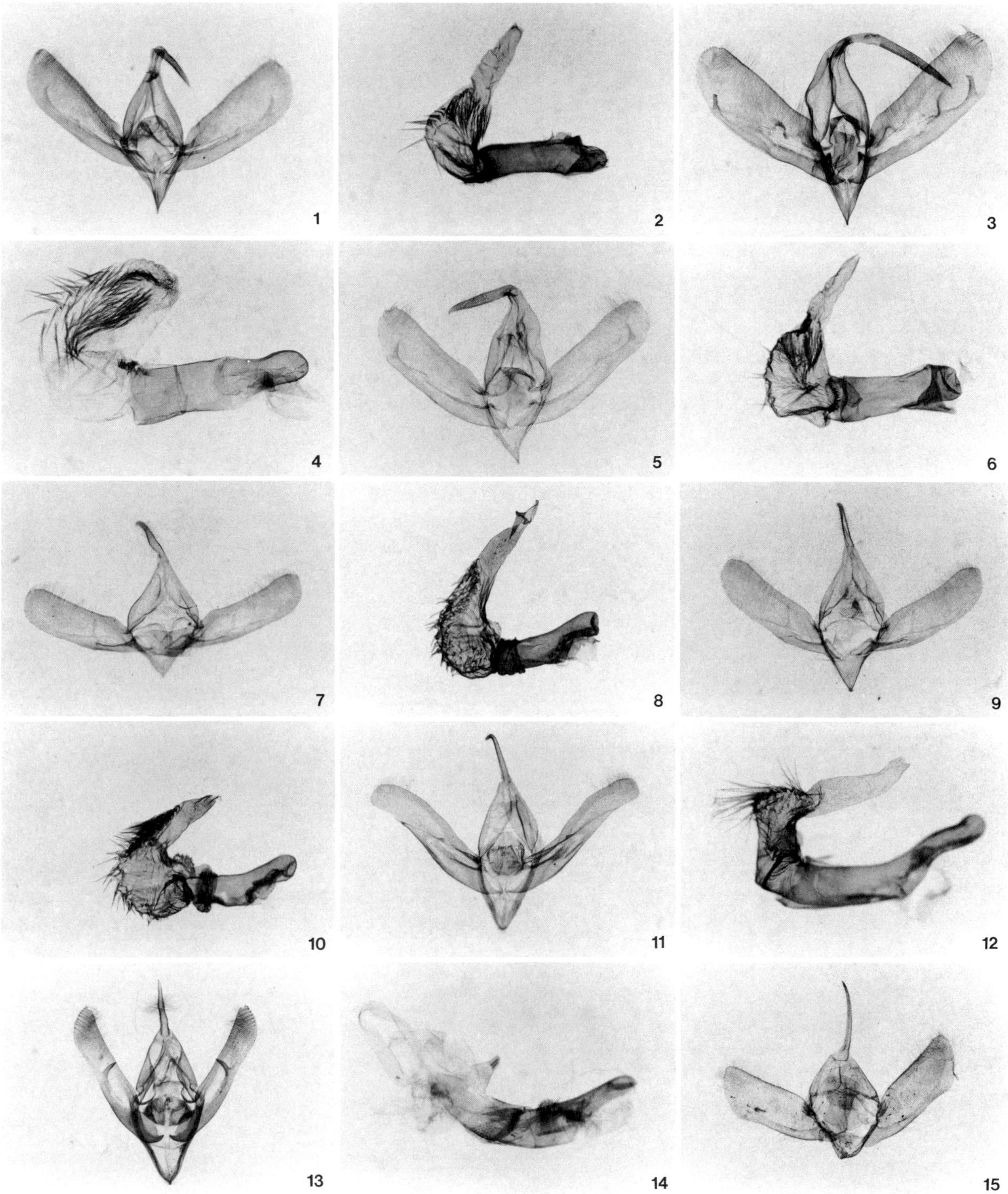

PLATE O: MALE GENITALIA OF STIRIINAE SPECIES

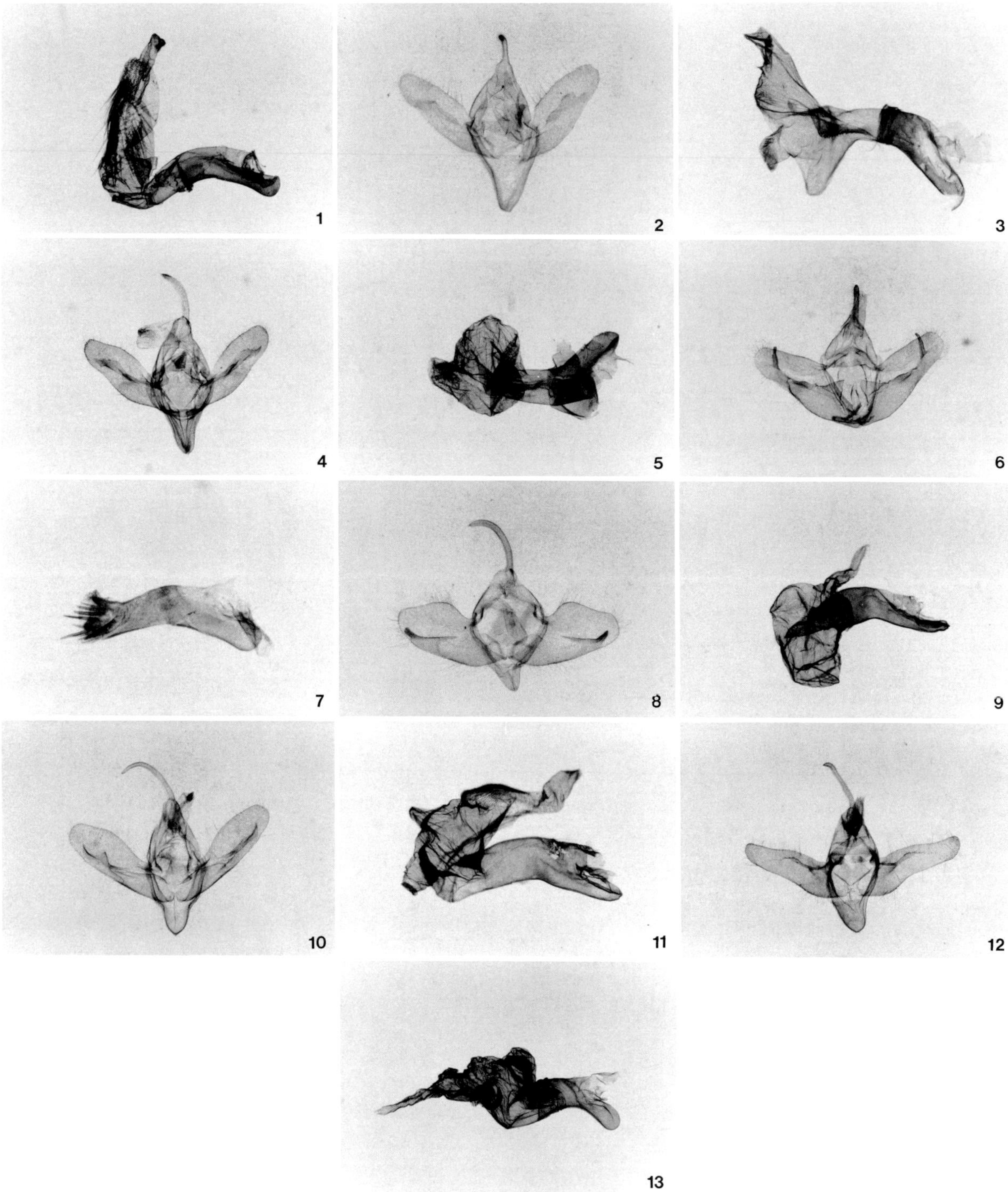

PLATE P: FEMALE GENITALIA OF STIRIINAE SPECIES

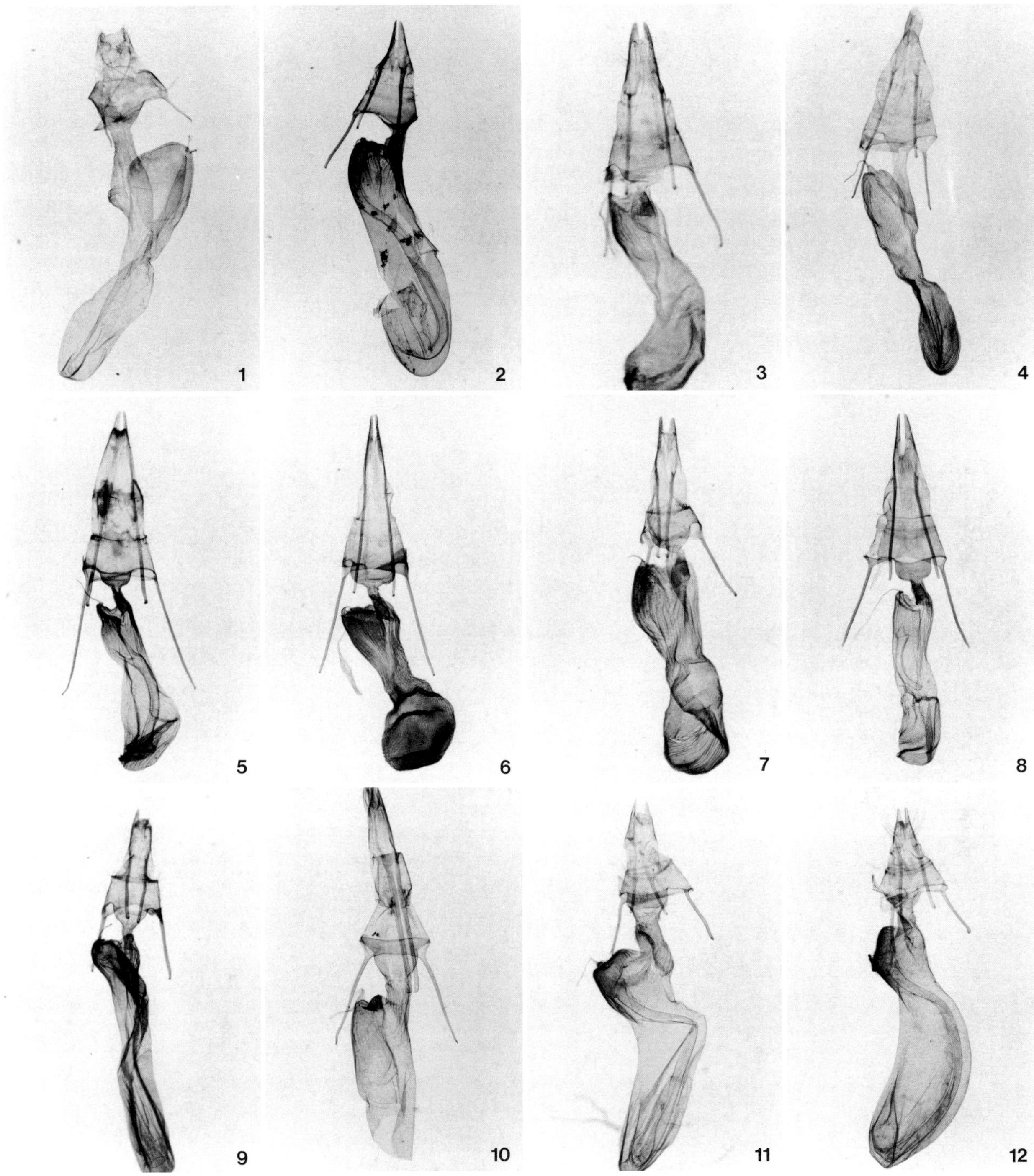

PLATE Q: FEMALE GENITALIA OF STIRIINAE SPECIES

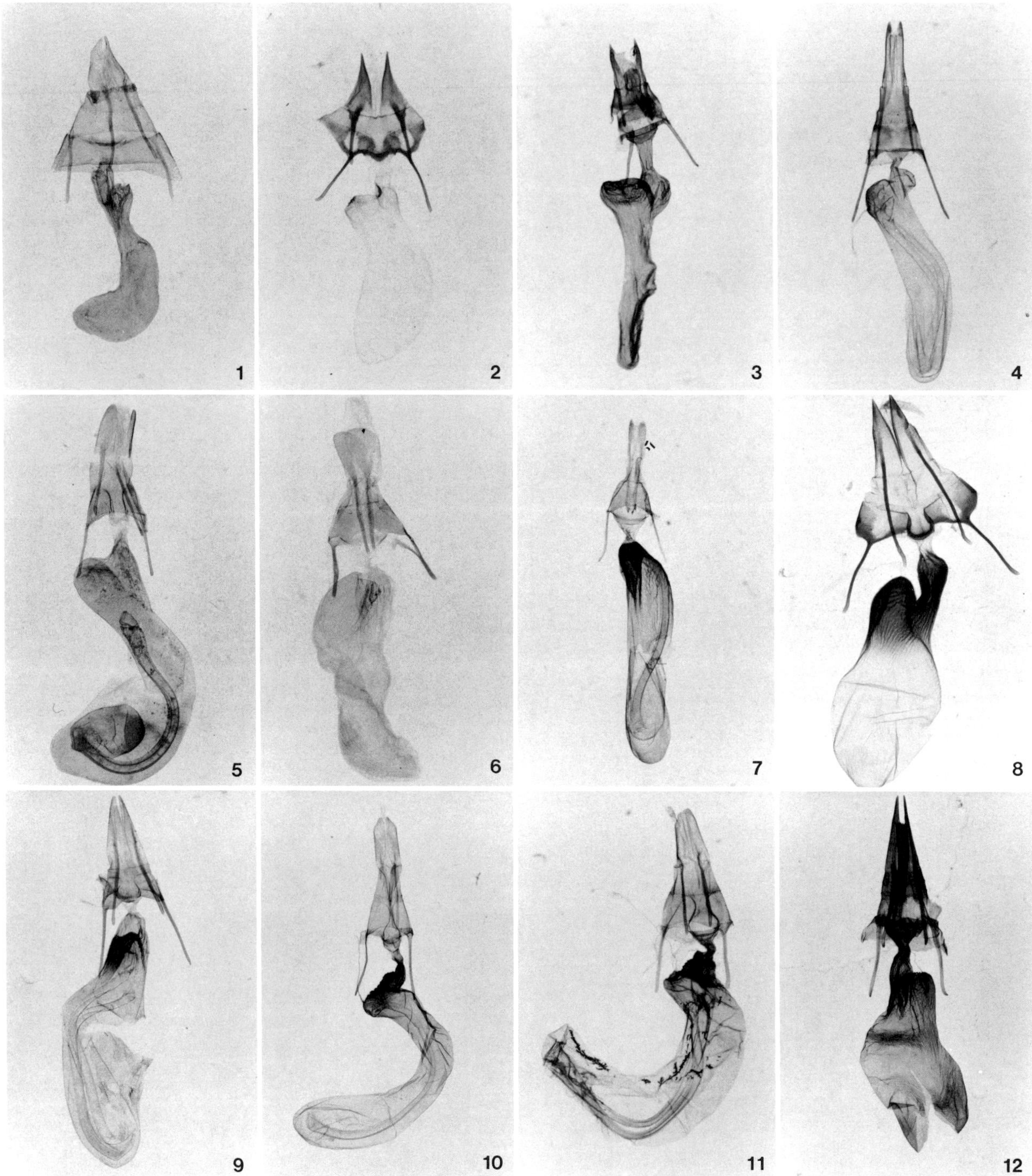

PLATE R: FEMALE GENITALIA OF STIRIINAE SPECIES

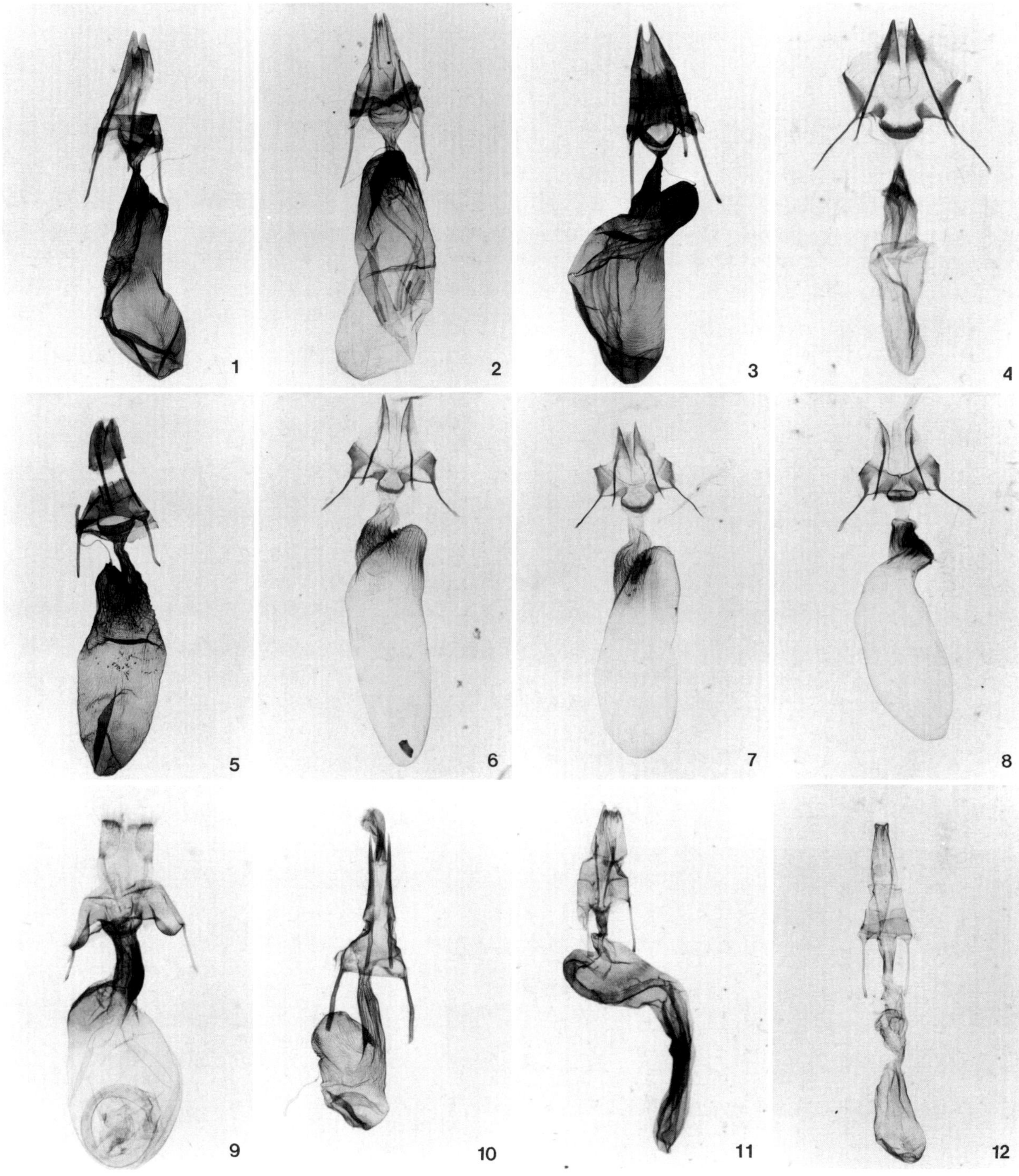

PLATE S: FEMALE GENITALIA OF STIRIINAE SPECIES MALE GENITALIA OF PSAPHIDINAE SPECIES

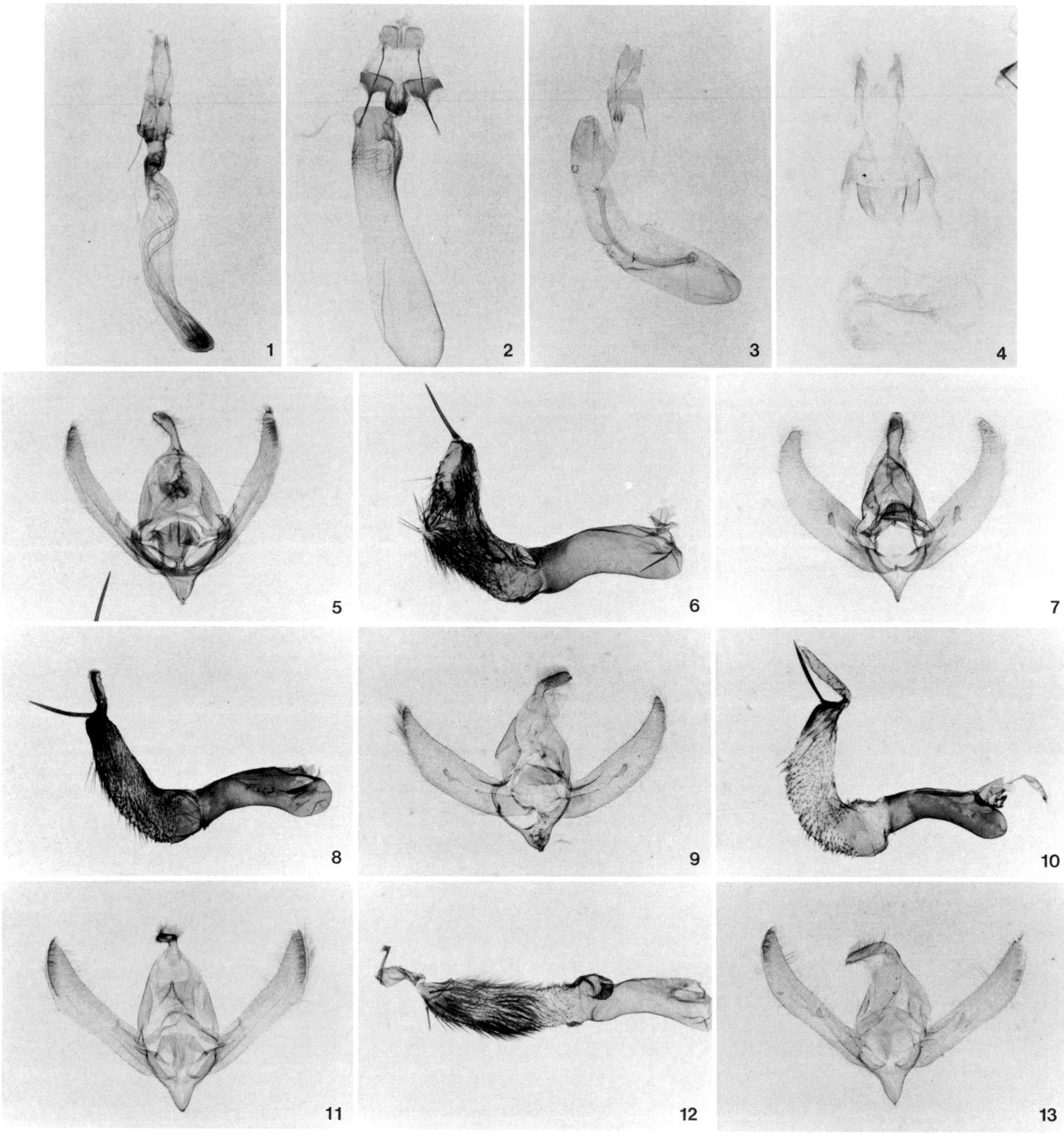

PLATE T: MALE GENITALIA OF PSAPHIDINAE SPECIES

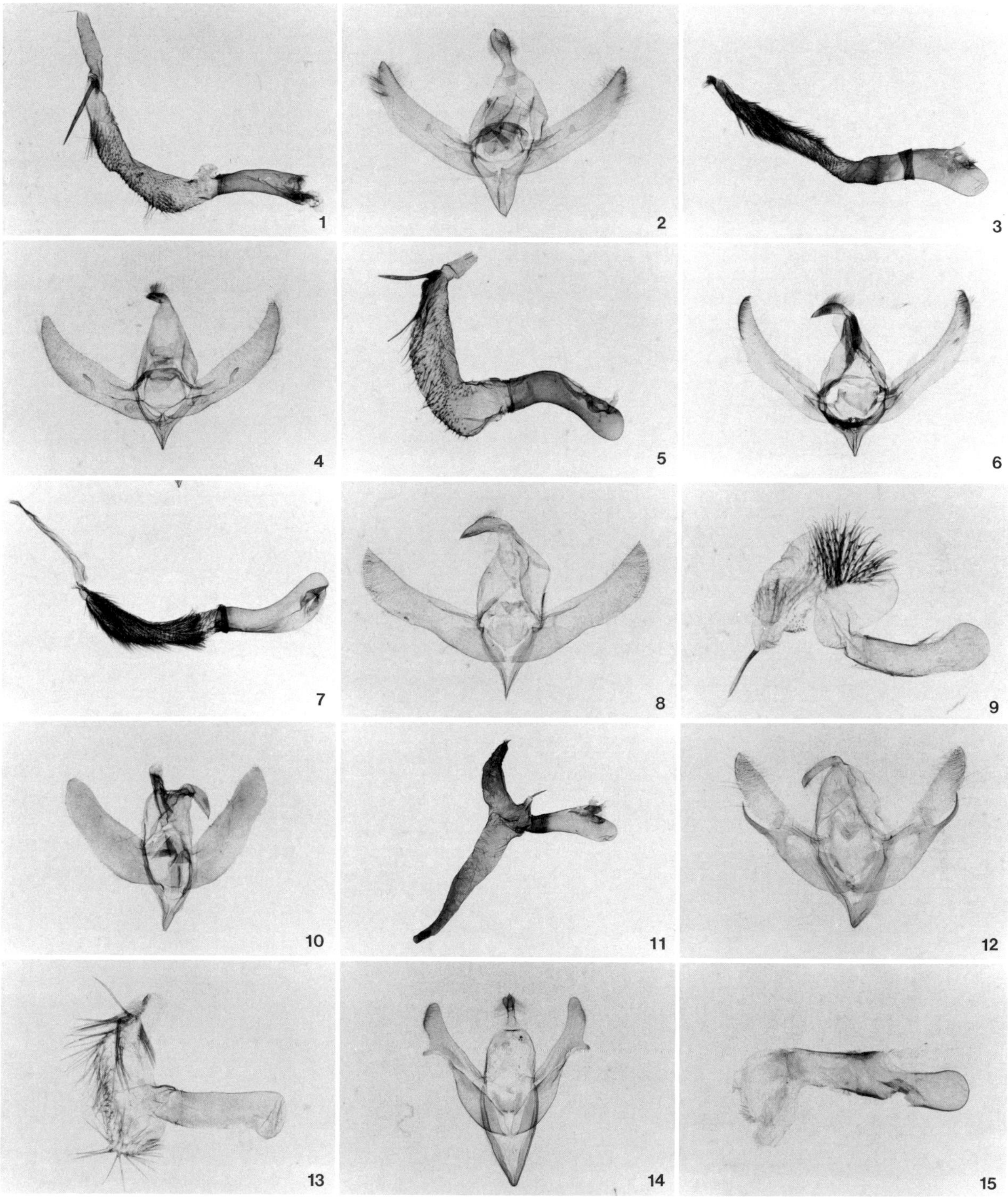

PLATE U: GENITALIA OF PSAPHIDINAE SPECIES

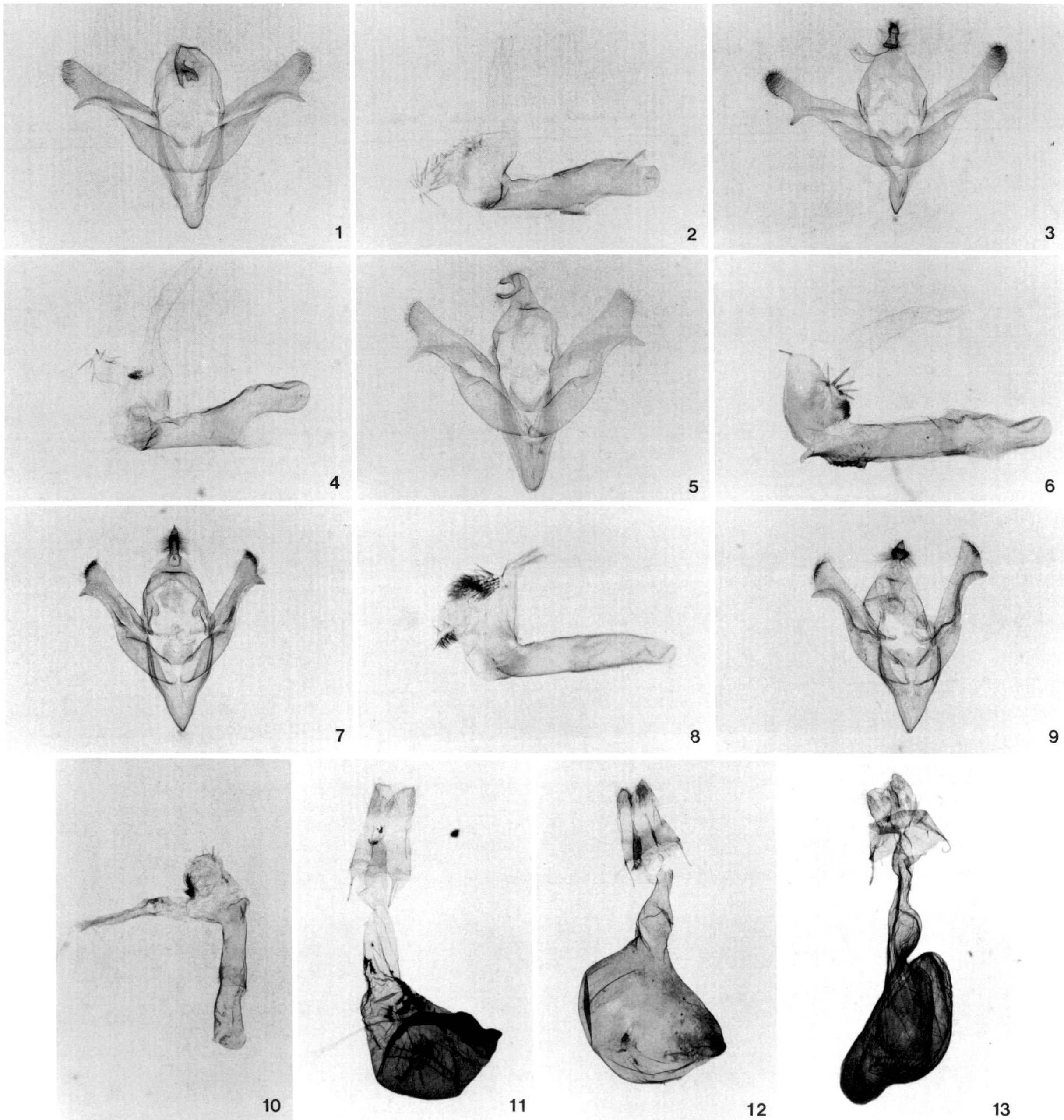

PLATE V: FEMALE GENITALIA OF PSAPHIDINAE SPECIES

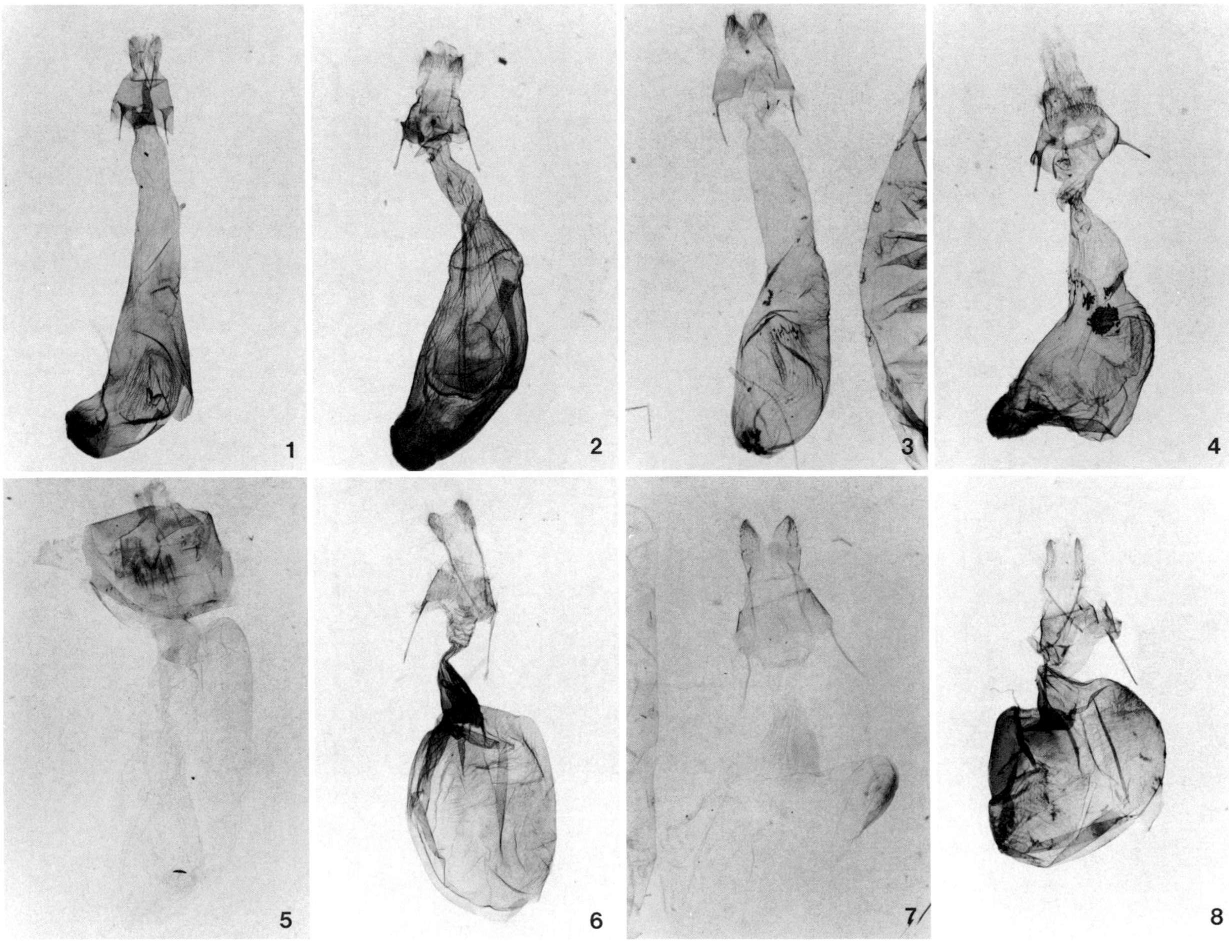

PLATE W: FEMALE GENITALIA OF PSAPHIDINAE SPECIES

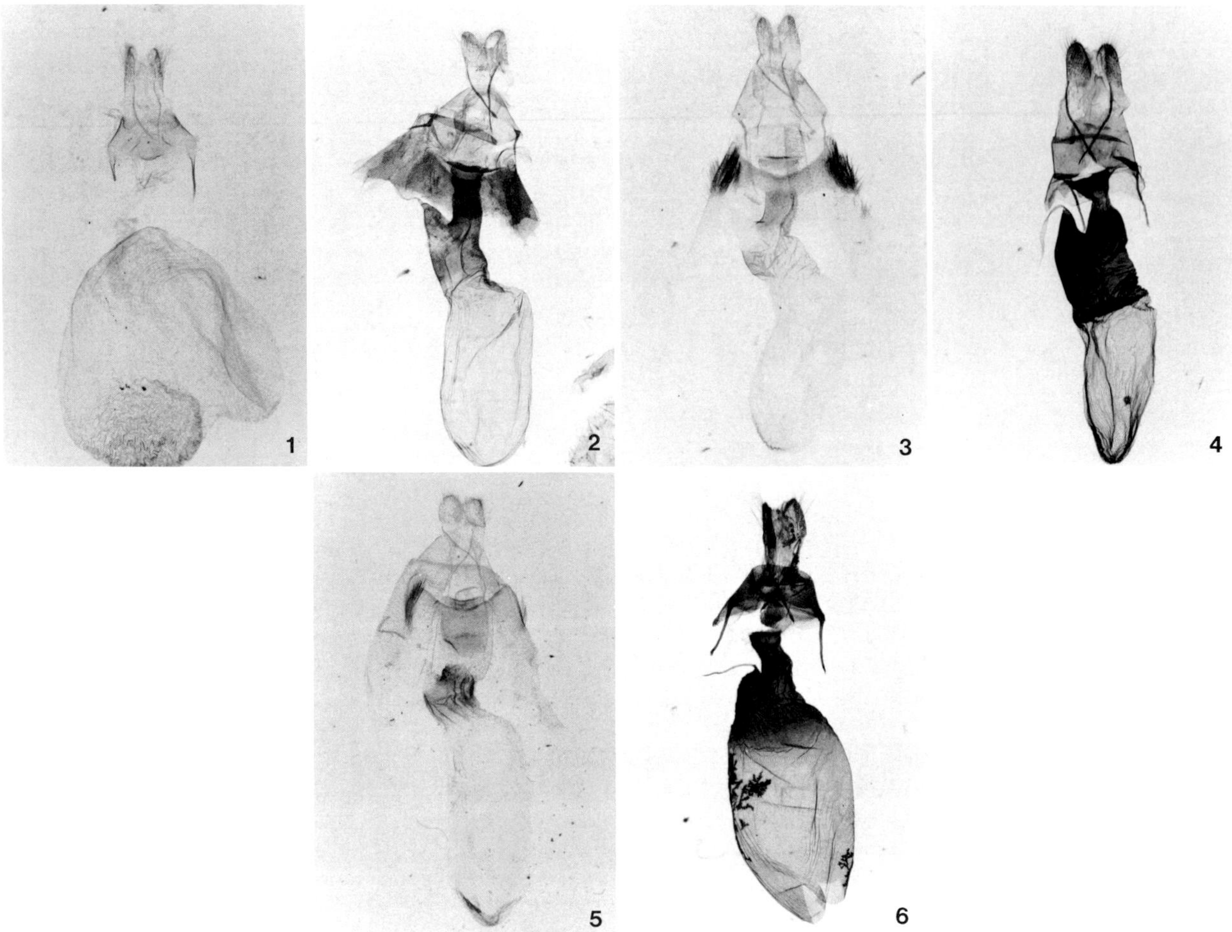

COLOR PLATES

Noctuoidea

PLATE 1

Noctuoidea

NOCTUIDAE

figs. 1–36

NATURAL SIZE 1:1

1. *Cucullia asteroides* Gn., ♂. Hackberry Lake, Valentine N.W.R., Cherry Co., Nebraska, 8 June 1983, D. C. Ferguson (USNM). (p. 34).
2. *Cucullia asteroides* Gn., ♀. Hackberry Lake, Valentine N.W.R., Cherry Co., Nebraska, 2 June 1983, D. C. Ferguson (USNM). (p. 34).
3. *Cucullia montanae* Grt., ♂. Estes Park, Larimer Co., Colorado, 16 July 1968, A. & M. E. Blanchard (USNM). (p. 35).
4. *Cucullia montanae* Grt., ♀. Hill City, Pennington Co., South Dakota, 2 August 1964, D. C. Ferguson (USNM). (p. 35).
5. *Cucullia similaris* Sm., ♂. Angel Creek, E Humboldt Mts., SSW of Wells, 7,000′, Elko Co., Nevada, 18 July 1971, D. C. Ferguson (USNM). (p. 36).
6. *Cucullia similaris* Sm., ♀. Angel Lake, road above Wells, Elko Co., Nevada, 7,000′, 20 July 1971, D. C. Ferguson (USNM). (p. 36).
7. *Cucullia omissa* Dod, ♂. Doyles, Codroy Valley, Newfoundland, 30 July 1962, D. C. Ferguson (USNM). (p. 36).
8. *Cucullia omissa* Dod, ♀. Doyles, Codroy Valley, Newfoundland, 31 July 1962, D. C. Ferguson (USNM). (p. 36).
9. *Cucullia postera* Gn., ♂. West Dover, Halifax Co., Nova Scotia, 30 June 1968, D. C. Ferguson (USNM). (p. 38).
10. *Cucullia postera* Gn., ♀. West Dover, Halifax Co., Nova Scotia, 24 July 1968, D. C. Ferguson (USNM). (p. 38).
11. *Cucullia postera* Gn., ♀. Annapolis Royal, Nova Scotia, 2 August 1967, D. C. Ferguson (USNM). (p. 38).
12. *Cucullia postera* Gn., ♀. Hereford, Arizona, C. R. Biedermann (USNM). (p. 38).
13. *Cucullia florea* Gn., ♂. Doyles, Codroy Valley, Newfoundland, 30 July 1962, D. C. Ferguson (USNM). (p. 37).
14. *Cucullia florea* Gn., ♀. Doyles, Codroy Valley, Newfoundland, 4 August 1962, D. C. Ferguson (USNM). (p. 37).
15. *Cucullia obscurior* Sm., ♂. Priest River Exp. For., Bonner Co., Idaho, 2,500′, 19 June 1979, D. C. Ferguson (USNM). (p. 39).
16. *Cucullia obscurior* Sm., ♀. Gisborne Mt., Bonner Co., Idaho, 5,000′, 29 June 1979, D. C. Ferguson (USNM). (p. 39).
17. *Cucullia florea-obscurior* (indeterminate), ♂. Major's Flat, nr. Ephraim, Sanpete Co., Utah, 7,100′, 12 July 1980, D. C. Ferguson (USNM). (p. 39).
18. *Cucullia lilacina* Schaus, ♂. Mount Locke, Davis Mountains, Texas, 6,700′, 10 June 1969, A. & M. E. Blanchard (USNM). (p. 40).
19. *Cucullia lilacina* Schaus, ♀. Basin, Big Bend National Park, Texas, 25 August 1965, A. & M. E. Blanchard (USNM). (p. 40).
20. *Cucullia eucaena* Dyar, ♂. Mount Locke, Davis Mountains, Texas, 6,700′, 25 August 1967, A. & M. E. Blanchard (USNM). (p. 40).
21. *Cucullia eucaena* Dyar, ♀. Fort Davis, Jeff Davis Co., Texas, 11 July 1969, A. & M. E. Blanchard (USNM). (p. 40).
22. *Cucullia oribac* Barnes, ♂. Lomas de Chapultepec, Distrito Federal, Mexico, 17 June 1939, T. Escalante (USNM). (p. 41).
23. *Cucullia oribac* Barnes, ♀. Lomas de Chapultepec, Distrito Federal, Mexico, 6 September 1940, T. Escalante (USNM). (p. 41).
24. *Cucullia convexipennis* G. & R., ♂. Debert, Colchester Co., Nova Scotia, 19 July 1961, D. C. Ferguson (USNM). (p. 42).
25. *Cucullia convexipennis* G. & R., ♀. Brooklyn, New York (USNM). (p. 42).
26. *Cucullia speyeri* Lint., ♂. Hackberry Lake, Valentine N.W.R., Cherry Co., Nebraska, 11 June 1983, D. C. Ferguson (USNM). (p. 46).
27. *Cucullia speyeri* Lint., ♀. Hackberry Lake, Valentine N.W.R., Cherry Co., Nebraska, 8 June 1983, D. C. Ferguson (USNM). (p. 46).
28. *Cucullia styx* Poole, ♂. Lomas de Chapultepec, Distrito Federal, Mexico, 6 June 1940, T. Escalante (USNM). (p. 48).
29. *Cucullia styx* Poole, ♀. Paratype. Paradise, Cochise Co., Arizona (USNM). (p. 48).
30. *Cucullia dorsalis* Sm., ♂. Eureka, Utah, 9 May 1910, T. Spalding (USNM). (p. 48).
31. *Cucullia dorsalis* Sm., ♀. The Castles, 8 mi E of Buena Vista, Chaffee Co., Colorado, 8,800′, 11 July 1982, D. C. Ferguson (USNM). (p. 48)
32. *Cucullia laetifica* Lint., ♂. Junction, Kimble Co., Texas, 18 September 1968, A. & M. Blanchard (USNM). (p. 49).
33. *Cucullia laetifica* Lint., ♀. Sierra Diablo, 20 mi NNW Van Horn, Culberson Co., Texas, 6,000′, 30 May 1973, D. C. Ferguson (USNM). (p 49).
34. *Cucullia lethe* Poole, ♂. Paratype. Sinton, Welder Wildlife Refuge, Texas, 11 October 1964, A. & M. E. Blanchard (USNM). (p. 50).
35. *Cucullia lethe* Poole, ♀. Kerrville, Texas, August 1909 (USNM). (p 50).
36. *Cucullia charon* Poole, ♂. Fort Davis, Jeff Davis Co., Texas, 27 March 1965, A. & M. E. Blanchard (USNM). (p. 51).

1
2
3
4
5
6
7
8
9
10
11
12
13
14
15
16
17
18
19
20
21
22
23
24
25
26
27
28
29
30
31
32
33
34
35
36

PLATE 2

Noctuoidea

NOCTUIDAE

figs. 1–41

NATURAL SIZE 1:1

1. *Cucullia charon* Poole, ♀. Panther Pass, Chisos Mountains, Brewster Co., Texas, 6,000′, 2 June 1973, D. C. Ferguson (USNM). (p. 51).
2. *Cucullia alfarata* Strecker, ♂. Eldora, Cape May Co., New Jersey, 7 August 1978, J. Muller (AMNH). (p. 53).
3. *Cucullia alfarata* Stkr., ♀. Philadelphia, Pennsylvania (USNM). (p. 53).
4. *Cucullia eccissica* Dyar, ♂. Verdugo Hills, Los Angeles Co., California, 24 March 1993, C. Henne (LACM). (p. 53).
5. *Cucullia eccissica* Dyar, ♀. East slope El Sequoia Dunes, Los Angeles Co., California, 7 March 1954, C. Henne (LACM). (p. 58).
6. *Cucullia intermedia* Speyer, ♂. Doyles, Codroy Valley, Newfoundland, 30 July 1962, D. C. Ferguson (USNM). (p. 43).
7. *Cucullia intermedia* Speyer, ♂. Ringwood Reserve, near Dryden, Tompkins Co., New York, 21 May 1965, D. C. Ferguson (USNM). (p. 43).
8. *Cucullia intermedia* Speyer. ♀. Staten Island, New York, July (USNM). (p. 43).
9. *Cucullia intermedia* Speyer. ♂. Angel Creek, E Humboldt Mountains, SSW of Wells, Elko Co., Nevada, 7,000′, 18 July 1971, D. C. Ferguson (USNM). (p. 43).
10. *Cucullia antipoda* Stkr., ♂. Eureka, Utah, 14 July 1909, T. Spalding (USNM). (p. 55).
11. *Cucullia antipoda* Stkr., ♂. Estes Park, Larimer Co., Colorado, 7,800′, 28 July 1967, A. & M. E. Blanchard (USNM). (p. 55).
12. *Cucullia antipoda* Stkr., ♂. Olancha, Inyo Co., California, 24–30 June (USNM). (p. 55).
13. *Cucullia antipoda* Stkr., ♂. Great Sand Dunes National Monument, Alamosa Co., Colorado, 8,200′, 25 June 1982, D. C. Ferguson (USNM). (p. 55).
14. *Cucullia antipoda* Stkr., ♂. Great Sand Dunes National Monument, Alamosa Co., Colorado, 8,200′, 25 June 1982, D. C. Ferguson (USNM). (p. 55).
15. *Cucullia antipoda* Stkr., ♀. Great Sand Dunes National Monument, Alamosa Co., Colorado, 8,200′, 28 June 1982, D. C. Ferguson (USNM). (p. 55).
16. *Cucullia antipoda* Stkr., ♀. Great Sand Dunes National Monument, Alamosa Co., Colorado, 8,200′, 28 June 1982, D. C. Ferguson (USNM). (p. 55).
17. *Cucullia antipoda* Stkr., ♂. Smokey Valley, Tulare Co., California, 4 June 1953, C. Henne (LACM). (p. 55).
18. *Cucullia eurekae* Poole, ♂. Holotype. Eureka, Utah, 4 June 1921, T. Spalding (USNM). (p. 56).
19. *Cucullia eurekae* Poole, ♀. Paratype. Eureka, Utah, 4 June 1921, T. Spalding (USNM). (p. 56).
20. *Cucullia incresa* (Sm.), ♂. Loma Linda, San Bernardino Co., California, 8–15 April (USNM). (p. 58).
21. *Cucullia incresa* (Sm.), ♀. Mint Canyon, Los Angeles Co., California, 19 June 1954, C. Hill (USNM). (p. 58).
22. *Cucullia incresa* (Sm.), ♂. Palmdale, California, 6 May 1946 (USNM). (p. 58).
23. *Cucullia incresa* (Sm.), ♀. Seaberg's Well, California, 12 April 1935, G. & J. Sperry (USNM). (p. 58).
24. *Cucullia heinrichi* B. & Benj., ♂. Borego Canyon, California, March 6 1940, J. Sperry (AMNH). (p. 59).
25. *Cucullia heinrichi* B. & Benj., ♀. Borego Desert Park, San Diego Co., 9 February 1941, D. Meadows (USNM). (p. 59).
26. *Cucullia astigma* (Sm.), ♂. Blue Ridge, San Gabriel Mountains, Los Angeles Co., California, 7,386′, 1 July 1964, C. Henne (LACM). (p. 60).
27. *Cucullia astigma* (Sm.), ♀. Big Rock Creek, San Gabriel Mountains, Los Angeles Co., California, 5,800′, 2 June 1962, C. Henne (LACM). (p. 60).
28. *Cucullia astigma* (Sm.), ♀. Glenwood Springs, Colorado, 1–7 June (USNM). (p. 60).
29. *Cucullia eulepis* (Grt.), ♂. Zapata Ranch, Alamosa Co., Colorado, 9,000′, 26 June 1982, D. C. Ferguson (USNM). (p. 61).
30. *Cucullia eulepis* (Grt.), ♀. Zapata Ranch, Alamosa Co., Colorado, 9,000′, 26 June 1982, D. C. Ferguson (USNM). (p. 61).
31. *Cucullia cucullioides* (B. & Benj.), ♂. Yermo, San Bernardino Co., California, 25 August 1942, D. Meadows (USNM). (p. 62).
32. *Cucullia cucullioides* (B. & Benj.), ♀. Olancha, Inyo Co., California, 8–15 June (USNM). (p. 62).
33. *Cucullia basipuncta* (B. & McD.), ♂. Christmas Tree Pass, Newberry Mountains, Clark Co., Nevada, 19 March 1988, S. McKown (USNM). (p. 62).
34. *Cucullia basipuncta* (B. & McD.), ♀. Clark Co., Nevada, 16–23 March (USNM). (p. 62).
35. *Cucullia mcdunnoughi* (Henne), ♂. Angel Lake, E Humboldt Mts., near Wells, Elko Co., Nevada, 8,200′, 23 July 1971, D. C. Ferguson (USNM). (p. 63).
36. *Cucullia mcdunnoughi* (Henne), ♀. Brewster, Washington, 14 July 1951, J. C. Hopfinger (USNM). (p. 63).
37. *Cucullia luna* Morr., ♂. Fort Niobrara, Cherry Co., Nebraska, 15 June 1983, D. C. Ferguson (USNM). (p. 64).
38. *Cucullia luna* Morr., ♀. Fort Niobrara, Cherry Co., Nebraska, 25 June 1983, D. C. Ferguson (USNM). (p. 64).
39. *Cucullia serraticornis* Lint., ♂. Pueblo del Sol, Huachuca Mts., Cochise Co., Arizona, 28 January 1986, R. Wielgus (USNM). (p. 65).
40. *Cucullia serraticornis* Lint., ♂. Palmdale, California, 28 February 1946 (USNM). (p. 65).
41. *Cucullia serraticornis* Lint., ♀. Burbank, Los Angeles Co., California, 2 February 1960, F. Sala (USNM). (p. 65).

1
2
3
4
5
6
7
8
9
10
11
12
13
14
15
16
17
18
19
20
21
22
23
24
25
26
27
28
29
30
31
32
33
34
35
36
37
38
39
40
41

PLATE 3

Noctuoidea

NOCTUIDAE

figs. 1–54

NATURAL SIZE 1:1

1. *Cucullia comstocki* (McD.), ♂. Vincent Station, Los Angeles Co., California, 11 February 1950 (USNM). (p. 67).
2. *Cucullia comstocki* (McD.), ♀. Pearblossom, Palmdale area, Los Angeles Co., California, 29 January 1962, F. Sala (USNM). (p. 67).
3. *Cucullia strigata* (Sm.), ♂. Glenwood Springs, Colorado, 1–7 April (USNM). (p. 67).
4. *Cucullia strigata* (Sm.), ♀. Brewster, Washington 14 April 1950, J. Hopfinger (USNM). (p. 67).
5. *Cucullia albida* Sm., ♂ Rounthwaite, Manitoba (USNM). (p. 68).
6. *Cucullia albida* Sm., ♀. Calgary, Alberta (USNM). (p. 68).
7. *Cucullia pulla* (Grt.), ♂. Pinyon Hills, San Bernardino Co., California, 6 February 1968 (USNM). (p. 69).
8. *Cucullia pulla* (Grt.), ♀. Mint Canyon, Los Angeles Co., California, 7 January 1950, F. Sala (USNM). (p. 69).
9. *Cucullia pulla* (Grt.), ♂. Eureka, Utah, 26 April 1922, T. Spalding (USNM). (p. 69).
10. *Cucullia dammersi* (McD.), ♂. near Perris, Riverside Co., California, 21 January 1950, C. Hill (USNM). (p. 70).
11. *Cucullia dammersi* (McD.), ♀. Paratype. Riverside, California, 1 January 1935, C. Dammers (USNM). (p. 70).
12. *Dolocucullia dentilinea* (Sm.), ♂. Fort Niobrara, Cherry Co., Nebraska, 14 June 1983, D. C. Ferguson (USNM). (p. 72).
13. *Dolocucullia dentilinea* (Sm.), ♀. Fort Niobrara, Cherry Co., Nebraska, 19 June 1983, D. C. Ferguson (USNM). (p. 72).
14. *Dolocucullia dentilinea* (Sm.), ♀. Upper Lundy Canyon, W of Mono Lake, California, 8,000′, 4 July 1987, D. C. Ferguson (USNM). (p. 72).
15. *Dolocucullia minor* (B. & McD.), ♂. Alpine, Brewster Co., Texas, 6 September 1964, A. & M. E. Blanchard (USNM). (p. 72).
16. *Dolocucullia minor* (B. & McD.), ♀. Basin, Big Bend National Park, Texas, 22 September 1963, A. & M. E. Blanchard (USNM). (p. 72).
17. *Opsigalea blanchardi* Todd, ♂. Fort Davis, Jeff Davis Co., Texas, 18 July 1971, A. & M. E. Blanchard (USNM). (p. 74).
18. *Opsigalea blanchardi* Todd, ♂. Fort Davis, Jeff Davis Co., Texas, 18 July 1971, A. & M. E. Blanchard (USNM). (p. 74).
19. *Emariannia cucullidea* Benj., ♂. Fort Davis, Jeff Davis Co., Texas, 29 August 1964, A. & M. E. Blanchard (USNM). (p. 76).
20. *Emariannia cucullidea* Benj., ♂. Fort Davis, Jeff Davis Co., Texas, 28 August 1964, A. & M. E. Blanchard (USNM). (p. 76).
21. *Fala ptychophora* Grt., ♂. Loma Linda, San Bernardino Co., California, 16–23 March (USNM). (p. 86).
22. *Plagiomimicus ochoa* (Barnes), ♀. Mount Locke, Davis Mountains, Texas, 25 August 1967, A. & M. E. Blanchard (USNM). (p. 87).
23. *Plagiomimicus olvello* (Barnes), ♂. Oak Spring, Big Bend National Park, Texas, 5 October 1967, A. & M. E. Blanchard (USNM). (p. 89).
24. *Plagiomimicus dimidiata* (Grt.), ♂. Snake Creek, 3 mi NW of Midway, Wasatch Co., Utah, 31 July 1971, D. C. Ferguson (USNM). (p. 90).
25. *Plagiomimicus aureolum* (Hy. Edw.), ♂. Ash Canyon Road, Huachuca Mountains, Cochise Co., Arizona, September 1980, N. McFarland (USNM). (p. 90).
26. *Plagiomimicus heitzmani* Poole, ♂. Holotype. East of Birmingham, Clay Co., Missouri, 2 September 1962, J. Heitzman (USNM). (p. 91).
27. *Plagiomimicus spumosum* (Grt.), ♂. Santa Ana Wildlife Refuge, Hidalgo Co., Texas, 6 May 1967, A. & M. E. Blanchard (USNM). (p. 92).
28. *Plagiomimicus spumosum* (Grt.). ♀. Paducah, Cottle Co., Texas, 8 September 1966, A. & M. E. Blanchard (USNM). (p. 92).
29. *Plagiomimicus spumosum* (Grt.), ♀. Brewster, Okanagan Co., ington, 900′, 10 July 1959, J. Hopfinger (USNM). (p. 92).
30. *Plagiomimicus spumosum* (Grt.), ♀. Loma Linda, San Bernardin California, 24–31 August (USNM). (p. 92).
31. *Plagiomimicus caesium* (Blanchard and Knudson). ♂. Welder W Refuge, San Patricio Co., Texas, 14 September 1985, E. Kn (USNM). (p. 94).
32. *Plagiomimicus caesium* (Blanchard and Knudson), ♀. Welder W Refuge, San Patricio Co., Texas, 14 September 1985, E. Kn (USNM). (p. 94).
33. *Plagiomimicus manti* (Barnes), ♂. Garner State Park, Uvalde Co., as, 1 October 1964, A. & M. E. Blanchard (USNM). (p. 95).
34. *Plagiomimicus curiosum* (Neum.), ♀. Paradise, Cochise Co., Ar (USNM). (p. 95).
35. *Plagiomimicus astigmatosum* (Dyar), ♂. Holotype. Arizona, A O. C. Poling (USNM). (p. 96).
36. *Plagiomimicus hilli* (B. & Benj.), ♀. Palm Springs, Riverside Co. ifornia, 4 November 1950 (USNM). (p. 96).
37. *Plagiomimicus hilli* (B. & Benj.), ♀. Holotype of *Antaplaga ca* Hill, Indio, California, 16 October 1921, C. Hill (USNM). (p. 9
38. *Plagiomimicus tepperi* (Morr.), ♂. Mount Locke, Davis Moun Texas, 6,700′, 27 August 1970, A. & M. E. Blanchard (USNM 97).
39. *Plagiomimicus tepperi* (Morr.), ♀. Stronghold Campground, Dra Mountains, Cochise Co., Arizona, 17 August 1974, R. Wielgus (US (p. 97).
40. *Plagiomimicus tepperi* (Morr.), Paratype. ♂. Satus Creek, Yakima Washington, 30 May 1949, E. C. Johnston [CNC]. (p. 97).
41. *Plagiomimicus mimica* Poole, ♂. Paratype. Paradise, Cochise Co izona (USNM). (p. 98).
42. *Plagiomimicus pityochromus* Grt., ♀. Laguna Park, Bosque Co., T 24 September 1970, A. & M. E. Blanchard (USNM). (p. 101).
43. *Plagiomimicus pityochromus* Grt., ♂. New Brighton, Pennsylvan August 1907, Merrick (USNM). (p. 101).
44. *Plagiomimicus triplagiatus* Sm., ♀. Deming, New Mexico, 1–7 Sep ber (USNM). (p. 101).
45. *Plagiomimicus expallidus* Grt., ♀. Canadian, Hemphill Co., Texa August 1971, A. & M. E. Blanchard (USNM). (p. 103).
46. *Cuahtemoca unicum* (B. & Benj.), ♀. Holotype. Paradise, Cochise Arizona, 1923, O. Duffner (USNM). (p. 105).
47. *Lineostriastiria hachita* (Barnes). ♀. Deming, New Mexico, 1–7 tember (USNM). (p. 106).
48. *Lineostriastiria sexseriata* (Grt.), ♂. Ash Canyon Road, Huac Mountains, Cochise Co., Arizona, September 1980, N. McFar (USNM). (p. 108).
49. *Lineostriastiria biundulalis* (Zell.), ♂. Artesia Wells, La Salle Co., T 28 September 1971, A. & M. E. Blanchard (USNM). (p. 108).
50. *Lineostriastiria biundulalis* (Zell.), ♂. Paducah, Cottle Co., Texa August 1971, A. & M. E. Blanchard (USNM). (p. 108).
51. *Lineostriastiria biundulalis* (Zell.), ♀. Sinton, Welder Wildlife Re Texas, 8 October 1964, A. & M. E. Blanchard (USNM). (p. 108)
52. *Lineostriastiria olivalis* (B. & McD.). ♀. Burbank, Los Angeles California, 4 April 1961, F. Sala (USNM). (p. 109).
53. *Lineostriastiria hutsoni* (Sm.), ♀. Newberry Springs, San Bernar Co., California, 4 April 1937, D. Meadows (USNM). (p. 111).
54. *Lineostriastiria hutsoni* (Sm.), ♂. White Tanks Regional Mountain F Maricopa Co., Arizona, 1 April 1976, R. Wielgus (USNM). (p. 1

1
2
3
4
5
6
7
8
9
10
11
12
13
14
15
16
17
18
19
20
21
22
23
24
25
26
27
28
29
30
31
32
33
34
35
36
37
38
39
40
41
42
43
44
45
46
47
48
49
50
51
52
53
54

PLATE 4

Noctuoidea

NOCTUIDAE

figs. 1–55

NATURAL SIZE 1:1

nthothrix *neumoegeni* Hy. Edw., ♂. Soledad Canyon, Los Angeles unty, California, 24 April 1946, L. Martin (USNM). (p. 112).

nthothrix *neumoegeni* Hy. Edw. ♀. California (USNM). (p. 112).

nthothrix *ranunculi* Hy. Edw. ♂. Lovejoy Buttes, Los Angeles Coun- California, 12 April 1943, L. Martin (USNM). (p. 113).

4. *Xanthothrix ranunculi* Hy. Edw., ♂. Lovejoy Buttes, Los Angeles County, California, 29 March 1945, L. Martin (USNM). (p. 113).

5. *Xanthothrix ranunculi* Hy. Edw., ♀. Little Rock, Los Angeles County, California, 3 March 1938, R. McElvare (USNM). (p. 113).

6. *Chrysoecia scira* (Druce), ♂. Ash Canyon Road, Huachuca Mountains,

Cochise County, Arizona, 8 August 1980, N. McFarland (USNM). (p. 115).

7. *Chrysoecia gladiola* (Barnes), ♂, Santo Domingo Valley, Baja California, Mexico, January 1983 (USNM). (p. 115).
8. *Chrysoecia atrolinea* (B. McD.), ♂. Paradise, Cochise Co., Arizona, September (USNM). (p. 117).
9. *Chrysoecia thoracica* (Hy. Edw.), ♂. Ash Canyon Road, Huachuca Mountains, Cochise County, Arizona, 17 August 1980, N. McFarland (USNM). (p. 119).
10. *Chrysoecia thoracica* (Hy. Edw.), ♀. Ash Canyon Road, Huachuca Mountains, Cochise County, Arizona, 19 August 1980, N. McFarland (USNM). (p. 119).
11. *Basilodes chrysopis* Grt., ♂. Government Springs, Big Bend National Park, Texas, 22 September 1971, A. & M. E. Blanchard (USNM). (p. 123).
12. *Basilodes chrysopis* Grt., ♀. Artesia Wells, La Salle County, Texas, 28 September 1971, A. & M. E. Blanchard (USNM). (p. 123).
13. *Basilodes pepita* Gn., ♂. Padre Island National Seashore, Texas, 24 September 1979, A. & M. E. Blanchard (USNM). (p. 120).
14. *Basilodes straminea* Poole, ♂. Holotype. Paradise, Cochise County, Arizona, (USNM). (p. 122).
15. *Hoplolythrodes arivaca* (Barnes), ♂. Holotype. Deming, New Mexico (USNM). (p. 125).
16. *Cirrhophanus dyari* Ckll., ♂. Fort Davis, Jeff Davis County, Texas, 24 August 1967, A. & M. E. Blanchard (USNM). (p. 126).
17. *Cirrhophanus dyari* Ckll., ♀. Ash Canyon Road, Huachuca Mountains, Cochise County, Arizona, 16 August 1980, N. McFarland (USNM). (p. 126).
18. *Cirrhophanus pretiosa* (Morr.), ♂. Eagle Lake, Colorado County, Texas, 27 April 1978, A. & M. E. Blanchard (USNM). (p. 129).
19. *Cirrhophanus triangulifer* Grt., ♀. Montgomcry County, Virginia (USNM). (p. 127).
20. *Cirrhophanus triangulifer* Grt., ♀. Green Village, New Jersey, 10 August (USNM). (p. 127).
21. *Eulithosia composita* Hy. Edw. ♂. Paradise, Cochise County, Arizona (USNM). (p. 130).
22. *Eulithosia composita* Hy. Edw., ♂. Paradise, Cochise County, Arizona (USNM). (p. 130).
23. *Eulithosia discistriga* (Sm.), ♂. Oak Spring, Big Bend National Park, Texas, 5 August 1964, A. & M. E. Blanchard (USNM). (p. 130).
24. *Eulithosia plesioglauca* (Dyar), ♂. Baboquivari Mountains, Pima County, Arizona, 15–30 August 1923, O. Poling (USNM). (p. 132).
25. *Eulithosia papago* (Barnes), ♀. Fort Davis, Jeff Davis County, Texas, 28 August 1964, A. & M. E. Blanchard (USNM). (p. 132).
26. *Eulithosia papago* (Barnes), ♀. 5 mi SE Livermore, Davis Mountains, Texas, 6,000′, 29 August 1970, A. & M. E. Blanchard (USNM). (p. 132).
27. *Angulostiria chrysochilus* (Dyar), ♂. Sanderson, Terel County, Texas, 28 September 1980, E. Knudson (USNM). (p. 134).
28. *Stiria rugifrons* Grt., ♂. Tall Timbers Research Station, Leon County, Florida, 26 September 1972, E. Todd (USNM). (p. 138).
29. *Stiria rugifrons* Grt., ♀. 6 mi NW Newcastle, Wyoming, 23 June 1965, R. Hodges (USNM). (p. 138).
30. *Stiria intermixta* Dyar, ♂. Alpine, Brewster County, Texas, 13 September 1963, A. & M. E. Blanchard (USNM). (p. 140).
31. *Stiria intermixta* Dyar, ♂. Alpine, Brewster County, Texas, 9 September 1963, A. & M. E. Blanchard (USNM). (p. 140).
32. *Stiria intermixta* Dyar, ♀. Mount Locke, Davis Mountains, 6,700′, 25 August 1967, A. & M. E. Blanchard (USNM). (p. 14
33. *Stiria dyari* Hill, ♂. Kingman, Mohave County, Arizona, 20 April H. Wilhelm (USNM). (p. 141).
34. *Stiria dyari* Hill, ♀. Kingman, Mohave County, Arizona, 23 April H. Wilhelm (USNM). (p. 141).
35. *Stiria blanchardi* (Hogue), ♂. Pine Canyon, Guadalupe Mountains as, 6,700′, 28 August 1967, A. & M. E. Blanchard (USNM). (p
36. *Stiria blanchardi* (Hogue), ♀. Government Spring, Big Bend Na Park, Texas, 6 October 1967, A. & M. E. Blanchard (USNM). (p
37. *Stiria sulphurea* Neum., ♂. Paradise, Cochise County, Arizona (U (p. 142).
38. *Stiria satana* Poole, ♂. Holotype. Baboquivari Mountains, Pima ty, Arizona, 1–5 September 1923, O. Poling (USNM). (p. 143).
39. *Stiria consuela* Stkr., ♀. Palm Springs, California, April, R. C (USNM). (p. 143).
40. *Stiria consuela* Stkr., ♂. Palm Springs, California, 18 March 19 Colidge (USNM). (p. 143).
41. *Chalcopasta territans* (Hy. Edw.), ♂. Paradise, Cochise County, A (USNM). (p. 145).
42. *Chalcopasta howardi* (Hy. Edw.), ♂. Artesia Wells, La Salle C Texas, 28 September 1971, A. & M. E. Blanchard (USNM). (p.
43. *Chalcopasta howardi* (Hy. Edw.), ♀. Chihuahuan Desert near N Mountain, Big Bend National Park, Texas, 21 September 1971 M. E. Blanchard (USNM). (p. 147).
44. *Chalcopasta fulgens* B. & McD., ♂. K-Bar Research Station, Big National Park, Texas, 23 September 1971, A. & M. E. Blar (USNM). (p. 147).
45. *Chalcopasta acema* (Druce), ♀. Huetama, Michoacan, Mexico, 2 gust 1933, T. Escalante (USNM). (p. 148).
46. *Neumoegenia poetica* Grt., ♂. Ash Canyon Road, Huachuca Mour Cochise County, Arizona, 20 July 1980, N. McFarland (USN 150).
47. *Chamaeclea pernana* (Grt.), ♂. Paradise, Cochise County, A (USNM). (p. 154).
48. *Chamaeclea pernana* (Grt.), ♀. Paradise, Cochise County, A (USNM). (p. 154).
49. *Chamaeclea basiochrea* B. & McD., ♂. Santa Ana Refuge, H County, Texas, 23 September 1973, A. & M. E. Blanchard (U (p. 155).
50. *Chamaeclea basiochrea* B. & McD., ♀. Santa Ana Refuge, H County, Texas, 23 September 1973, A. & M. E. Blanchard (U (p. 155).
51. *Bistica noela* (Druce), ♀. Madera Canyon, Santa Rita Mountains, Cruz County, Arizona, 8 September 1959, J. Franclemont (JG 156).
52. *Argentostiria koebelei* (Riley), ♀. Mecca, California, March, K. Co (USNM). (p. 152).
53. *Thurberiphaga diffusa* (Barnes), ♂. Redington, Arizona (USN 160).
54. *Hemioslaria pima* B. & Benj., ♂. Holotype. Baboquivari Mour Pima County, Arizona, 15–30 August 1923, O. Poling (USN 158).
55. *Heminocloa mirabilis* (Neum.), ♂. Baboquivari Mountains, Pima C Arizona, 10 October 1923 (USNM). (p. 159).

1 2 3 4 5 6 7
8 9 10 11 12
13 14 15 16 17
18 19 20 21 22
23 24 25 26 27
28 29 30 31
32 33 34 35 36
37 38 39 40 41
42 43 44 45 46
47 48 49 50 51 52
53 54 55

PLATE 5

Noctuoidea

NOCTUIDAE

figs. 1–49

NATURAL SIZE 1:1

1. *Psaphida resumens* Wlk., ♂. Forest Home, Ithaca, New York, 3 May 1965, D. Ferguson (USNM). (p. 166).
2. *Psaphida resumens* Wlk., ♂. Colesville, Montgomery County, Maryland, 27 March 1976, D. Ferguson (USNM). (p. 166).
3. *Psaphida resumens* Wlk., ♀. The Wedge, McClellanville, South Carolina, 21 March 1968, D. Ferguson (USNM). (p. 166).
4. *Psaphida thaxterianus* (Grt.), ♂. T5N R1W Sec 14, Clinton County, Michigan, 12 April 1972, M. Nielsen (USNM). (p. 167).
5. *Psaphida thaxterianus* (Grt.), ♂. Oakland County, Michigan, 19 April 1944, J. Newman (USNM). (p. 167).
6. *Psaphida thaxterianus* (Grt.), ♀. T5N R1W Sec 14, Clinton County, Michigan, 13 April 1971, M. Nielsen (USNM). (p. 167).
7. *Psaphida grandis* (Sm.), ♂. Belleview, Marion County, Florida, 11 February 1989, J. Kutis (USNM). (p. 168).
8. *Psaphida grandis* (Sm.), ♂. Oakland County, Michigan, 15 April 1941, P. Moody (USNM). (p. 168).
9. *Psaphida grandis* (Sm.), ♂. T5N R1W Sec 14, Clinton County, Michigan, 12 April 1972, M. Nielsen (USNM). (p. 168).
10. *Psaphida grandis* (Sm.), ♀. Mouth of Seneca, Pendleton County, West Virginia, 4 April 1981, E. Metzler (USNM). (p. 168).
11. *Psaphida electilis* (Morr.), ♂. Forest Home, Ithaca, New York, 3 May 1965, D. Ferguson (USNM). (p. 169).
12. *Psaphida electilis* (Morr.), ♂. Forest Home, Ithaca, New York, 1 May 1965, D. Ferguson (USNM). (p. 169).
13. *Psaphida electilis* (Morr.), ♂. Forest Home, Ithaca, New York, 23 April 1961, D. Ferguson (USNM). (p. 169).
14. *Psaphida electilis* (Morr.), ♂. Bend, San Saba County, Texas, 11 March 1982, R. Poole (USNM). (p. 169).
15. *Psaphida electilis* (Morr.), ♀. Forest Home, Ithaca, New York, 1 May 1965, D. Ferguson (USNM). (p. 169).
16. *Psaphida rolandi* (Grt.), ♂. Bethany, New Haven County, Connecticut, 13 April 1968, D. Ferguson (USNM). (p. 168).
17. *Psaphida rolandi* (Grt.), ♂. Forest Home, Ithaca, New York, 1 May 1965, D. Ferguson (USNM). (p. 168).
18. *Psaphida rolandi* (Grt.), ♂. 3 mi NW Warsaw, Benton County, Missouri, 7 March 1976, J. Heitzman (USNM). (p. 168).
19. *Psaphida rolandi* (Grt.), ♂, Spring, Harris County, Texas, 27 February 1963, A. & M. E. Blanchard (USNM). (p. 168).
20. *Psaphida rolandi* (Grt.), ♀. Houston, Harris County, Texas, 23 February 1962, A. & M. E. Blanchard (USNM). (p. 168).
21. *Psaphida damalis* (Grt.), ♂. Sonoma County, California, 8–15 February (USNM). (p. 171).
22. *Psaphida styracis* (Gn.), ♂. Bear Mountains, New York, 25–31 October, H. Erb (USNM). (p. 170).
23. *Psaphida styracis* (Gn.), ♂. Yale Preserve, New Haven, Connecticut, 25 March 1964, D. Ferguson (USNM). (p. 170).
24. *Psaphida styracis* (Gn.), ♀. Bellmore, Long Island, New York, 6 April 1913, G. Englehardt (USNM). (p. 170).
25. *Psaphida styracis* (Gn.), ♀. Central Park, Long Island, New York, 17 April 1912, G. Englehardt (USNM). (p. 170).
26. *Psaphida styracis* (Gn.), ♀. Central Park, Long Island, New York, 11 April (USNM). (p. 170).
27. *Psaphida styracis* (Gn.), ♂. Lake Houston, Harris County, Texas, 30 January 1983, E. Knudson (USNM). (p. 170).
28. *Pseudocopivaleria sonoma* (McD.), ♂. Laytonville, Mendocino County, California, 29 March 1948, R. Sternitzky (USNM). (p. 175).
29. *Pseudocopivaleria sonoma* (McD.), ♂. Spring Crest, Riverside County, California, March 11 1966, R. Leuschner (RHL). (p. 175).
30. *Pseudocopivaleria anaverta* Buckett and Bauer, ♂. Holotype. 24 mi SSW of Valyermo, Los Angeles County, California, 4,800′, 14 April 1964, N. McFarland (USNM). (p. 176).
31. *Pseudocopivaleria anaverta* Buckett and Bauer, ♀. 2.5 mi SSW of Valyermo, Los Angeles County, California, 12 April 1964, N. McFarland (USNM). (p. 176).
32. *Copivaleria grotei* (Morr.), ♂. Six Mile Creek, Ithaca, New York, 23 April 1961, D. Ferguson (USNM). (p. 176).
33. *Brachionycha borealis* (Sm.), ♂. Luzerne County, Pennsylvania, April 1903 (USNM). (p. 178).
34. *Pleromella opter* Dyar, ♂. Mint Canyon, Los Angeles County, California, 18 March 1950 (USNM). (p. 179).
35. *Provia argentata* B. & McD., ♂. Singing Springs, Los Angeles County, California, 7 April 1950, C. Hill (USNM). (p. 181).
36. *Feralia jocosa* (Gn.), ♂. Big Indian Lake, Halifax County, Nova Scotia, 28 May 1959, D. Ferguson (USNM). (p. 183).
37. *Feralia jocosa* (Gn.), ♂. Mt. Uniacke, Nova Scotia, 20 May 1955, D. Ferguson (USNM). (p. 183).
38. *Feralia jocosa* (Gn.), ♂. Williams' Lake, Purcell's Cove, Halifax County, Nova Scotia, 17 May 1963, D. Ferguson (USNM). (p. 183).
39. *Feralia jocosa* (Gn.), ♀. Forest Hills, Massachusetts, 12 April 1913, W. Reiff (USNM). (p. 183).
40. *Feralia deceptiva* McD., ♂. Barton, Clackamas County, Oregon, 21 March 1961, C. Nelson (USNM). (p. 184).
41. *Feralia major* Sm., ♂. Fall Creek, Ithaca, New York, 3 May 1965, D. Ferguson (USNM). (p. 185).
42. *Feralia major* Sm., ♂. Colesville, Montgomery County, Maryland, 3 April 1980, D. Ferguson (USNM). (p. 185).
43. *Feralia major* Sm., ♀. Lake Kejimkujik, Queens County, Nova Scotia, 19 May 1962, D. Ferguson (USNM). (p. 185).
44. *Feralia comstocki* Grt., ♂. Montague Mines, Halifax County, Nova Scotia, 29 May 1954, D. Ferguson (USNM). (p. 185).
45. *Feralia comstocki* Grt., ♂. Mount Uniacke, Nova Scotia, 20 May 1952, D. Ferguson (USNM). (p. 185).
46. *Feralia februalis* Grt., ♂. McMinnville, Oregon, 24 March 1940, K. Fender (USNM). (p. 186).
47. *Feralia februalis* Grt., ♂. Sunland, California, 22 February 1946 (USNM). (p. 186).
48. *Feralia februalis* Grt., ♂. Lake Pillsburg, Lake County, California, 1936, O. Scott (USNM). (p. 186).
49. *Feralia meadowsi* Buckett, ♂. Holotype. Avalon, Santa Catalina Island, California, 23 February 1933, D. Meadows (USNM). (p. 187).

1
2
3
4
5
6
7
8
9
10
11
12
13
14
15
16
17
18
19
20
21
22
23
24
25
26
27
28
29
30
31
32
33
34
35
36
37
38
39
40
41
42
43
44
45
46
47
48
49

NOTES

1. ABBREVIATIONS FOR COLLECTORS AND COLLECTIONS (Institution(s) receiving private collection are indicated in square brackets, e.g., [USNM])

AEB	A. E. Brower, Augusta, ME [USNM]
AMNH	American Museum of Natural History, New York
ANSP	Academy of Natural Sciences, Philadelphia
BM	Bryant Mather, Clinton, MS
BMNH	Natural History Museum, London
CAS	California Academy of Sciences, San Francisco
CFDA	California Department of Food and Agriculture, Sacramento
CM	Carnegie Museum of Natural History, Pittsburgh
CNC	Canadian National Collection, Ottawa
CU	Cornell University, Ithaca
ECK	Edward C. Knudson, Houston, TX
EIHU	Entomological Institute, Hokkaido University, Sapporo
ERL	Entomology Research Laboratory, Burlington, Vermont
FMNH	Field Museum of Natural History, Chicago
FSCA	Florida State Collection of Arthropods, Gainesville
GB	George Balogh, Portage, MI
HUMB	Museum Alexander Humboldt, Berlin
INHS	Illinois Natural History Survey, Champaign
ITZ	Institute voor Taxonomische Zöologie, Amsterdam
JBH	John B. Heppner, Gainesville, FL
JGF	John G. Franclemont, Ithaca, NY [CU, USNM]
JRH	J. Richard Heitzman, Independence, MO
JSK	John S. Kutis, Belleview, FL
KWP	K. W. Philip, Fairbanks, AK
LACM	Natural History Museum, Los Angeles County
LEM	Lyman Entomological Museum, Montreal
LS	Linnaean Society, London
MCZ	Museum of Comparative Zoology, Cambridge
MEM	Mississippi Entomological Museum, Mississippi State
MNSA	Museu Argentino de Ciencias Naturales "Bernardino Rivadavia," Buenos Aires
MSU	Michigan State University, East Lansing
MZC	Museo Zoologico Cubano, Havana
NCSU	North Carolina State University, Raleigh
NHMV	Naturhistorisches Museum, Vienna
NSM	Nova Scotia Museum, Halifax
NYSM	New York State Museum, Albany
PMBC	Provincial Museum of British Columbia, Victoria
RHL	Ronald H. Leuschner, Manhattan Beach, CA
RN	Naturhistoriska Riksmuseet, Stockholm
ROK	Roy O. Kendall, San Antonio, TX [TAMU]
ROM	Royal Ontario Museum, Toronto
SDNH	San Diego Natural History Museum, San Diego
SIUC	Southern Illinois University, Carbondale
TAMU	Texas A.&M. University, College Station
UA	University of Alberta, Edmonton
UAF	University of Arkansas, Fayetteville
UBC	University of British Columbia, Vancouver
UCB	University of California, Berkeley
UCD	University of California, Davis
ULK	University of Louisville, Louisville
UM	University of Michigan, Ann Arbor
UMC	University of Missouri, Columbia
UMO	University Museum, Oxford
USNM	National Museum of Natural History, Washington
UWM	University of Wisconsin, Madison
VAB	Vernon A. Brou, Abita Springs, LA
WEM	William E. Miller, Saint Paul, MN
WPC	Wedge Plantation Collection, McClellanville

YPM	Yale Peabody Museum, New Haven
ZMC	Zoologisk Museum, Copenhagen
ZMH	Zoologiska Museum, Helsinki
ZSBS	Zoologische Sammlung des Bayerischen Staates, Munich

2. COMMON NAMES

The use of an asterisk "*" in the text denotes a name listed in Stoetzel, M. B., et al., *Common Names of Insects & Related Organisms 1989* published by the Entomological Society of America.

French-language common names have been taken from Benoit, P., et al., 1975, *French Names of Insects of Canada,* 4th edition, published for the Quebec Society for the Protection of Plants by Department of Agriculture of Quebec. The abbreviation "m." after a name indicates that it is masculine, "f." that it is feminine.

3. CITATIONS OF AUTHORITIES

Authors' names without parentheses indicate that the specific name is associated with the genus in which it was described.

Authors' names in parentheses indicate that the specific name has been transferred from the genus in which it was described to another genus.

4. WING LENGTH

Wing length is the measurement in millimeters from the base to the apex of the forewing.

5. LOCATION OF TYPE SPECIMEN

The current location of the type specimen is given by the appropriate abbreviation in square brackets immediately following the type locality. The words "type lost" indicate that it no longer exists. If no information is given, a type may exist; but its present location was not determined.

6 NOMENCLATURE FOR LARVAL SETAE

Hinton's (*Trans. Royal Ent. Soc. London,* **97**: 1–37, 1946) terminology as modified by Stehr (Order Lepidoptera, *in* Stehr, F. W., *Immature Insects,* 288–304, 1987) is used to refer to larval setae.

CHECK LIST

CUCULLIINAE

CUCULLIA Schr., 1802
EUDERAEA Hbn., [1821]
EUCALIMIA Hbn., [1821]
CALLAENIA Hbn., [1821]
EMPUSA Hbn., [1821], preocc. by Illiger, 1798
ARGYRITIS Hbn., [1821]
TRIBUNOPHORA Hbn., 1822
LOPHIA Sodoffsky, 1837
LATHOSEA Grt., 1881
NYCTEROPHAETA Sm., 1882
EPINYCTIS Grt., 1882
RANCORA Sm., 1892
COPICUCULLIA Sm., 1894
CHELIGALEA Hamp., 1906
ARGYROGALEA Hamp., 1906
ARGYROMATA Hamp., 1906
EMPUSADA Hamp., 1906
PSEUDOCOPICUCULLIA Dumont, 1928
PSEUDONYCTEROPHAETA Berio, 1934
CALOCUCULLIA Ronkay & Ronkay, 1967
SHARGACUCULLIA Ronkay and Ronkay, 1992
asteroides Gn., 1852
montanae Grt., 1882
montana Hamp., 1906, emend.
similaris Sm., 1892
omissa Dod, 1916
florea Gn., 1852
indicta Sm., 1904
postera Gn., 1852
obscurior Sm., 1892
lilacina Schaus, 1898
agua Barnes, 1905
eucaena Dyar, 1919
oribac Barnes 1904
strigata Schaus, 1898, preocc. by Sm., 1892
arizona Sm., 1905
perstrigata Hamp., 1906
convexipennis G. & R., 1868
intermedia Speyer, 1870
cinderella Sm., 1892
speyeri Lint., 1874
styx Poole, 1995
dorsalis Sm., 1892
laetifica Lint., 1875
cita Grt., 1883
hartmanni French, 1888
lethe Poole, 1995
charon Poole, 1995
alfarata Stkr., 1898
phila Sm., 1908
eccissica Dyar, 1919
antipoda Stkr., 1877
propinqua Sm., 1894, preocc. by Evers., 1842
luteodisca (Sm., 1909)
adopitna (B. & Benj., 1927)
eurekae Poole, 1995
incresa (Sm., 1910)
heinrichi B. & Benj., 1924
astigma (Sm., 1894)
jemezensis (Dyar, 1922)
eulepis (Grt., 1875)
bistriga Sm., 1892
mala (Sm., 1908)
cucullioides (B. & Benj., 1923)
basipuncta (B. & McD., 1918)
mcdunnoughi (Henne, 1940)
luna Morr., 1875
magdalena (Hlst., 1882)
notatella (Grt., 1882)
serraticornis Lint., 1874
solidaginis Strk., 1874
matricariae Strk., 1874
comstocki (McD., 1937)
strigata (Sm., 1892)
brucei (Sm., 1903)
ketchikana (B. & Benj., 1922)
albida Sm., 1894
albicinerea (Sm., 1903)
pulla (Grt., 1881)
pullata (Grt., 1890), emend.
spauldingi (B. & Benj., 1922)
dammersi (McD., 1935)

DOLOCUCULLIA Poole, 1995
dentilinea (Sm., 1899)
mexicana (Draudt, 1924)
minor (B. & McD., 1913)

OPSIGALEA Hamp., 1906
blanchardi Todd 1966

EMARIANNIA Benj., 1933
cucullidea Benj., 1933

STIRIINAE

FALA Grt., 1875
ptycophora Grt., 1875

PLAGIOMIMICUS Grt., 1873
STIBADIUM Grt., 1874
ANTAPLAGA Grt., 1877
POLENTA Morr., 1880
NEOPHAEUS Dyar, 1918
ochoa (Barnes, 1904)
dollii Sm., 1908
olvello (Barnes, 1907)
dimidiata (Grt., 1877)
aureolum (Hy. Edw., 1882)
heitzmani Poole, 1995
spumosum (Grt., 1874)
mavina (B. & McD., 1910)
caesium (Blanchard & Knudson, 1984)
manti (Barnes, 1904)
curiosum (Neum., 1883)
astigmatosum (Dyar, 1921)
hilli (B. & Benj., 1923)
caliente (Hill, 1923)

tepperi (Morr., 1875)
richii Grt., 1886
mimica Poole, 1995
pityochromus Grt., 1873
media (Morr., 1875)
triplagiatus Sm., 1890
expallidus Grt., 1882

CUAHTEMOCA Hogue, 1963
unicum (B. & Benj., 1926)

LINEOSTRIASTIRIA Poole, 1995
hachita (Barnes, 1904)
sexseriata (Grt., 1881)
biundulalis (Zell., 1872)
olivalis (B. & McD., 1916)
hutsoni (Sm., 1907)
fuliginosa (Sm., 1907)

Xanthothrix Hy., Edw., 1878
EUEDWARDSIA Grt., 1882
neumoegeni Hy. Edw., 1881
ranunculi Hy. Edw., 1878

CHRYSOECIA Hamp., 1908
GORGORA Dyar, 1914
CHICHIMECA Hogue, 1963
scira (Druce, 1889)
benjamini (Hill, 1924)
gladiola (Barnes, 1907)
atrolinea (B. & McD., 1912)
thoracica (Hy. Edw., 1884)

BASILODES Gn., 1852
DEOBRIGA Wlk., 1869
pepita Gn., 1852
chrysopasa (Wlk., 1869)
straminea Poole, 1995
chrysopis Grt., 1881
catharops Dyar, 1911

HOPLOLYTHRODES Poole, 1995
arivaca (Barnes, 1907)

CIRRHOPHANUS Grt., 1872
dyari Ckll., 1899
triangulifer Grt., 1872
pretiosa (Morr., 1875)

EULITHOSIA Hy. Edw., 1884
HOPLOLYTHRA Hamp., 1910
composita Hy. Edw., 1884
discistriga (Sm., 1903)
plesioglauca (Dyar, 1912)
comstocki (Hill, 1924)
papago (Barnes, 1907)

ANGULOSTIRIA Poole, 1995
chryseochilus (Dyar, 1909)

STIRIA Grt., 1874
rugifrons Grt., 1874
intermixta Dyar, 1918
dyari Hill, 1924
blanchardi (Hogue, 1965)
sulphurea Neum., 1882
satana Poole, 1995
consuela Stkr., 1900

CHALCOPASTA Hamp., 1908
RODRIGUESIA Dyar, 1912
territans (Hy. Edw., 1884)
arianda (Druce, 1889)
arizona (French, 1889)
anopis Dyar, 1918
chalcophanis Dyar 1918
howardi (Hy. Edw., 1877)
ornata (Ottol., 1898)
houardi Hamp., 1910, emend.
fulgens B. & McD., 1912
acema (Druce, 1889)

INCERTAE SEDIS

NEUMOEGENIA Grt., 1882
KALLITRICHIA Ottol., 1898
TRICHOCALA Dyar, [1903]
poetica Grt., 1882

ARGENTOSTIRIA Poole, 1995
koebelei (Riley, 1893)
coebelei (Hamp., 1910), emend.

CHAMAECLEA Grt., 1883
pernana (Grt., 1881)
basiochrea B. & McD., 1916
mapensa Dyar, 1919

BISTICA Dyar, 1912
noela (Druce, 1892)
noela Dyar, 1912, preocc. by Druce, 1892

HEMIOSLARIA B. & Benj., 1924
pima B. & Benj., 1924

HEMINOCLOA B. & Benj., 1924
mirabilis (Neum., 1884)

THURBERIPHAGA Dyar [1920]
diffusa (Barnes, 1904)
catalina Dyar [1920]

PSAPHIDINAE (part)
DICHOPINAE

Psaphidini

PSAPHIDA Wlk., 1865
DICOPIS Grt., 1874
COPIPANOLIS Grt., 1874
EUTOLYPE Grt., 1874

resumens Wlk., 1865
viridescens (Wlk., 1865)
muralis (Grt., 1874)
thaxterianus (Grt., 1874)
rolandi (Grt., 1874)
vernalis (Morr., [1875])
grandis (Sm., 1898)
electilis (Morr., 1875)
depilis (Grt., 1881)
bombyciformis (Sm., 1892)
styracis (Gn., 1852)
cubilis (Grt., 1874)
stigma (Sm., 1890)
borealis (Sm., 1892)
fasciata (Sm., 1892)
damalis (Grt., 1879)

PSEUDOCOPIVALERIA Buckett and Bauer, 1966
sonoma (McD., 1941)
anaverta Buckett and Bauer, 1966

COPIVALERIA Grt., 1883
grotei (Morr., 1875)

BRACHIONYCHA Hbn., [1819]
ASTEROSCOPUS Boisduval, 1829
PETASIA Steph., 1829
SELENOSCOPUS Heinemann, 1859
borealis (Sm., 1899)

PLEROMELLA Dyar, 1921
opter Dyar, 1921

PROVIA B. & McD., 1910
argentata B. & McD., 1910

Feraliini

FERALIA Grt., 1874
MOMOPHANA Grt., 1875
ARTHROCHLORA Grt., 1875
jocosa (Gn., 1852)
furtiva Sm., 1909
deceptiva McD., 1920
major Sm., 1890
comstocki Grt., 1874
columbiana Sm., 1903
comstocci (Hamp., 1906), emend.
februalis Grt., 1874
meadowsi Buckett, [1968]

INDEX TO ANIMAL NAMES

Principal entries are given in bold face
Plate references are given as (1:5)

Generic names cited only in combination with specific names, whether in synonymy or text, are not given in the index. Look for such entries under the specific name. For example, *Cucullia postera* will be found under *postera* but not under *Cucullia.*

INDEX TO PLANT NAMES